International Conference on Recent Advances in Mechanical Engineering and Nanomaterials (ICRAMEN 2021)

IOP Conference Series: Materials Science and Engineering
Volume 1206

Pune, India
16 – 17 October 2021

ISBN: 978-1-7138-4665-9
ISSN: 1757-8981

Printed from e-media with permission by:

Curran Associates, Inc.
57 Morehouse Lane
Red Hook, NY 12571

Some format issues inherent in the e-media version may also appear in this print version.

Institute of Physics
Dirac House, Temple Back
Bristol BS1 6BE UK

Phone: 44 1 17 929 7481
Fax: 44 1 17 920 0979

techtracking@iop.org

Additional copies of this publication are available from:

Curran Associates, Inc.
57 Morehouse Lane
Red Hook, NY 12571 USA
Phone: 845-758-0400
Fax: 845-758-2633
Email: curran@proceedings.com
Web: www.proceedings.com

TABLE OF CONTENTS

IOP Conf. Series: Materials Science and Engineering 1206 (2021) 011001 doi:10.1088/1757-899X/1206/1/011001

International Conference on
Recent Advances in Mechanical Engineering and Nanomaterials
ICRAMEN 2021
16-17 October 2021

Preface

Research Association of Masters of Engineering is committed to providing a peer reviewed platform to outstanding researchers to exhibit their findings for the furtherance of technology to provide a research forum. We are devoted to continual innovation to better support the wishes of our communities, making sure the integrity of the research we publish, and championing the importance of open research.

ICRAMEN 2021 was a two days event that aimed to showcase state-of-the-art methodologies and technologies in Mechanical Engineering and Nanomaterials. It focused on new ideas and paves the way to disseminate the latest innovations and practices. It facilitates opportunities to network, collaborates, and exchange ideas with renowned leaders, scientists, and researchers in Mechanical Engineering and Nanomaterials. It acts as a crucial platform for industry and academia to foster innovative ideas, theories, frameworks, and applications. ICRAMEN 2021 was encouraged recent and futuristic advancements, challenges, and new strategies in the frontiers of Mechanical Engineering and Nanomaterials.

Dr. C. C. Handa, Professor and Head, Department of Mechanical Engineering, KDK College of Engineering, Nagpur, Maharashra, India was the Chief Guest of the conference and delivered inaugural address. Dr. S. A. Khan, Professor, Department of Mechanical Engineering, International Islamic University Malaysia, Kuala Lumpur, Malaysia was the Guest of Honor and Keynote Speaker of the conference and delivered a key-note address on "Recent Advances in Aerospace Engineering". Several students/research scholar and academicians have presented their research work at ICRAMEN 2021 in the areas of design engineering, thermal engineering and fluid science, material science and nanomaterials, production and industrial engineering. We believe that the knowledge exchanged at ICRAMEN 2021 immensely helps to the researchers working on relevant fields and contributes to the growth of science and technology.

As the coronavirus (COVID-19) spreads worldwide, taking into account the various circumstances of the participant, ICRAMEN was adopted all available social media methods, such as online presentation besides the regular presentation styles (oral & poster). The online presentation session was arranged via. ZOOM platform. All authors were participated via in-person presentation at the time of their session on the online platform. Our sincere thanks to the keynote speakers, reviewers, technical and advisory committee members, organizing committee, and all authors for their contribution and making ICRAMEN 2021 a grand success.

Dr. Manoj A. Kumbhalkar
Conference Chair, ICRAMEN 2021
President,
Research Association of Masters of Engineering, India

Organizing Committee

Conference Chair

Dr. Manoj A. Kumbhalkar, President, Research Association of Masters of Engineering, India

Convener

Dr. Radheshyam H. Gajghat, Member, Research Association of Masters of Engineering, India

Dr. R. Kishor, Member, Research Association of Masters of Engineering, India

Co-Convener

Nand Jee K., Member, Research Association of Masters of Engineering, India

Coordinator

Dr. Jagdish D. Kene, Member, Research Association of Masters of Engineering, India

Mr. Vivekanand Thakare, Member, Research Association of Masters of Engineering, India

Keynote Speaker

Dr. S. A. Khan, Professor, Department of Mechanical Engineering, International Islamic University Malaysia, Kuala Lumpur, Malaysia

Session Chair

Dr. Vivek Pandey, School of Mechanical, Chemical and Materials Engineering, Adama Science and Technology University, Adama, Ethiopia

Dr. Surajkumar Mukti, Department of Mechanical Engineering, National Institute of Technology (NIT), Raipur

International Advisory Committee

Dr. S. A. Khan, Professor, Department of Mechanical Engineering, International Islamic University Malaysia, Kuala Lumpur, Malaysia

Dr. Mehmet Serkan KIRGIZ, Professor, İstanbul Sabahattin Zaim University, Istanbul, Turkey

Dr. Wadhah Hussein Al Doori, Tikrit University, Tikrit, Iraq

Dr. Gyanendra Kumar Singh, Associate Professor, Adama Science and Technology University, Adama, Ethiopia

Dr. Kamil Kahveci, Department of Mechanical Engineering, Trakya University, Edirne/Turkey

Dr. Te Fu Chen Lunghwa University of Science and Technology, Taiwan

Dr. S. Rajkumar, School of Mechanical and Electromechanical Engineering, Hawassa University, Ethiopia

Dr. Fuat Kara, Düzce University, Konuralp Campus, DÜZCE Turkey

Dr. Engin NAS, Dr. Engin PAK Cumayeri Vocational School, Duzce University, Cumayeri, Duzce (Turkey)

Dr. Menderes KAM, Dr. Engin PAK Cumayeri Vocational School, Duzce University, Cumayeri, Duzce (Turkey)

Dr. Mustaffa, AYYILDIZ Duzce University, Faculty of Technology, Konuralp Campus, Duzce (Turkey)

Dr. Swathi Sunkara, Process Engineer, Intel, Portland, Oregon

Dr. Omer ERKAN, Duzce University, Faculty of Technology Konuralp Campus, Duzce (Turkey)

Dr. Sunada Chakravarthy, Research Scientist, GSL Solar Materials, US

Dr. Mustafa Turkyilmazoglu, Hacettepe University, Ankara, Turkey

Dr. Jaejin Jang, University of Wisconsin, Milwaukee, WI, United States

Dr. Nallamothu Ramesh Babu, SoMCME, ASTU, Adama

Dr. Shuva Ghosh, Bangladesh University of Engineering and Technology, Bangladesh

Dr. Vivek Pandey, School of Mechanical, Chemical and Materials Engineering, Adama Science and Technology University, Adama, Ethiopia

National Advisory Committee

Dr. T Ram Prabhu, Assistant Director/Scientist (Materials and Manufacturing Processes), Defence R & D Organization (DRDO), Bangalore, India

Dr. Chirra Kesava Reddy, Professor & Principal, Newton Institute of Science and Technology, Guntur, Andhrapradesh, India

Dr. Mrinal Kanti Manik, Director, LDC Institute of Technical Studies, Allahabad, Prayagraj, India

Dr. P. V. Walke, Deputy Director and Dean IQAC, G. H. Raisoni College of Engineering, Pune, India

Dr. G. Venkatachalam, Professor Grade 1, School of Mechanical and Building Sciences, VIT University, Vellore, India

Dr. Chandrahas C. Handa, Professor & Head, Karmavir Dadasaheb Kannamwar College of Engineering, Nagpur, India

Dr. R. S. Shelke, Professor & Head, G. H. Raisoni College of Engineering, Nagpur, India

Dr. (Mrs.) Shiena Shekhar, Professor & Head, Bhilai Institute of Technology, Durg, India

Dr. Rakesh L. Himte, Professor & Head, Rungta College of Engineering & Technology, Bhilai, Chhattisgarh

Dr. P. T. Saravankumar, Professor & Head, Dept of Automobile Engineering, Hindusthan Institute of Technology, Coimbatore, India

Dr. Begori Venkatesh, Professor, Vardhaman College of Engineering, Shamshabad, Hyderabad, India

Dr. Ajith Ramesh, Associate Professor, Amrita School of Engineering, Amrita Vishwa Vidyapeetham (University), Coimbatore, India

Dr. D. B. Jani, Associate Professor, Government Engineering College, Dahod, Gujrat, India

Dr. Suraj Kumar Mukti, Assistant Professor, National Institute of Technology, Raipur, India

Dr. Srinivasu Gangi Setti, Assistant Professor, Department of Mechanical Engineering, National Institute of Technology, Raipur, India

Dr. Pulivarti S. Rao, Professor & Head, Christian College of Engineering & Technology, Bhilai, India

Dr. Shitalkumar A. Rawandale, Dean - Industry Institute Interaction, Pimpri Chinchwad College of Engineering, Pune, India

Dr. Joji Thomas, Associate Professor, Saveetha Engineering College, Chennai, India

Dr. Manish D. Pasarkar, Associate Professor, Bajaj Institute of Technology, Wardha, India

Dr. Kiran More, Associate Professor, D.Y. Patil Institute of Technology, Ambi, Pune, Maharashtra, India

Dr. Pavan Kumar Manvikar, Assistant Professor, Shri Madhwa Vadiraja Institute of Technology Bantakal, Udupi, India

Dr. Sanjay Sharma, Assistant Professor, Ajay Kumar Garg Engineering College, Gaziabad, UP, India

Dr. Sandip Joshi, Assistant Professor, Ramdeobaba college of engineering and management, Nagpur, India

Dr. Vivek Khond, Assistant Professor, G. H. Raisoni College of Engineering, Nagpur, India

Dr. Ashish S. Raut, Assistant Professor, G. H. Raisoni College of Engineering, Nagpur, India

Dr. Shashi Kant Verma, Institute of Plasma Research, Ahmedabad, Gujarat, India

Nikhil Jumade, CFD Engineer, Fiat Automobiles, Pune, India

Vednath Kalbande, Assistant Professor, G. H. Raisoni College of Engineering, Nagpur, India

IOP Conf. Series: Materials Science and Engineering 1206 (2021) 011001 doi:10.1088/1757-899X/1206/1/011001

Call for Paper

ICRAMEN aims to bring together leading academic scientists, researchers, and research scholars to exchange and share their experiences and research results in the field of Mechanical Engineering and Nanomaterial.

Major Themes

1. Design Engineering
2. Thermal Engineering
3. Production and Industrial Engineering
4. Material Science and Nanomaterials

Topics including but not limited to the following:

Mathematical formulation and modelling
Computer aided modelling
Experimental analysis
Finite element analysis
Applied Mechanics
Automotive Engineering
Computer aided manufacturing
Composite and Smart Materials
Thermal Engineering
Heat and Mass Transfer
Industrial and Systems Engineering
Machinery and Machine Design
Manufacturing and Production Processes
Material Science and Processing
Composite Materials
Production Technology
Robotic Automation and Control
Solid Mechanics
System Dynamics and Simulation
Mechanical Power Engineering
Computational Fluid Dynamics
Fluid Mechanics and Machinery
Additive Manufacturing
Supply Chain Management
Functionally Graded Materials
Mechanics of Biological Materials
Nanomaterials and Alloys
Nano and Microscale Engineering
Nanocomposites
Smart Materials
Shape Memory Alloys

IOP Conf. Series: Materials Science and Engineering 1206 (2021) 011001 doi:10.1088/1757-899X/1206/1/011001

LIST OF ACCEPTED PAPERS

Track 1 - Design Engineering

Paper Id	*Author(s)*	*Article title*
paper_35	Neetirajsinh Chhasatia, Jitendra Chaudhari and Amit Patel	Ridge Gap Waveguide based band pass filter for Ku-band Application
paper_36	Shrutika Dahake, Nilesh Awate and Rupesh Shelke	Design And Analysis of Release Mechanism Using Shape Memory Alloy for Spacecraft
paper_37	Deepesh Makhija, Sanjay Jain, Madhusudan Achari and Kunal Ghosh	Aerodynamic analysis of aircraft model using indigenously developed wind tunnel facility
paper_38	Dhairya Raval, Sanjay Jain, Madhusudan Achari and Kunal Ghosh	Design and analysis of smoke flow visualization apparatus for wind tunnel
paper_40	Saurabh Kumar	Free vibration analysis on axially graded beam resting on variable Pasternak foundation
paper_44	Nitin Sawarkar, K S Zakiuddin, Roshan Umate, Rajkumar Chagde	Application of Jaw Type Flexible Clutches In Human Powered Machines: A Literature Review
paper_46	Surojit Biswas and Priyankar Datta	Finite Element Model for Free Vibration Analyses of FG-CNT Reinforced Composite Beams using Refined Shear Deformation Theories
paper_51	Devismita Sanjay, Neeraj Kumar and Souptick Chanda	Stress-strain distribution in intact L4-L5 vertebrae under the influence of physiological movements: A finite element (FE) investigation
paper_54	Prakash Kumar Sen and Mahesh Bhiwapurkar	Numerical Simulation and Parametric Analysis of Fatigue Crack in UIC60 Rail Thermite Welded Joint

Track 2 - Material Science and Nanomaterials

Paper Id	*Author(s)*	*Article title*
paper_12	Satadru Kashyap and Debanil Das	Study of effects of weathering on natural fiber composites
paper_28	Abhishek Singh and S.C Jayswal	Mechanical characterization of TiO2 nanoparticles based on glass fiber reinforced polymer composite
paper_30	Sk Faruque Ahmed, Mohibul Khan and Nillohit Mukherjee	Enhancing electrical properties of carbon nanotubes thin films by silicon incorporation
paper_33	Mukesh Thakur and Rohan Senanayake	Experimental Analysis using an Innovative Catalytic Converter coated with Nano-particles for Pollution Control from Automobiles
paper_39	Mohd. Anas and Anurag Sharma	2-Dimensional inline system for peppermint oil extraction
paper_45	Sourabh Shinde, Taukir Momin, Vispi Karkaria and Parshuram Karandikar	Investigation of substitute jar materials for Laboratory grade ball milling machine to process electrode materials for energy storage devices
paper_47	Ujjwal Ks and Dr. Sharath Chandra N	A review of wear resistance materials used in power-screw mechanism for aerospace applications

paper_48	Siva Chakra Avinash Bikkina	Analysis of Electromagnetic Reflection Loss for Mesh Structure with Al6061 MMC for Aerospace Applications
paper_52	Srinu Budumuru	Electromagnetic shielding effectiveness for Al6061 metal matrix composite based mesh wire reinforced with Flyash for oblique incidence of EM wave
paper_53	Sumit Kumar, Praveen Kumar and Rudra Pratap	Reliability Failure in Microelectronic Interconnects by Electric Current Induced Chemical Reaction

Track 3 - Thermal Engineering and Fluid Science

Paper Id	*Author(s)*	*Article title*
paper_18	Shyamal Chakrabarty, Uday Wankhede and Rupesh Shelke	Impact of Non-convective Zone and Lower Convective Zone Thickness on the Performance Characteristics of Salinity Gradient Solar Pond
paper_24	N. Golden Stepha and D. Kavin Jacob	Analysis on physical properties of micropolar nano fluid past a continuously moving permeable plate
paper_27	Ujwal Sontakke and Santosh Jaju	Green Hydrogen Economy and Opportunities for India
paper_31	Ritwik Bhattacharya and Pranab Samanta	Simulation-based Study of Graphene-water Nanofluid flow through Microchannel Heatsink
paper_50	Sunil Gupta, Bb Arora and Akhilesh Arora	Economics-Based Payback and Life Cycle Cost Savings Assessment of Inverter Type Air Conditioners

Track 4 - Production and Industrial Engineering

Paper Id	*Author(s)*	*Article title*
paper_4	Umesh Kumar Vates, Nand Jee Kanu, Eva Gupta, Gyanendra Kumar Singh, Naveen Anand Daniel and Bhupendra Prakash Sharma	Optimisation Of FDM 3d Printing Process Parameters on ABS Based Bone Hammer Using RSM Technique
paper_16	Vijay Kalbande, Chandrahas Handa, Radheshyam Gajghat and Shitalkumar Rawandale	Development of Online Platform for Checking Placement Capability of Engineering Students
paper_23	Vijay Kalbande, Dr Nitin Manadavgade, Radheshyam Gajghat and Shitalkumar Rawandale	New Approach to Identify Industry Institute Skill Gap of Engineering Students
paper_25	Avez Shaikh, Ajinkya Shinde, Satish Chinchanikar, Guruprasad Zagade and Sonia Pardeshi	Comparative assessment of hard turning under dry and minimum quantity lubrication
paper_34	Rohit Sharma and Ubaid Ahmad Khan	Applying analytical hierarchy process for addressing the agile manufacturing drivers
paper_49	Tanmay Pati, Shruti Kabra and Utkarsh Chadha	Statistical Quality Study of the Parts Produced in an Automobile Industry: A Daimler India Case Study

IOP Conf. Series: Materials Science and Engineering 1206 (2021) 011001 doi:10.1088/1757-899X/1206/1/011001

Dr. C. C. Handa
Professor & Head,
Department of Mechanical Engineering,
KDK College of Engineering,
Nagpur, Maharashtra, India.

Message of Chief Guest

It gives me great pleasure to extend my greetings and warmest wishes to the Research Association of Masters of Engineering for organizing an **International Conference on Recent Advances in Mechanical Engineering and Nanomaterials (ICRAMEN 2021) during** October 16-17, 2021.

Advances in Mechanical Engineering and nanomaterials are the main area covered by the conference. The growth and transformations of colloidal nanocrystals are a major issue for the creation of functional nanomaterials. In practical terms, knowledge of how to selectively synthesis the required sizes and forms of nanocrystal that would benefit various applications. Interdisciplinary character of research enthuses the current activity in the field of nanomaterials. People from varied backgrounds combine hands to succeed in establishing a simple way of synthesizing and applying heterogeneous nanomaterials.

Despite extensive efforts and evidence that the tuning of various nanomaterials to specific applications may be extremely advantageous, many basic features of complex nanomaterial systems are still poorly understood and empiricism is still widespread.

I am quite sure that this conference will ignite the creative ideas and exchange of knowledge to fill up the gaps in ongoing research on multifunctional nanomaterials.

I once again commend the core organizing team and wish the conference a resounding success.

Dr. C. C. Handa

IOP Conf. Series: Materials Science and Engineering 1206 (2021) 011001 doi:10.1088/1757-899X/1206/1/011001

Dr. S. A. Khan
Professor,
Department of Mechanical Engineering,
International Islamic University Malaysia, Kuala Lumpur, Malaysia

Message of Guest of Honor

I feel truly delighted to learn that the Research Association of Masters of Engineering e- platform for the scientific community to share their ideas and research by organizing an **International Conference on Recent Advances in Mechanical Engineering and Nanomaterials (ICRAMEN 2021)** during October 16-17, 2021.

Its subject is fascinating since it focuses on the latest developments in mechanical and nanomaterial engineering. Through the range of discoveries and technologies produced using nanomaterials, we see a silent revolution. Now, nanomaterials and products, including paints, filters, insulation, and lubricant additives, are utilized in several manufacturing processes and medical treatments. There can be no exaggeration that multidisciplinary research now constitutes the foundation of any country's economic success and that humankind directly benefits. The finest examples of multidisciplinary research are nanotechnology and nanoscience.

It is also vital to grasp the underlying fundamental concepts of science, apart from developing applicable technology. ICRAMEN 2021, therefore, takes on particular relevance. I think the attendees will profit immensely.

My congratulations to the organizing team, and I extend my best wishes for its successful completion.

Dr. S. A. Khan

IOP Conf. Series: Materials Science and Engineering 1206 (2021) 011001 doi:10.1088/1757-899X/1206/1/011001

Dr. M. A. Kumbhalkar
President,
Research Association of Masters of Engineering

Message of Conference Chair

I am very happy to share that Research Association of Masters of Engineering is organising an **International Conference on Recent Advances in Mechanical Engineering and Nanomaterials (ICRAMEN 2021)** during October 16-17, 2021.

The fast advancements in technology make the promotion of technical competence in the workforce highly essential nowadays. One of the main problems for scientists is the implementation of their highly specialized research in many domains of human knowledge and interdisciplinary study. This problem must be seen as an opportunity for human resources development in a variety of areas. The other significant problem is to discover how these material advancements may be utilized to better the everyday job and the common person's surroundings. In these recent years there has been a lot of focus to promoting cross-disciplinary research with specific issues in nanoscience and nanotechnology.

I'm confident there will be many intellectually engaging contacts and the presentation of constructive ideas which would benefit a broad spectrum of participants in the conference on current achievements in mechanical engineering and nanomaterials.

I congratulate the team of RAME and extend my best wishes for its success.

Dr. Manoj A. Kumbhalkar

IOP Conf. Series: Materials Science and Engineering 1206 (2021) 011001 doi:10.1088/1757-899X/1206/1/011001

Dr. R. H. Gajghat
Member,
Research Association of Masters of Engineering

Message of Convener

I am pleased to know that the Research Association of Masters of Engineering is organizing an **International Conference on Recent Advances in Mechanical Engineering and Nanomaterials (ICRAMEN 2021)** during October 16-17, 2021.

The present requirement of our society is to develop the abilities to address different challenges of individuals with a scientific temperament. Research in nanotechnology covers several fields, which get different information from multidisciplinary sources. Nano research is getting more and more inclusive. I'm delighted that the invited speakers for the event are well-known speakers from around the world.

I am convinced that at this magnificent conference on knowledge the mainstay of nanoscience and nanotechnology will put together the insight and wisdom appropriate to human growth.

I congratulate the Research Association of Masters of Engineering for organizing such event and convey my best wishes for all success.

Dr. R. H. Gajghat

IOP Conf. Series: Materials Science and Engineering 1206 (2021) 011001 doi:10.1088/1757-899X/1206/1/011001

Dr. Kishor S. Rambhad
Member,
Research Association of Masters of Engineering

Message of Convener

It gives me immense pleasure to know that Research Association of Masters of Engineering is organising an **International Conference on Recent Advances in Mechanical Engineering and Nanomaterials (ICRAMEN 2021)** during October 16-17, 2021.

The subject of the conference is intriguing since it includes a wide variety of mechanical and nanomaterial characteristics. The emergence of huge potential uses of nanomaterials constantly leads to major societal transformations. The wall of traditional subjects collapses in the modern world and we see the spread of one domain's ideas into the next. The coagulation of thoughts can lead us all to the crossroads when the unity of knowledge from many fields can be noticed.

I am confident that debates and conversations at this e-conference will encourage the younger brains to understand the science and technology of nanomaterials.

My congratulations to the whole organizing team and best wishes for the grand success of the event.

Dr. K. S. Rambhad

IOP Conf. Series: Materials Science and Engineering 1206 (2021) 011001 doi:10.1088/1757-899X/1206/1/011001

Mr. Nand Jee K.
Member,
Research Association of Masters of Engineering

Co-Convener Message

As a co-convener of the conference, I take great pride in welcoming the great scientists, academicians, young researchers, and students from all over the world to attend the **International Conference on Recent Advances in Mechanical Engineering and Nanomaterials (ICRAMEN 2021**) during **October 16-17, 2021**.

In ICRAMEN 2021 you may obtain insights into the latest technology and research that attracts great interest via the gigantic and enthusiastic presence of experts, young, brilliant researchers and outstanding student populations.

ICRAMEN 2021 aims to bring together a multidisciplinary group of scientists and engineers from across the world to present and exchange insights on the conference's primary subject. It encourages top-tier study and globalizes research on quality in general, thereby increasing the worldwide competition for debates, presentations and attention to recent excellent accomplishments in the field of scientific materials.

We are looking forward to an excellent meeting with great scientists from different countries around the world and sharing new and exciting results. *I express pleasure to complement my team of RAME for their sincere hard work in formulating an excellent programme.*

Nand Jee K.

IOP Conf. Series: Materials Science and Engineering 1206 (2021) 011002 doi:10.1088/1757-899X/1206/1/011002

Peer review declaration

All conference organisers/editors are required to declare details about their peer review. Therefore, please provide the following information:

- **Type of peer review: Single-blind / Double-blind / Triple-blind / Open / Other**
  All papers have been reviewed by at least two reviewers and follow a double-blind peer review process. For the peer review, an editor has been assigned to the submitted paper as per the major theme of the conference, i.e., design engineering, thermal engineering and fluid science, material science and nanomaterials, production and industrial engineering. Four editors have been handling the review process by assigning two domain expert reviewers to each paper. The papers have been evaluated as per the quality criteria of IOP Publishing and the decisions were sent to the authors, such as: strong accept, accept, weak accept, borderline paper, and reject. The weakly accepted and borderline papers were sent back to authors for revision and resubmission was done through EasyChair. The resubmitted paper was again evaluated by the editor, and a final decision has been conveyed to the author.

- **Conference submission management system:** Easychair
  Easychair was chosen as the conference paper submission and review management system for **ICRAMEN 2021**. All papers have been submitted and also reviewed by the Easychair account only. A revised submission was also accepted only by the easychair account. All email correspondence with the author was done through easychair email and also from conference@rame.org.in.

- **Number of submissions received:** 52
  We have received a total of 52 papers, out of which accepted and registered papers have been considered for presentation at the conference and publication in a journal.

- **Number of submissions sent for review:** 50
  Two papers were not considered for review due to late submission and was rejected by the editor.

- **Number of submissions accepted:** 28

- **Acceptance Rate (Number of Submissions Accepted / Number of Submissions Received X 100):** 54%

- **Average number of reviews per paper:** 2.2

- **Total number of reviewers involved:** 23

- **Any additional info on review process:**

  1. The Plagiarism Checker X software has been used to check the similarity index of the papers. The similarity index of all the papers is not more than 20%.

2. ICRAMEN 2021 was a virtual conference, with all presentations delivered via the Zoom online platform.
3. All papers have been evaluated by the editor and reviewed by a minimum of two reviewers. Also, some papers are sent back to the author for further revision after the presentation at the conference to amend the corrections suggested by the session chair.

- **Contact person for queries:** Dr. Manoj A. Kumbhalkar, Conference Chair, ICRAMEN 2021. President, Research Association of Masters of Engineering (RAME), India. manoj.kumbhalkar@rame.org.in, manoj.kumbhalkar@rediffmail.com

Optimization of FDM 3D Printing Process Parameters on ABS based Bone Hammer using RSM Technique

Umesh Kumar Vates[1], Nand Jee Kanu[2,3,*], Eva Gupta[4,5], Gyanendra Kumar Singh[6], Naveen Anand Daniel[1], Bhupendra Prakash Sharma[1]

[1]Associate Professor, Mechanical Engineering, Amity University, Noida, India
[2]PhD Research Scholar, Mechanical Engineering, SVNIT, Surat, India
[3]Assistant Professor, Mechanical Engineering, JSPM Narhe Technical Campus, Pune, India
[4]PhD Research Scholar, ASET, Amity University, Noida, India.
[5]Assistant Professor, Electrical Engineering, TSSM's Bhivrabai Sawant College of Engineering and Research, Pune, India
[6]Faculty, Mechanical Design and Manufacturing Engineering, Adama Science and Technology University, Adama, Ethiopia

[*]Corresponding Author Email: nandssm@gmail.com

Abstract. Rapid prototyping (RP) uses a cycle where a real model is made by explicitly adding material as thin cross-sectional layers. Fused deposition modelling (FDM) 3D printer is being use for synthesis of ABS based bone hammer. Response surface methodology (RSM) based L27 design of experiment were adopted to perform the experiment using four influencing parameters such as layer thickness, infill percentage, orientation and nozzle temperature for the three responses deflection, hardness and weight. Response surface methodology was used for modelling and optimization of considered process parameters. In present investigation, it is evident that bone hammer fabrication process parameters have been optimized on data such as bone hammer weight 19.8091g, hardness 104.5921 BHN, and force of 15-degree deflection 36.0681 N has been produced with RSM prediction with influence of process parameters such as layer thickness 0.250 mm, infill percentage 63.3333, orientation 60-degree, nozzle temperature 240°C.

Keywords. RSM, RP, FDM, ABS, Additive Manufacturing, Bone Hammer Optimization

1. Introduction

1.1 An insight

It is well known that additive manufacturing technologies (AMT) become popular in manufacturing sectors since 1980s. Previous technology involving in layered construction of physical models based on a digital 3D solid model was stereo lithography. One of the other layered manufacturing methods invented in the 1980s is the technology of the deposition of molten thermoplastic material. In this recent technology, material feed in the form of a very thin solid plastic wire is generally passed through the nozzle where it is heated to a temperature at which it becomes an intermittent phase of solid and liquid interface. The material is distributed as per design cross-section of the current building model. Rapid prototyping (RP) licenses engineers to make considerable models quickly rather than straightforward two-dimensional pictures, such models are even using for various huge purposes from passing on contemplations to partners and customers to testing of different pieces [1]. Other than this, RP offers a lot of various advantages, for instance, unambiguous data managing and limit, ability to make complex shapes and interlocking plans, freed from contraption/work piece garbage, nonappearance of molds, kicks the pail, establishments and models, mass customization. As of now, the prevalent RP estimates open market are entwined declaration illustrating (FDM), stereo lithography (SL), and explicit laser sintering (SLS), overlaid object creating 3D printing and solid ground mitigating (SGC). Regardless, the introduction of any RP association is assessed with respect to manufacture time, quality ascribes like surface disagreeableness and dimensional accuracy, mechanical and tribological properties, cost of creation and energy use. These display extents of RP measures are through and through influenced by their participation limits. We explored the effect of cooperation limits on the perimetric exactness of the FDM models. Also we investigated the effect of cycle limits on the physical properties of the FDM models. Definite conditions for versatility, various strengths of the FDM models were made using response surface technological way of thinking and alluring quality limit approach was used to expect the ideal blend of cycle limits [2, 3]. Mathematical models were made and ideal structure course was settled using a graphical strategy. The paper presents exploratory delayed consequences of the models on picked mechanical properties like force at 15 degrees evasion, Brinell hardness, and weight at consistent volume 25. Tests were made using the estimation and arranged a lot using SolidWorks 2013 [4]. They were made on a 1200es machine understanding the joined testimony exhibiting advancement (FDM) in light of plan of investigation masterminded using MATLAB considering all likely outcomes with changing of variable components to get the best strong hammer printed. Added substance manufacturing measurements are used to make genuine models from three-dimensional (3D) PC upheld plan (CAD) math data. The article is based on layer-by-layer stacking, with supporting developments for overhanging math and sabotages being made where essential (measure subordinate). The circuit of reconfigurable parametric inside matrix structures subject to rough segments will change the mechanical properties, the material uses and the structure time. The models were arranged on the machine stage at different focuses to the printing course. Similarly, what's more we changed genuine components of the printer to improve progressed results like the layer thickness, Infill rate, Orientation, and Nozzle temperature. The material used for the advancement of tests was ABS [5-7]. All the testing part was done truly on machines under evaluated bearing from specialists. Wide genuine testing is done as the fragments worked by the FDM cooperation have anisotropic properties. The mechanical properties and dimensional-shaped accuracy of elements manufactured by the aforementioned technology depend on the direction of its position on the building platform, the layer thickness and the heat of the working chamber [8].

There are papers in which authors analysed both selected mechanical properties, as well as the accuracy of elements manufactured using FDM. There are no studies concerning rheological tests in spite of current fact concerning to FDM, i.e., relaxation, especially in relation to process parameters such as printing direction. Samples were produced using FDM Technology. It is very important to determine the critical parameters of FDM and their values in regards to tensile strength. The impact of the degree of filling of the sample by model material was tested. Biodegradable material acrylonitrile butadiene styrene (ABS) was used to build the model. The study included 27 types of samples, which were printed along the three main axes: X, Y, Z and selected thickness of the combined layers. The amount of material wastage was kept in mind. The study confirmed the dependence of the mechanical properties of the selected printing direction during manufacturing [9-11]. Depending on the location of models on the working platform, different type of filling of support material was used. Based on the fact, support thickness was increased or decreased. Due to the fact that most of the elements of machines and mechanisms manufactured by additive technologies, they are subjected to a constant load, the research presented in this paper can be regarded as well founded. Additive manufacturing (AM) is a suitable name to portray the advances that assemble 3D protests by adding endless supply of material, regardless of whether the material is plastic, metal, concrete or human tissue and is an extraordinary way to deal with modern creation that empowers the production of lighter, more grounded parts and frameworks [3].

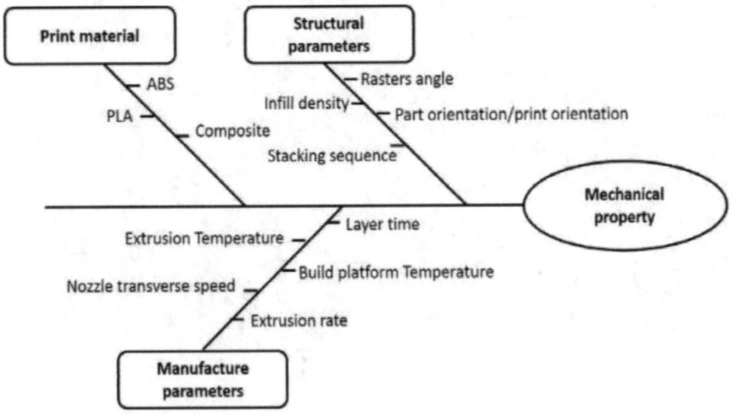

Figure 1: Input - output of 3D printing

It gives expert looking item, on which the structure layers are hard to see, with noteworthy exactness and extensive assortment of materials than FDM. Fig. 1 is the 3D printing fishbone diagram for process. Metal additive manufacturing, components with complex plans, or those profiting by being light, for example, parts for airplanes, vehicles, trucks or clinical applications, will discover points of interest, while segments with basic plans, or where weight isn't an issue, are better delivered utilizing customary, subtractive manufacturing. The aviation and clinical businesses are especially appropriate to see development. Fig. 2 indicates the 3D printing setup with nozzle.

Figure 2: Printing plant

1.2. AM materials

Materials for additive manufacturing are mentioned below:
Thermoplastics: Thermoplastic polymers remain the most famous class of additive manufacturing. Polylactic acid (PLA), acrylonitrile butadiene styrene (ABS), and polycarbonate (PC) each offer unmistakable favourable circumstances in various applications. Water-solvent polyvinyl alcohol (PVA) is normally used to make impermanent help structures.
Metals: Many metals and its compounds are utilized in additive manufacturing, as valuable metals such as silver, gold, treated steel and titanium. Fig. 3 indicates the traditional and additive manufacturing.
Ceramics: An assortment of earthenware production has additionally been utilized in additive manufacturing, including zirconia, alumina and tricalcium phosphate. Similarly, substitute layers of powdered glass and glue are prepared together to make altogether new classes of glass items.
Biochemical: Biochemical medical care applications incorporate the utilization of solidified material from silicon, calcium phosphate and zinc to help bone structures as new bone development happens. Analysts are additionally investigating the utilization of bio-inks created from undifferentiated cells to frame everything from veins to bladders and past.

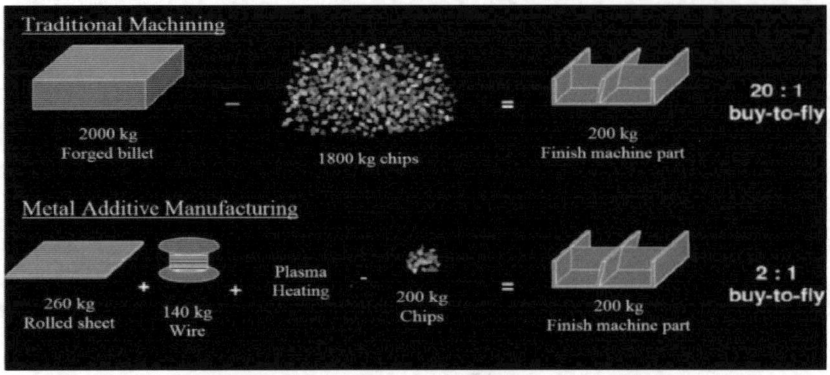

Figure 3: Traditional v/s additive manufacturing

The impediment organizations face while examining additive manufacturing is the expense. New advances have made the expense considerably moderate, for a solid modern machine. Since additive manufacturing diminishes capital it is expected to scale up creation without rolling out significant enhancement, makers might have the option to push up and benefit of their plan of action. Revised version of product easily available: AM is something beyond making an actual item, but is tied in with carrying plan and development to the front line. In manufacturing, altering a plan during creation increases critical cost or time delays as tooling on a creation line is changed out. Additive manufacturing takes care of this issue by moving ceaselessly from static plans and empowering specialists to deliver various renditions of a solitary plan in a financially savvy way [12-14]. AM will not intrude on business tasks as there are wide assortment of preparing programs accessible for originators and makers at all degrees of comprehension. Reduction in waste production: AM creates less waste than customary manufacturing. The eliminated material is typically fuel or shavings that can't be reused and end up as waste. Contrasted and conventional manufacturing measures, additive manufacturing can essentially decrease energy use by utilizing less material and disposing of steps in the creation cycle.

1.3 Response surface methodology

Response surface methodology technique (RSMT) researches the associations between a couple of educational variables and in any event one response factors. The standard considered RSM is to use a gathering of arranged investigations to obtain an ideal response. RSM could be used to increase the formation of a phenomenal substance by smoothing out of operational segments. Maybe than customary methodologies, the association among measure variables can be directed by authentic techniques.

1.4 Fused Filament Fabrication (FFF)

It is also known fused deposition modeling (FDM), sometimes referred as filament freeform fabrication, and is a 3D printing measure that uses a constant fiber of a thermoplastic material. Fiber is dealt with from a colossal spool through a moving, warmed printer extruder head, and is kept on the creating work. The speed of the extruder head may similarly be controlled to respite and start declaration and design a barged in on plane without hanging or spilling between sections. "Consolidated fiber fabricate" was created by the people from the RepRap errand to give an articulation that would be genuinely unconstrained in its usage, given brand names covering "entwined declaration showing". The 3D printer head or 3D printer extruder is a part in material ejection added substance creating at risk for rough material mellowing and outlining it into a relentless profile [15, 16]. A wide arrangement of fiber materials are removed, including thermoplastics.

2 Materials and Methods

2.1. Introduction

ABS is an impact-resistant engineering thermoplastic and amorphous polymer. ABS is made up of three monomers: acrylonitrile, butadiene and styrene as structure of molecules given in Fig. 4.
Acrylonitrile: It is a synthetic monomer produced from propylene and ammonia. The component contributes to ABS chemical resistance & heat stability.
Butadiene: It is produced as a by-product of ethylene production from steam crackers. The component delivers toughness and impact strength to ABS polymer.
Styrene: It is manufactured by dehydrogenation of ethyl benzene. It provides rigidity and process ability to ABS plastic, ABS is produced by emulsion or continuous mass technique. The chemical formula of acrylonitrile butadiene styrene is (C8H8·C4H6·C3H3N)n. The natural material is an opaque ivory color and is readily colored with pigments or dyes.

ABS is a strong and durable, chemically resistant resin but gets easily attacked by polar solvents. It offers greater impact properties and slightly higher heat distortion temperature than HIPS [17-19]. Acrylonitrile butadiene styrene has a broad processing window and can be processed on most standard machinery. It can be injection-molded, blow-molded, or extruded. It has a low melting temperature making it particularly suitable for processing by FDM based 3D printing machine.
ABS falls between standard resins and often meets all the desire properties at a reasonable price-cost effectiveness. Key suppliers of ABS plastics include: SABIC, RTP Company, LG Chem, Ineos etc. 3D Printing ABS grades are readily available from 3D Systems, Stratasys, Techmer Engineered Solution.

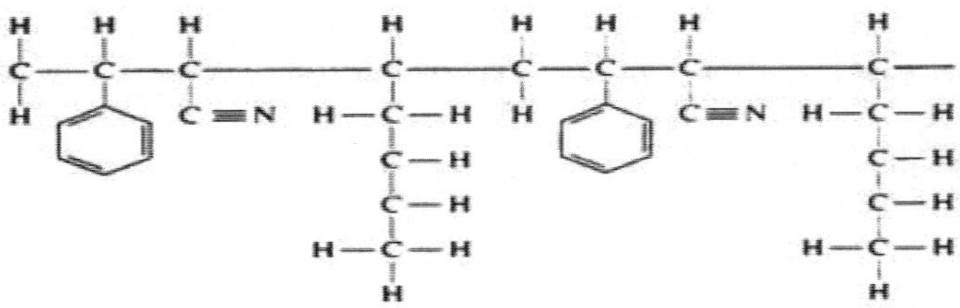

Figure 4: Molecular structure of ABS

2.2 Properties of ABS

Mechanical properties:

- High rigidity
- Good impact resistance, even at low temperatures
- Good insulating properties
- Good weldability
- Good abrasion and strain resistance
- High dimensional stability (Mechanically strong and stable over time)
- High surface brightness and excellent surface aspect

Chemical properties:

- Very good resistance to diluted acid and alkalis
- Moderate resistance to aliphatic hydrocarbons
- Poor resistance to aromatic hydrocarbons, halogenated Hydrocarbons and alcohols

Mechanical and electrical properties of ABS plastics have been explored in Table 1 as standard to enable understanding bone hammer requirement.

Table 1: Properties of ABS

Mechanical properties	Elongation at break	10 - 50 %
	Flexibility (flexural modulus)	1.6 - 2.4 GPa
	Hardness Shore D	100
	Strength at yield (tensile)	29.6 - 48 MPa
	Toughness (Notched Izod Impact at room temperature)	200 - 215 J/m
Electrical properties:	Arc resistance	60 - 120 sec
	Dielectric constant	2.7 - 3.2
	Dielectric strength	15.7 - 34 kV/mm
	Volume resistivity	14 - 16 x 10^{15} Ohm.cm

2.3 Printing process

- Fused filament fabrication (FFF) is the additive manufacturing bases technology; a fine plastic material filament is generally fed in to the influence of heated and moving head that melts and extrudes it to depositing layer- by- layer and form desired shape. Each layer is deposited over the layer with help of moving platform. Additional vertical support structures are required to sustain overhanging part during printing.
- FFF operated with software but requires STL file (stereolithography file format). It is also requiring support structures for holding the part [20].
- Nozzle is capable to move in both vertical and horizontal directions with given mechanical range which moves in the XY plane. Process: 1 – 3D Printer Extruder, 2 – deposited material (modelled part), 3 – controlled movable table
- Thin bead of deposited extruded plastic metal (called road) generally solidifies quickly and get rigid collapse with substrate just after movement of nozzle on table
- Servo motors are employed for extrusion head movement as the specified geometry.
- Platform is lowered in the z direction just after the completion of one layer in order to start the next layer. Such process is generally continuing till the complete fabrication of substrate.
- Thermal environment is very important control parameter for adequate bonding between the successive layers. It should be maintaining below the melting point [21].
- Materials such as Polylactic acid (PLA), Acrylonitrile Butadiene Styrene (ABS), Polycarbonate (PC), Polyamide (PA), Polystyrene (PS), and lignin, rubber may be applied on FDM also for fabrication of substrate as per design geometrical consideration. [22]. In addition, even the colour of a given thermoplastic material may affect the strength of the printed object
- During the synthesis of substrate, it is verdict that inert gas like nitrogen or argon significantly impacts on layer adhesion capabilities and mechanical properties of the 3D printed objects.

2.4 Process Parameters

The following are the most critical parameters generally found in literatures such as orientation, nozzle temperature (degree Celsius), printing speed (mm/s), infill density, shell thickness and layer thickness etc. In the present investigation, the process parameters are orientation, nozzle temperature, infill percentage and layer thickness.

2.5. Design of experiment

- Nozzle temp is measured as degree centigrade (ranging from 220-230-240) represented by -1, 0, 1.
- Steel ball intender for brinell hardness test was 1.58mm and constant load was 100 kg
- Layer thickness is the layer height of each successive addition of material in AM (ranging from 0.15-0.2-0.25) represented by -1, 0, 1.
- The density of the pattern to be printed is known as infill percentage (ranging from 30-60-90) represented by -1, 0, 1.
- Orientation directly governs the productivity, part quality and cost of manufacturing (ranging from 50-55-60) represented by -1, 0, 1.

2.6 CAD model of bone hammer

Fig. 5 is the CAD model of different parts of bone hammer. It has been designed as per requirement in biomedical field. Lot of questioner's float entire group of patient and doctors then after design factors were decided.

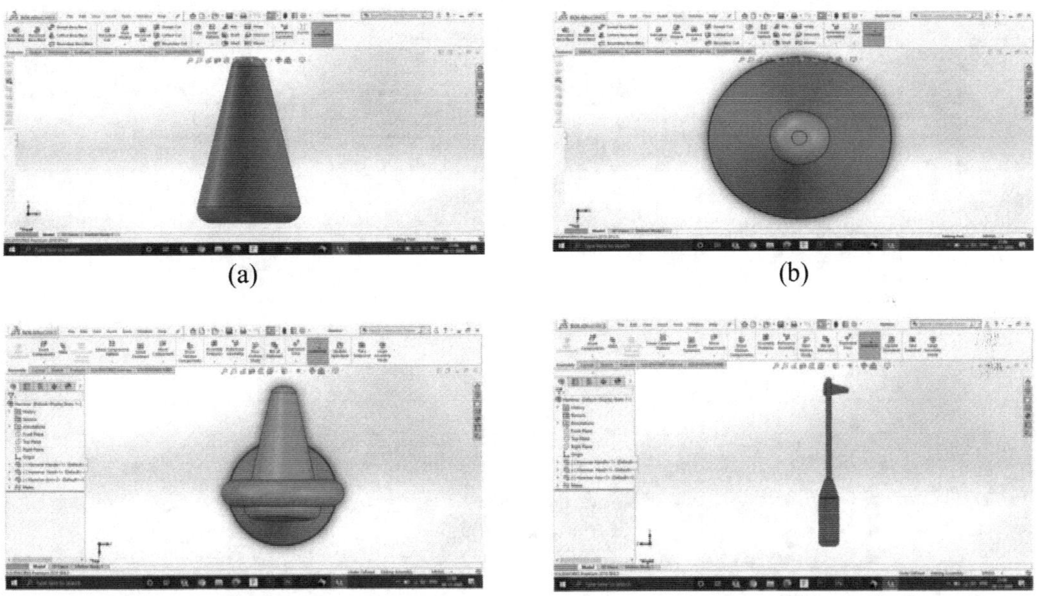

(a) (b) (c) (d)

Figure 5: Different portion of Bone hammer

3. RSM Modelling

- RSM based design of experiments (DOE) was conducted at three level input parameters using FDM 3D printing in different grade of ABS materials.
- Strength and bend test will be conducted on 3D printed bone hammer in each of the specimen. [2, 3]
- Using the experimental data responses have been predicted using RSM modelling.
- RSM is the technique which works based on integration of algorithms [23].
- Predicted and experimental responses were compared for error removal.

Table 2: Experimental Data

Sample No	Layer Thickness (mm)	Infill (%)	Orientation (degree)	Nozzle Temp. (^{0}C)	Deflection on 15 degree	Brinell Hardness (BHN)	Weight (g)
1	0.250	60	60	230	42	96	16.4331
2	0.200	30	50	220	37	86	10.0866
3	0.150	90	50	230	41	83	20.2629
4	0.200	30	60	230	35	16	16.3535
5	0.250	60	50	220	52	86	15.8180
6	0.200	90	50	220	46	34	19.3400
7	0.150	60	50	240	53	97	16.3102
8	0.200	90	55	230	55	100	17.0400
9	0.200	60	50	230	52	100	14.8900
10	0.150	60	50	220	30	97	16.0900
11	0.250	30	50	230	32	57	11.9700
12	0.200	90	50	240	58	89	11.9700
13	0.200	30	55	220	33	105	20.3900
14	0.200	60	55	230	48	96	11.2208
15	0.200	90	60	230	46	101	20.4858
16	0.250	60	50	240	44	104	20.1800
17	0.150	30	50	230	46	103	16.5224
18	0.250	90	50	230	44	87	11.4500
19	0.200	60	50	230	33	90	19.5600
20	0.200	60	60	220	53	90	16.0140
21	0.200	30	55	240	43	89	15.8000
22	0.150	60	60	230	48	88	12.1243
23	0.200	60	55	220	28	93	16.3400
24	0.250	60	60	230	42	100	16.2400
25	0.200	60	50	230	47	89	16.0081
26	0.150	60	55	230	55	100	16.0200
27	0.200	60	60	230	36	81	16.5000

Figure 6 3D printed bone hammer

Bone hammers have been manufactured through FDM 3D printer as per the design of experiment listed in Table 2 which is being expressed in Fig. 6. It has been fabricated on FDM 3D printed bone hammer.

It is very clear from Fig. 7 that bone hammer weight increasing with increase of layer thickness, infill percentage and nozzle temperature. Hardness is also proportional to the nozzle temperature. Deflection of bone hammer is also proportional to the infill percentage and nozzle temperature. All the interaction plots (Fig. 7) are self-explanatory and verifying the main effect plots. It is evident from Fig. 7 that high nozzle temp.; infill percent and low layer thickness are responsible to increase weight of bone hammer. However, hardness increase with low thickness of layer at high nozzle temperature also high deflection occurs at low infill percentage [24-30].

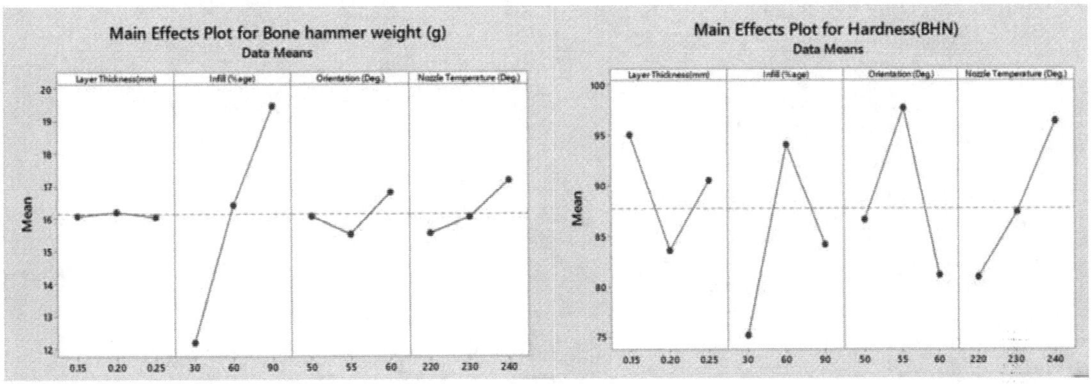

(a)

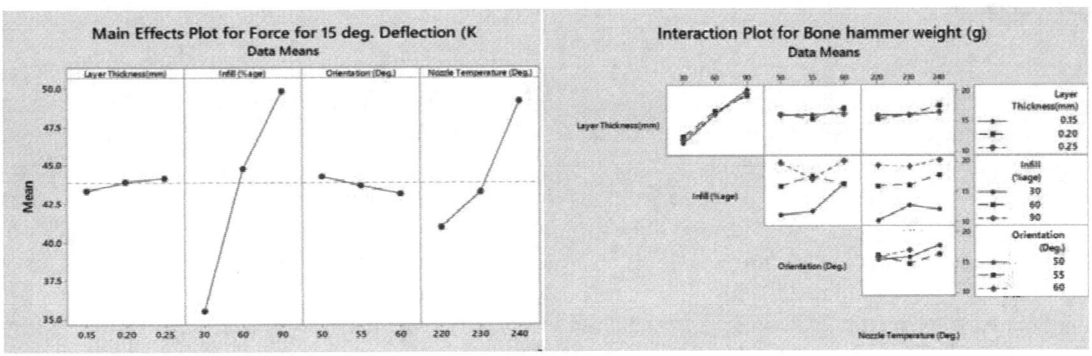

(b)

Figure 7: Main effect plot of bone hammer process parameters

4. Optimization of process parameters

Fig. 8 indicates that optimal process parameters layer thickness 0.250, infill percent 63.3333, orientation 60 and nozzle temperature 240 are the best suited to produce quality bone hammer with economical consideration. Present optimization plot has been plotted on composite desirability 'D' 0.8795 which indicates the probability to achieve such responses. It was again tested with conducting experiment by considering optimal parameters; the result was almost same as having bone hammer weight 20.3045g, hardness 105.5653 BHN, and force at 15 degree deflection 35.4471 N with 1.64 percent error.

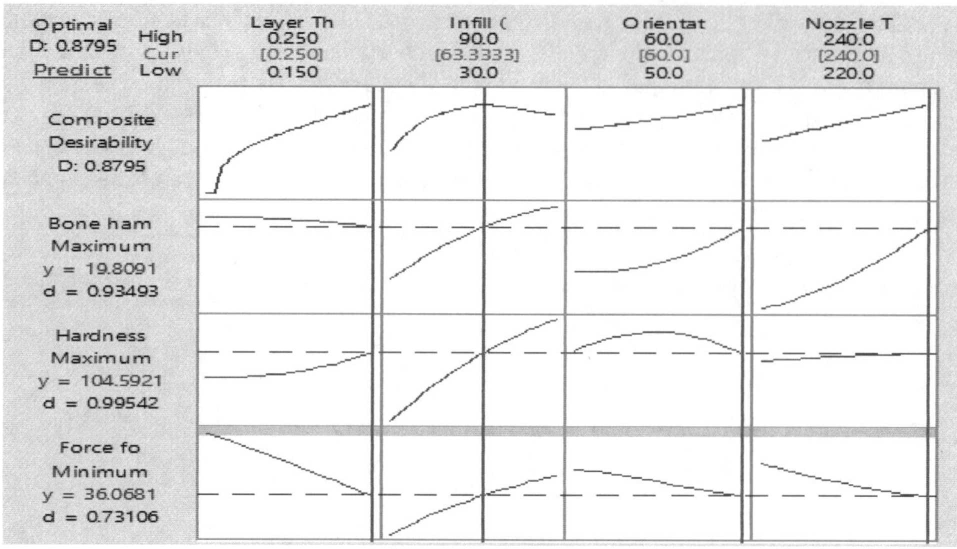

Figure 8: Optimization plot

5. Conclusions

It is evident that 3D printing technology is the alternate way throughout product development which can reduce design risk and, result in better quality of products. Main competitor materials of 3D printing are ABS and PLA. Unlike PLA, ABS is a renewably derived plastic. It is therefore biodegradable and biocompatible whereas PLA is only biocompatible. However, ABS is 100% recyclable, non-toxic and harmless. Best quality of bone hammer responses (having bone hammer weight 19.8091g, hardness 104.5921 BHN, and force at 15 degree deflection 36.0681 N has been produced with RSM prediction with influence of process parameters (layer thickness 0.250, infill percentage 63.3333, orientation 60 degree, nozzle temperature 240°C). RSM is the best suitable technique through which further prediction may be done for biomedical equipment manufacturing. Response surface methodology (RSM) allows researcher to approximate the behavior of a process in the vicinity of the optimum. Rapid prototyping, on the same platform as production, aids quick progression of novel designs and features. Economically feasible to print multiple custom designs in one build [31-35].

References

[1] El-Gizawy AS, Cardona J, Graybill B 2010. An integrated approach for characterization of properties and meso-structure of fused deposition modeling ULTEM 9085 products. *International SAMPE Symposium and exhibition.*

[2] Kanu,N.J; Gupta,E; Vates, U.K; and Singh,G.K; 2019 An insight into biomimetic 4D printing, *RSC Adv.*, 9, 38209

[3] Ahn SH, Montero M, Odell D, Roundy S, Wright PK. 2002 Anisotropic material properties of fused deposition modelling ABS. *Rapid Prototyping Journal;* p. 248–257.

[4] Daniel, N.A; Singh,N.K; Vates,U.K; Sharma, B.P; and Sivaraos 2019 Optimization of Critical Parameters of EDD Steel in Die Cavity Manufacturing, *Advances in Industrial and Production Engineering, Springer Nature Singapore Pte Ltd., Lecture Notes in Mechanical Engineering*, pp 357-363.

[5] Saqib S, Urbanic, R.J.2011 An experimental study to determine geometric and dimensional accuracy impact factors for fused deposition modelled parts. *4th International Conference on Changeable Agile, Reconfigurable and Virtual Production;* p. 293-298.

[6] Alexander, P., Allen, S., Dutta, D., 1998. Part orientation and build cost determination in layered manufacturing. *Computer-Aided Design, Vol 30, No 5,pp,343-356.*
[7] Cz. Kundera, T. Kozior 2014, Research of the Elastic Properties of Bellows Made in SLS Technology, *Advanced Materials Research,* Volume 874 pp. 77-81.
[8] Bonten C. Kunststofftechnik: Einfuehrung und Grundlagen 2021, Characterization of FDM printed product: A review, polymers, 124 (2) p. 252-264.
[9] Khangarot H, Sharma, S, Vates,U.K, Singh. GK, & Kumar, V; 2018. "Optimization of Surface Grinding Process Parameters Through RSM" *Proceedings of the International Conference on Modern Research in Aerospace Engineering* pp 291-302, Print ISBN: 978-981-10-5848-6, Online ISBN: 978-981-10-5849-3, Springer, Singapore.
[10] Turner BN, Strong B, Gold SA. 2014 A review of melt extrusion additive manufacturing processes: I. Process design and modeling. *Rapid Prototyping Journal*;20/3:192-204.
[11] Rodriguez JF, Thomas JP, Renaud JE.2001 Mechanical behavior of acrylonitrile butadiene styrene fused deposition materials. Experimental Investigation. *Rapid Prototyping Journal;* p.148-158.
[12] Cz. Kundera, 2013, Seals of rotating systems, *Monograph M53, Kielce University of Technology, Kielce,* p. 286-294.
[13] Cz. Kundera, J. Bochnia 2014, Investigating the stress relaxation of photopolymer O-ring seal models, *Rapid Prototyping Journal,* 20 (6) 533-540.
[14] Inspekt Mini, 2011 Universal Testing Machine Inspekt mini 3kN - *User's Guide, Hegewald & Peschke, Meß- und Prüftechnik GmbH 5- p.1-478.*
[15] Villalpando L.2013 Characterization of parametric internal structures for components built by fused deposition modeling. Electronic Theses and Dissertations: University of Windsor.
[16] Agarwala MK, Jamalabad VR, Langrana NA, Safari A 1996, Whalen PJ, Danforth SC. Structural quality of parts processed by fused deposition.Rapid Prototyping Journal p. 4-19.
[17] L. Koch, A. Deiwick and B. Chichkov, 2018 Laser-Based Cell Printing, 3D Printing and Bio fabrication, Springer International Publishing, Cham, pp. 303–329.
[18] Hopkinson, N., Dickens, P., 2003. Analysis of rapid manufacturing using layer manufacturing processes for production. Proc.Instn Mech.Engrs Vol.217 Part C.
[19] Anitha R, Arunachalam S, Radhakrishnan P. 2001. Critical parameters influencing the quality of prototypes infused deposition modelling.Journal of Materials Processing Technology.p. 385-388.
[20] J. Skrzypek,1986, High Strain Rate Thermo-Inelasticity of Damaged AISI 316H, Plasticity and creep, PWN, Warsaw p. 628-637.
[21] Sahu, K.L; Kumar,N; Singh,A; Daniel, N A; and Vates, U.K; 2020 'Investigation of Laser- MIG hybrid welding performances in Al alloys with influence of Ar-He-Ne mix shielding' Review' 2nd International Conference on Future Learning Aspects of Mechanical Engineering (FLAME) Department of Mechanical Engineering Amity University Uttar Pradesh (India)
[22] ISO 3384: 2002, Rubber, vulcanized or thermoplastic – Determination of stress relaxation in compression at ambient and at elevated temperatures.
[23] UK Vates, N.A.Daniel, Dr.N.K.Singh, Parametric Optimization & Simulation of Mild Steel Cup in Deep Drawing using LS-Dyna, International Journal of Applied Engineering Research 10 (9), 24479-24489, 2015.
[24] Ayushi, UK Vates, S Mishra, NJ Kanu, Biomimetic 4D printed materials: A state-of-the-art review on concepts, opportunities, and challenges. Materials Today: Proceedings. 47 (2021) 3313-3319. https://doi.org/10.1016/j.matpr.2021.07.148.
[25] NJ Kanu, Modeling of stress wave propagation in matrix cracked laminates. AIP Advances. 11 (2021) 085217. https://doi.org/10.1063/5.0057749.
[26] NJ Kanu, D Patwardhan, E Gupta, UK Vates, GK Singh, Numerical investigations of stress-deformation responses in fractured paediatric bones with prosthetic bone plates. IOP Conference Series: Materials Science and Engineering. 814 (2020) 012038. https://doi.org/10.1088/1757-899X/814/1/012038.

[27] A Lal, NJ Kanu, The nonlinear deflection response of CNT/nanoclay reinforced polymer hybrid composite plate under different loading conditions. IOP Conference Series: Materials Science and Engineering. 814 (2020) 012033. https://doi.org/10.1088/1757-899X/814/1/012033.
[28] NJ Kanu, E Gupta, UK Vates, GK Singh, Self-healing composites: A state-of-the-art review. Composites Part A: Applied Science and Manufacturing. 121 (2019) 474-486. https://doi.org/10.1016/j.compositesa.2019.04.012.
[29] N Jain, E Gupta, NJ Kanu, Plethora of Carbon nanotubes applications in various fields–A state-of-the-art-review. Smart Science. (2021). https://doi.org/10.1080/23080477.2021.1940752.
[30] NJ Kanu, S Bapat, H Deodhar, E Gupta, GK Singh, UK Vates, GC Verma, V Pandey, An Insight into Processing and Properties of Smart Carbon Nanotubes Reinforced Nanocomposites. Smart Science. (2021). https://doi.org/10.1080/23080477.2021.1972913.
[31] CM Asre, VK Kurkute, NJ Kanu, Power generation with the application of vortex wind turbine. Materials Today: Proceedings. (2021). https://doi.org/10.1016/j.matpr.2021.08.228.
[32] MA Kumbhalkar, KS Rambhad, NJ Kanu, An insight into biomechanical study for replacement of knee joint. Materials Today: Proceedings. 47 (2021) 2957-2965. https://doi.org/10.1016/j.matpr.2021.05.202.
[33] M.A. Kumbhalkar, D. T. Rangari, R. D. Pawar, R. A. Phadtare, K. R. Raut, A. N. Nagre, "Finite Element Analysis of Knee Joint with Special Emphasis on Patellar Implant", Trends in Mechanical and Biomedical Design" in book series of Lecture Notes in Mechanical Engineering, pp 319-333, 2020. ISBN 978-981-15-4487-3. https://doi.org/10.1007/978-981-15-4488-0_29
[34] M A Kumbhalkar, P H Jaiswal, H M Bansod, "Design and Manufacturing of Knee Joint using CAD/CAM and Rapid Prototyping" published in Journal of The Institution of Engineers (India), Mechanical Engineering, 25-28, Volume 92, April 18, 2011.
[35] M. A. Kumbhalkar, Umesh Nawghare, Rupesh Ghode, Yogesh Deshmukh, Bhushan Armarkar "Modeling and Finite Element Analysis of Knee Prosthesis with and without Implant", published in Universal Journal of Computational Mathematics, Vol. 1(2), Horizon Research Publishing, USA, pp. 56–66, 2013. https://doi.org/10.13189/ujcmj.2013.010204 24.

Study of effects of weathering on natural fiber composites

Satadru Kashyap[1], Debanil Das[1]

[1]Department of Mechanical Engineering, Tezpur University, Sonitpur, Assam - 784028 (India)

Email: satadru@tezu.ernet.in

Abstract. Composite materials are known to have many advantages and they can be used in the manufacture of a wide variety of products which make them commercially valuable. However, they may come into contact with various conditions that might lead to weathering. Weathering might lead to discoloration, decrease in durability, decline in mechanical properties and decrease in mass in composites. Hence, it is paramount to study the effects of weathering on composites before they are used in service. For this study, wood polyester composite (WPC), jute polyester composite (JPC) and coir polyester composite (CPC) were fabricated and subjected to four weathering tests - water ageing test at room temperature, water immersion test at 100°C, exposure to external environment and exposure to UV rays. It was seen that CPC samples subjected to water ageing at room temperature and immersion at 100°C showed comparatively higher water absorption rate than the WPC and JPC samples as coir fibres in CPC are hollow at the centremost region. JPC and CPC also turned whitish after 24 hours of water ageing. However, exposing the samples to the environment for a period of 144 hours did not show any significant change in the sample. It was also found that the tensile strength of all the composites decreased after being exposed to UV rays in a QUV spray weathering tester. Additionally, the mechanical properties of the composites were also modelled with analytical techniques (Halphin Tsai method) and FEM analysis and the results were found to be analogous.

Keywords – natural fibre, composites, polyester resin, weathering, mechanical properties.

1. Introduction

Composite materials are combination of two materials which can be custom tailored as per requirement. They have high fatigue life, high strength to weight ratio, high specific strength and stiffness, durability, [1-2] etc. Hence, they are widely used in the manufacture of many modern products making them commercially valuable.

However, composite materials on coming in contact with certain conditions like harsh climatic condition, exposure to ultraviolet radiation [3-4], absorption of large quantity of water, exposure to chemicals such as acids, bases [4-7], etc may experience weathering. Due to weathering, various harmful effects are seen in composites such as decrease in durability and life span, discolouration [8-9], decrease in tensile strength, flextural strength, hardness, interlaminar shear strength of composites [3,9-11], loss of mass [12-13], etc. Furthermore weathering tends to cause more harm in natural fiber based composites than artificial fiber based composites.

Natural fibers such as coir, jute etc are some of the most affordable fibers and are abundantly available in most parts of India. They are composed primarily of cellulose and lignin. They are usually used in the manufacture of cloth, carpets, bags, ropes, wrapping clothes, mats, vehicle seats, etc.

Numerous researchers have investigated and contributed their valuable findings on the ill effects of weathering on composite materials. Butylina et al. [8] studied the effects of external weathering on wood-polypropylene composites. The surface changes of the composites, colour fading, higher moisture absorption and decrease in charpy impact strength were the various effects that were observed in the study. Seki et al. [14] observed that the water aging test decreases the inter-laminar shear strength on jute polyester composite. Dash et al. [15] found that the flexural strength, tensile strength and inter-laminar shear strength of the composites lowered extensively upon their exposure to external weathering and water. Yan, L. et al. [11] found that accelerated weathering of flax-FRP composites decreases the tensile modulus and strength of the weathered composites by 34.9% and 29.9%, respectively. Amaro et al. [6] studied the effects of alkaline and acid solutions like sodium hydroxide (NaOH) and hydrochloric acid (HCl) on glass/epoxy composites and found that they decreased the flexural strength and modulus of the composite with the exposure time. Peng et al. [16] showed that the surface properties of polypropylene composites reinforced with cellulose, lignin and wood flour on being exposed to UV weathering, accelerated the discoloration of composites due to presence of lignin and decreased the flexural properties. Vates, U. K et al. [17] performed a study which involved composites of aluminium alloy metal matrix composites reinforced with reinforced with nitinol particulates (NiTip) ranging from 0%, 4%, 8%, and 12%. The results showed that the surface roughness and hardness of the AAMMC increased with an increase in the percentage of NiTip. This was attributed due to the superior matrix-reinforcement interfacial bonding. Devi et al. [19] in their study which included ageing of pineapple leaf fiber–reinforced polyester composites, reported that the absorption of water in fiber reinforced composite is due to the capillary phenomenon. Furthermore it was also reported that there was decrease in tensile strength of the composite.

In this study, three different types of composites, *viz.* wood polyester composite (WPC), jute polyester composite (JPC) and coir polyester composite (CPC) were fabricated to study their mechanical properties. Thereafter, the ill effects of weathering on the fabricated composites was studied using different techniques. Additionally, analytical and FEM models were developed to model the mechanical properties of the natural fiber composites.

2. Materials and Methodology

2.1. Materials

Three types of composite *viz.* wood polyester composite, jute polyester composite and coir polyester composites were prepared for this research. Polyester resin was used as the matrix material. Polyester resin was procured from Carbonblack Composites, Mumbai. Methyl ethyl ketone peroxide (MEKP) was used as hardener and cobalt accelerator was used as a catalyst during fabrication. Reinforcements such as wood dust, jute and coir fiber were procured locally. The reinforcements were cleaned and dried under sunlight for 4 days. The fibers (jute, coir) were chopped into smaller fragments of 14 mm each. The wood dust was ground and sieved to obtain finer quality wood dust.A MS mold used for fabrication process was prepared having dimensions of 140 mm x 140 mm x 5 mm and samples were cut from the bigger composite plate. Composites were fabricated by hand-layup technique and a weight of 150 kg was used over the mold during the curing process. Images of fabricated composites are shown in figure 1. For fabricating the composites, 15% wt. fraction of reinforced materials (jute, coir, wood dust) was used.

Figure 1. Composite panels of WPC, JPC and CPC.

2.2. Mechanical measurements

Composites were subjected to tensile testing by Universal Testing Machine UTM (model: Hounsfield H100K-S) of capacity 100 kN. The dimensions of the specimen were taken as 115 mm x 10 mm x 5 mm as per ISO 527 standards. Tensile testing was done both before and after exposing the samples to QUV weathering. The strain rate used was 2 mm/min.

2.3. Weathering tests

2.3.1. Water ageing in normal distilled water. For this study the fabricated composite samples were initially dried under sunlight for a period of 96 hours or 4 days. The samples from each type of composites were then immersed in three different containers containing distilled water for a period of

336 hours or 14 days. The initial weight of each sample was noted. The change in weight and change in surface of the samples was monitored every 24 hours. Photographs of the samples were also taken every 24 hours to monitor the surface and colour changes.. Percentage of water content (W.C.) is calculated using Eq. (1):

$$W.C = \frac{W_w - W_D}{W_D} \times 100\% \quad (1)$$

W_W= Weight of the wet specimen (gm)
W_D= Weight of the dry specimen (gm)

2.3.2. Water ageing at boiling temperature and gradual cooling. For this study the fabricated composite samples were initially dried under sunlight for a period of 96 hours or 4 days. Samples were taken and initial weight was measured. The samples were then submerged in a container with distilled water at 100°C. After 9 hours,samples were taken out, wiped with a clean cloth and weight was measured again. Percentage of water content (W.C) is calculated using Eq. (2):

$$W.C = \frac{W_w - W_D}{W_D} \times 100\% \quad (2)$$

W_W= Weight of the wet specimen (gm)
W_D= Weight of the dry specimen (gm)

2.3.3. Exposure to external environment. The fabricated composite samples were placed outdoors in a region where they could easily come into contact with different environmental conditions like rainfall, sunlight, dew, moisture, etc. The weight of these samples were recorded every 24 hours on a weighing scale. Photographs of the samples were also taken every 24 hours to monitor the surface and colour changes.

2.3.4. Exposure of samples to QUV spray weathering tester. The composite samples were exposed to QUV spray weathering tester. The test was performed under ASTM G154 standards wherein the samples were exposed to 8 hours of UV radiation at 70° ± 3° black panel temperature followed by condensation for a period of 4 hours at 50° ± 3°C black panel temperature in a cyclic pattern. The fabricated samples were exposed in the QUV weathering tester for a total period of 500 hours (21 days). QUV spray weathering tester simulates the outdoor weathering by exposing the samples of composite materials to alternating UV light and moisture at controlled temperature. UV light simulates effects of sunlight and rainfall and dew is simulated by means of condensing humidity or spray of water.

2.4. Analytical analysis

Analytical analysis was carried for JPC and CPC samples. Table 1 shows the elastic properties of the natural fibers and polyester resin that were used for the purpose of fabrication. Analytical analysis was carried out by a semi empirical method called Halphin-Tsai method [2]. The longitudinal Young's modulus, E_1, is calculated as per rule of mixture given by the Eq. (3):

$$E_1 = E_{1(f)}V_f + E_m V_m \quad (3)$$

In Eq.(3), E_m and $E_{1(f)}$ are the modulus of elasticity of matrix and fiber respectively and V_m and V_f are volume fraction of matrix and fiber respectively. The transverse Young's modulus, E_2 and E_3 is calculated from the Eq.(4):

$$\frac{E_2 or E_3}{E_m} = \frac{1+\xi\eta V_f}{1-\eta V_f} \quad (4)$$

Where, $\eta = \frac{\gamma - 1}{\gamma + \xi}$, $\gamma = \frac{E_{f2} or E_{f3}}{E_m}$

In Eq.(4), ξ is known as the reinforcing factor and depend on packing geometry, fiber geometry, loading conditions. In transverse direction, for circular or rectangular fiber, ξ is given as:

$\xi = 2\times(wt) + 40\times(V_f)^{10}$.

Table 1. Elastic properties of jute, coir and polyester resin.

Property	Jute [18]	Coir[18]	Polyester resin
Longitudinal Young's Modulus, E_1 (MPa)	39618	4600	3300
Transverse Young's Modulus, E_2, E_3 (MPa)	5500	4700	3300
Axial Shear Modulus, G_{12}, G_{13} (MPa)	7124.1	1760	1253
Transverse Shear Modulus, G_{23} (MPa)	2137	1760	1253
Major Poisson's Ratio, ϑ_{12}, ϑ_{13}	0.11	0.30	0.316
Minor Poisson's Ratio, ϑ_{23}	0.35	0.30	0.316

2.5. FEM Analysis

A 3D FEM modelling of the fiber composite was done in ANSYS APDL 18. 8 noded SOLID 185 was chosen as the element type for modelling of the fabricated fiber reinforced composites.

The element SOLID 185 is defined by eight nodes and has 3 degrees of freedom at each of the nodes, translations in the nodal x, y and z direction. The model is considered to be orthotrotropic in nature. The length, width and thickness of the specimen are 115 mm, 10 mm and 5 mm respectively. Two number of ply with thickness 2.5 mm each are taken with orientation of (0°/0°). One end of the sample is fixed in all direction; the other end of the sample is also fixed in all direction except in longitudinal direction where loading is applied.The applied boundary condition on the composite is shown in figure 2(a), and the meshing is shown in figure 2(b).

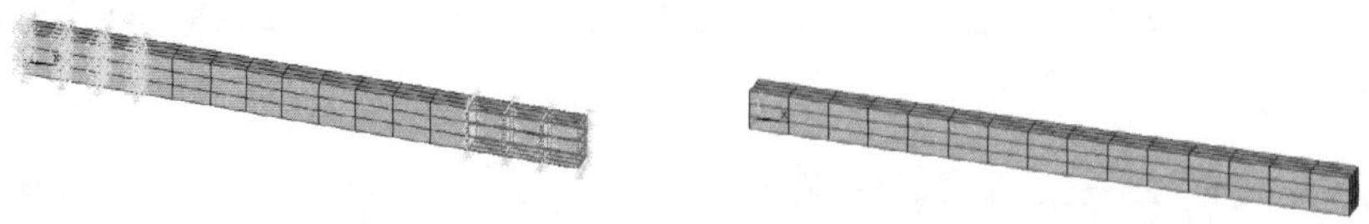

(a) (b)

Figure 2. (a) Boundary condition for tensile test; (b) Meshing on composite material.

3. Results and Discussion

3.1. Weathering tests

Four different types of weathering tests were conducted to investigate the effects of various environments on the three different types of composites. It was observed that weathering adversely affected the mechanical properties of the composites in addition to causing visible physical changes in the composites.

3.1.1. Water ageing at room temperature. Containers were filled with distilled water at room temperature and the fabricated composite samples were submerged in them for a period of 336 hours or 14 days. The water absorption rate was calculated every 24 hours. Table 2 and Figure 3 shows the rate of water absorption of the fabricated. It was observed that change in rate of water absorption is highest for the initial twenty four hours and then the rate steadily increases to a point of saturation. CPC samples shows the highest water absorption rate followed by JPC and WPC respectively. The reason for higher water absorption in CPC and JPC can be explained by the fact that coir and jute fibers are hydrophilic in nature i.e. they have a tendency to absorb water. Furthermore, coir and jute fibers are cellulose based fibers with hollow central regions which can hold water. Thus, rate of water absorption for CPC and JPC was higher. Slight damage to the surface of the composite was seen for the fiber reinforced composite but the same was not the case for WPC. Deposition of slight whitish material was seen for CPC and JPC samples however, no noticeable change in dimension of the samples was observed. Devi et al.reported the absorption of water in fiber reinforced composite is due to the capillary phenomenon [19]. These fibers were reported to swell with absorption of water (increasing the weight with increasing fiber loading) which resulted in leading to delamination and debonding.

Table 2. Change in weight of fabricated samples due to water ageing.

Material	Sample	Initial Weight (gms)	Final Weight (gms)	% Water Content
Jute Polyester Composite	JPC 1	7.01	-	-
	JPC 2	6.87	7.36	7.13
	JPC 3	7.1	7.43	4.64
Wood Polyester Composite	WPC 1	7.26	-	-
	WPC 2	6.88	7.11	3.63
	WPC 3	7.19	7.42	3.47
Coir Polyester Composite	CPC 1	8.27	-	-
	CPC 2	7.69	8.46	10.01
	CPC 3	7.3	8.06	10.41

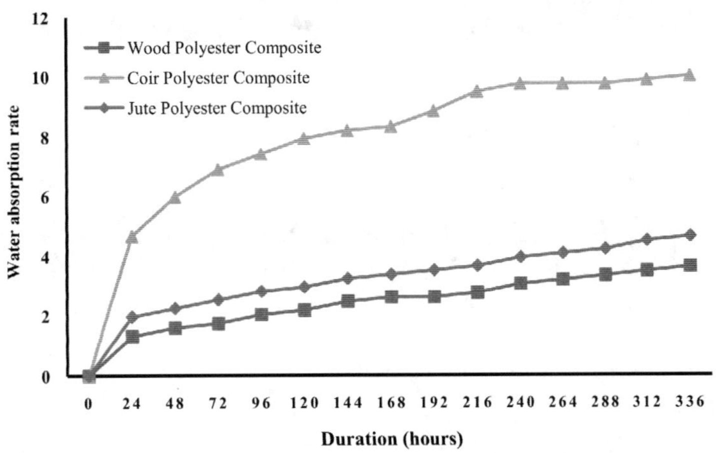

Figure 3. Water absorption rate of fabricated samples due to water ageing.

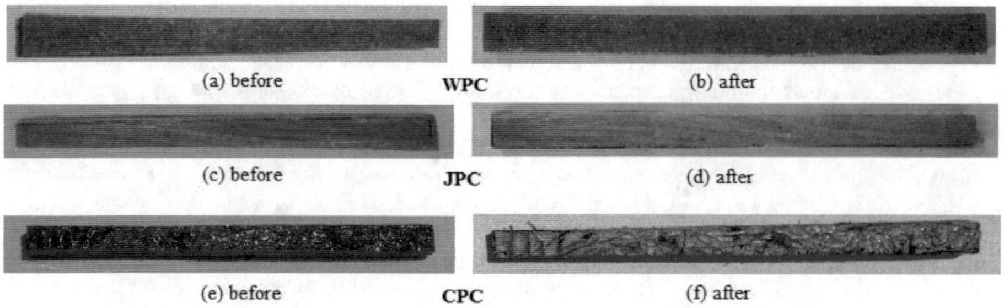

Figure 4. Before water ageing and after water ageing for 14 days images of (a-b) WPC, (c-d) JPC, (e-f) CPC samples.

Figure 4(a) and 4(b) shows the WPC samples before and after water ageing in distilled water for 14 days. It is clearly seen that there was no visible and dimensional changes after water ageing. In case of JPC samples (figures 4(c) and 4(d)), it was observed that there was a slight whitish deposition on the composite upon water ageing. However, in case of CPC samples (figures 4(e) and 4(f)), a high deposition of a whitish material takes place. The fibers are also seen to come out of the laminate in certain areas along with some swelling in certain areas.

3.1.2. Water ageing at boiling temperature and gradual cooling. To study the effects of hot and moist environmental conditions on the composite samples, boiling water ageing tests of the composite samples was conducted. This test was carried on for a period of 9 hours in distilled water at 100°C and the samples were allowed to gradually cool down.

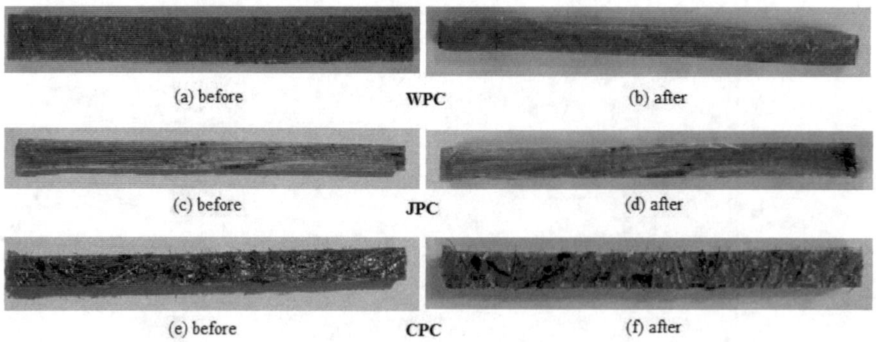

Figure 5. Before boiling water ageing and after boiling water ageing images of (a-b) WPC, (c-d) JPC, (e-f) CPC samples.

Figures 5(a) and 5(b) shows WPC samples before and after being exposed to water ageing at boiling temperature. It was observed that the samples after the end of the test showed surface damage. Swelling of the wood polyester composite sample was also seen. Figures 5(c), 5(d), 5(e) and 5(f) shows JPC and CPC samples respectively before and after being exposed to water ageing at boiling temperature for a period of 9 hours. It is seen that the samples after the end of exposure period showed surface damage along with swelling and white powdery deposition on the surface.

Table 3. Change in weight of fabricated samples subjected to distilled water at 100°C.

Composite	Weight (gms) 0 hours	9 hours	% Water Absorption
Wood Polyester Composite	7.83	7.91	1.02
Jute Polyester Composite	6.98	7.12	2.005
Coir Polyester Composite	7.68	8.11	5.59

Table 3 shows the water absorption rate of the samples of composites on being immersed in boiling water. It is seen that in this case too, rate of water absorption was highest for coir polyester composite followed by jute polyester composite and wood polyester composite but unlike in normal water the rate of water absorption of samples immersed in boiling water was comparatively higher.

3.1.3. Environmental weathering. To study the effects of environmental weathering on the fabricated composite materials, the samples are placed outside in a region where it can get proper exposure to sunlight, rainfall, moisture, dew etc. No significant changes were observed in the composite samples after environmental weathering of 144 hours. This is probably because a longer amount of time will be required to properly note the effects of environment on the samples.

Table 4. Change in weight of fabricated samples subjected to environmental weathering.

Material	Weight (gm) 0 hours	24 hours	48 hours	72 hours	96 hours	120 hours	144 hours
Jute Polyester Composite	7.03	7.03	7.04	7.02	7	7.03	7.02
Wood Polyester Composite	8.11	8.12	8.12	8.1	8.11	8.13	8.11
Coir Polyester Composite	7.63	7.58	7.57	7.56	7.56	7.57	7.57

3.1.4. Tensile strength before QUV weathering. The fabricated samples of JPC, CPC and WPC were subjected to tensile test to investigate their mechanical properties. Table 5 shows the results of the tensile testing. It was seen that the tensile strength is maximum for JPC, followed by CPC and WPC respectively. JPC and CPC are both fiber reinforced composites with stronger tougher which are able to withstand maximum loading in the direction of their fibres, thereby improving the property of the composite. However, the same is not the case for WPC as it is a particulate reinforced composite and depends mostly on the interfacial adhesion at the particulate-matrix boundary. Therefore, it was observed to have a lower tensile strength than JPC and CPC respectively.

Table 5. Tensile strength of the composites before weathering.

SI No.	Tests Method	Units	Results Obtained WPC	JPC	CPC
1	Tensile Strength (before weathering)	N/mm^2	22.46	81.7	15.66
			12.29	72.9	18.07
			10.46	71.8	16.86
			AVG: 15.07	AVG: 75.47	AVG: 16.86

3.1.5. Tensile test after exposure to QUV Weathering. After performing the QUV weathering test, JPC, CPC and WPC samples were again subjected to tensile test to investigate the effects of weathering on the mechanical properties (refer table 6). It was revealed that after 500 hours (21 days) of continuous exposure to ultraviolet/condensation cycle in the QUV spray weathering testing machine, the tensile strength of all the samples decreased. The decrease in tensile strength was seen to be maximum for JPC followed by CPC and WPC respectively. The decrease in tensile strength after performing QUV weathering test may be due to the degradation of the fibres as well as damage in the polyester matrix of the composite. Furthermore, due to this exposure, the bonding between the matrix and reinforcements also weakened. Thus the load carrying capacity of the composites decreased which in turn led to reduction in tensile strength (mechanical properties).

3.2. Analytical results.
The elastic constants of the fabricated 2 ply single fiber composite samples *viz.* JPC and CPC have been calculated by means of the semi empirical Halphin-Tsai method. The details of the results obtained have been discussed below.

3.2.1. Effect of fiber loading on composite materials. The elastic constants of the JPC and CPC samples *viz.* Longitudinal Young's Modulus (E_1), Transverse Young's Modulus (E_2, E_3), Axial Shear Modulus (G_{12}, G_{13}), Transverse Shear Modulus (G_{23}) and major Poisson's Ratio (ϑ_{12}, ϑ_{13}) were calculated by Halphin-Tsai method. Fiber weight fraction was varied from 15% to 50% and the change in elastic constants was observed. The results were plotted graphically from Figures 6 (a-e).

Table 6. Tensile strength of the composites after 500 hours of QUV weathering.

Sl No.	Tests Method	Units	Results Obtained		
			WPC	JPC	CPC
1	Tensile Strength (After 500 hours UV Exposure) (at 2mm/min)	N/mm^2	15.81	59.8	9.6
			13.36	65.7	9.83
			15.05	75.5	12.46
			AVG: 14.74	AVG: 67.0	AVG: 10.63

The value of longitudinal young's modulus, E_1 has been calculated by application of rule of mixture method while the other elastic constants were calculated through Halphin Tsai method. It is seen that for the fabricated 2 ply fiber reinforced composites, the elastic properties (E_1, E_2, E_3, G_{12}, G_{13}, G_{23}) increases with the increase in fiber loading for both JPC and CPC. Furthermore, it is also seen that the elastic constants other than major poisson's ratio, for jute polyester composite are better than coir polyester composites. This is due to the fact that jute fibers have inherently better elastic properties than coir fibers. Thus using jute fiber instead of coir fiber as reinforcement material for fabrication of composite will give better elastic properties to the composite. The major Poisson's ratio was however found to decreases with increases in fiber weight fraction for both JPC and CPC.

3.3. FEM modelling
3D solid models of the JPC and CPC samples were prepared in ANSYS APDL 18.1. The dimension of the specimen was taken as per the size of the samples. The FEM models of tensile testing for JPC and CPC samples are shown in Figure 7 (a-b). The experimental results of the tensile test (refer table 6) for the composites and the FEM analysis in ANSYS were found to be approximately similar. However a slight variation in the results can be seen. The variation is shown in Table 7.

Table 7. Comparison of the values of tensile test experimentally and via FEM.

Material	Experimental (N/mm^2)	FEM (N/mm^2)
JPC	75.47	76.39
CPC	16.86	17.39

4. Conclusion
Composite materials are constituted of two or more materials and they have better physical and chemical properties than either of their constituent materials. As such, they have many advantages and thus they find their application in a wide range of field. Due to excessive use in different conditions, these materials may come into contact with various natural as well as artificial agents that may lead to

weathering in composites which in turn might lead to their degradation, discoloration and further deterioration in elastic and mechanical properties.

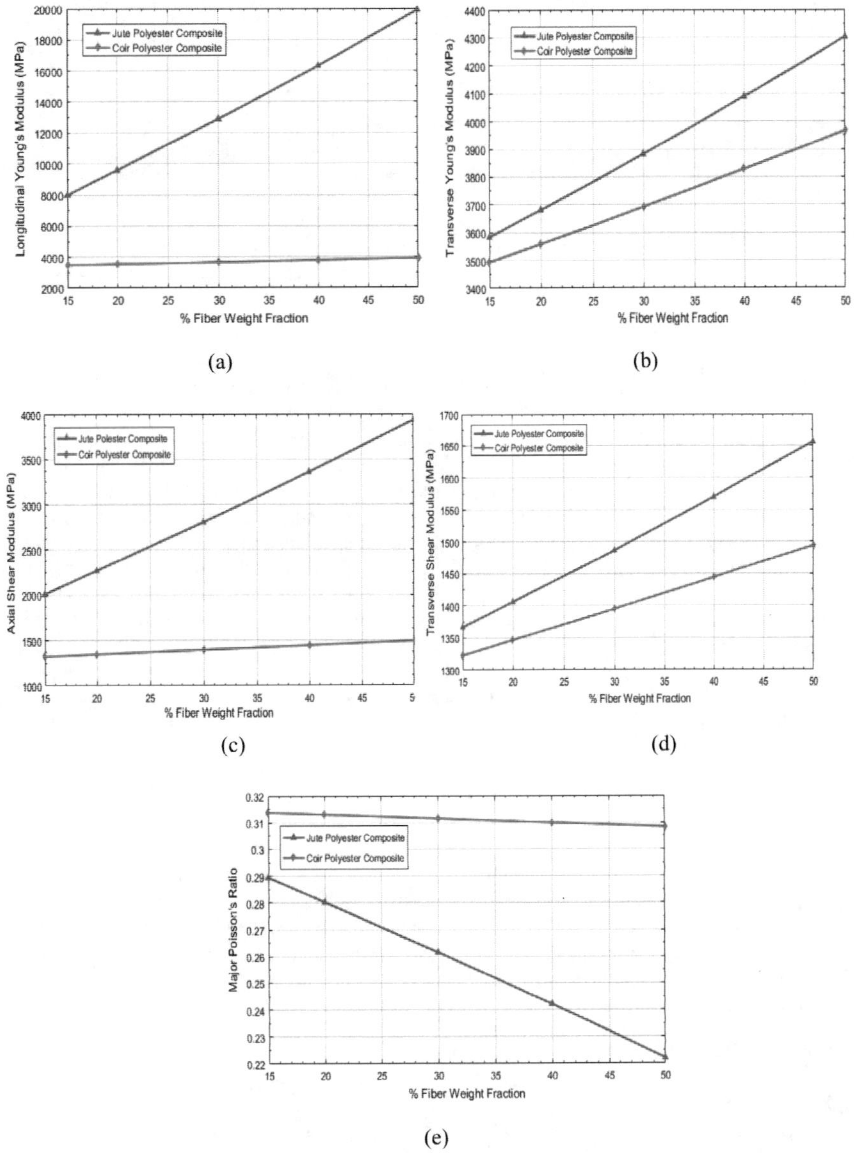

Figure 6. Graphical representation of the elastic constants of JPC and CPC (a) longitudinal Young's Modulus (E_1), (b) Transverse Young's Modulus (E_2, E_3), (c) Axial Shear Modulus (G_{12}, G_{13}), (d) Transverse Shear Modulus (G_{23}) and (e) major Poisson's Ratio(ϑ_{12}, ϑ_{13}) calculated by Halphin Tsai method.

In this study, three different composites viz. WPC, JPC and CPC were fabricated and exposed to various weathering tests like water ageing, immersion and subsequent cooling of samples in boiling water, QUV accelerated weathering, etc. to test their durability, longevity, change in mechanical properties and to understand their behavior under different weathering conditions.

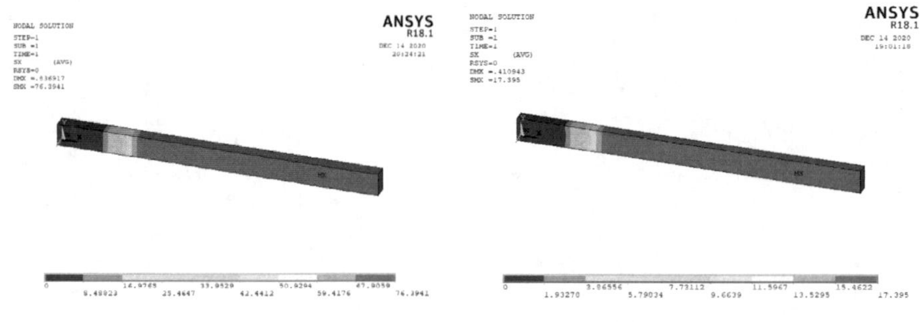

(a) (b)

Figure 7. FEM model of tensile testing (a) JPC and (b) CPC.

It was observed that the properties of the composites degraded upon weathering. Due to weathering ill effects like swelling, increased water absorption rate, whitish surface deposition was seen in the composites samples. Water ageing test at room temperature revealed that rate of water absorption was highest in CPC which was followed by JPC and WPC. Furthermore the samples exposed to accelerated weathering showed a significant drop in tensile strength, this can be due to degradation of the fibres as well as damage in the polyester matrix of the composite as a result of exposure to accelerated weathering. The bonding between the matrix and reinforcements also weakened and therefore the load carrying capacity of the composites decreased which in turn led to reduction in tensile strength (mechanical properties).

Composite materials have lots of advantages and thus they find their application in manufacture of a large number of products but they have certain limitations. Composites on being exposed to certain conditions like water ageing, prolonged exposure to UV radiation, hot and humid environmental conditions etc. may eventually experience deterioration in mechanical properties and physical damage in the surface of the composite thereby decreasing longitivity and durability. Analysis of the different elastic constants of the composites have shown that with the increase in fiber loading of the composites, the elastic constant of the fiber composites also improve.

This can be ascertained due to the fact that fibers impart further strength to the matrix materials after being fabricated as a composite. The elastic properties of jute being better than coir, results in better elastic properties of the jute polyester composite than coir polyester composite.

A 3D FEM model of the fabricated sample was done in ANSYS 18.1. 8 noded SOLID 185 defined by eight nodes and has 3 degrees of freedom at each of the nodes, translations in the nodal x, y and z direction was chosen as the element type The model is considered to be orthotrotropic in nature and the dimensions were taken as per specimen. One end of the sample is fixed in all direction; the other end of the sample is also fixed in all direction except in longitudinal direction where loading is applied. FEM modeling showed that the model satisfactorily predicts with the experimental results.

5. Reference

[1] Dr. Mohite, P M 2019 *Composite Introduction*. Retrieved on 23 Feb. 2019 from http://home.iitk.ac.in/~mohite/ae681.html

[2] Kaw A K 2006 *Mechanics of Composite Materials*, Second Edition 1-58

[3] Kumar B G, Singh R P and Nakamura T 2002 *Journal of Composite Materials* **36** 2713-33.

[4] Mahmoudi N, Tayeb N and Bourdim M 2017 Effect of sulfuric acid on the behavior of a composite based polymer reinforced with glass fibers. *Revue des matériauxeténergiesre nouvelables* **2**

[5] Agarwal A, Garg S, Rakesh P K, Singh I and Mishra B K 2010 *Indian Journal of Engineering & Material Science* **17** 471-76
[6] Amaro A M, Reis P N B, Neto M A and Louro C 2013 *Polymer degradation and stability* **98** 853-62
[7] Kanerva M, Jokinen J, Sarlin E, Pärnänen T, Lindgren M, Järventausta M. and Vuorinen J 2019 *Composite Structures* **212** 524-34.
[8] Butylina S, Hyvärinen M and Kärki T 2012 *Polymer Degradation and Stabili,* **97** 337-45.
[9] Heinrick M, Crawford B and Milani A S 2017 *MOJ Polymer Science* **1** 18–24.
[10] Abdullah H, Siddiqui R A and Araimi, S A 2000 *IIUM Engineering Journal* **1**.
[11] Yan L, Huang L and Braunschweig H 2015 *12th International Symposium on Fiber Reinforced Polymers for Reinforced Concrete Structures*
[12] Cesarino I, Dias O A T, Negrão D R, Rocha L L N and Leão AL 2019 *Journal of Composites Science*, **3** 24
[13] Liao K, Schultheisz C R and Hunston D L 1999 *Composites Part B: Engineering*, **30** 485-93.
[14] Seki Y K, Sever M, Sarıkanat I Sen, and A Aral 2011 Jute/polyester composites: The effect of water aging on the interlaminar shear strength (ILSS) *6th International Advanced Technologies Symposium* 368-71
[15] Dash B N, Rana A K, Mishra H K, Nayak S K and Tripathy S S 2000 *Journal of applied polymer science* **78** 1671-79
[16] Peng Y, Liu R, Cao J and Chen Y 2014 *Applied Surface Science* **317** 385-92
[17] Chauhan A, Vates U K, Kanu N J, Gupta E, Singh G K, Sharma B P, Gorrepati S R 2021 *Materials Today: Proceedings*, **38**, Part 5, 3027-34,
[18] Kim H J and Seo D W 2006 *International Journal of Fatigue* **28** 1307-14
[19] Devi U L, Joseph K, Nair K C M and Thomas S 2004 *Journal of applied polymer science* **94** 503-10

Impact of Non-convective Zone and Lower Convective Zone Thickness on the Performance Characteristics of Salinity Gradient Solar Pond

S G Chakrabarty[1], U S Wankhede[2] and R S Shelke[3]

[1] Research Scholar, Department of Mechanical Engineering, G. H. Raisoni College of Engineering, Nagpur, 440016, India.
[2] Associate professor, Department of Mechanical Engineering, Government College of Engineering, Chandrapur, 442403, India
[3] Head & Associate Professor, Department of Mechanical Engineering, G. H. Raisoni College of Engineering, Nagpur, 440016, India

Corresponding author's e-mail address: shyamal2706@gmail.com

Abstract. A solar pond technology employs a layer of salinity gradient to prevent heat loss due to convection from the lower convective zone. Thus, the energy received from solar radiation is stored in a lower convective zone. The thickness of various zones significantly affects the behaviour of solar pond temperature. In this present study, a transient numerical investigation is conducted to evaluate the impact of depths of different zones on the performance characteristics of solar pond. The variation in maximum temperature and maturation period under the influence of non-convective zone and lower convective zone thickness is discussed. The energy obtained from a solar pond significantly depends on various losses associated with the zones. Thus, an assessment of conduction and ground heat loss is presented for the variation in thickness of zones. An attempt is also made to study the effect of thickness of zones on the temperature of the lower convective zone. It is found that the configuration of a smaller thickness of LCZ and a higher thickness of NCZ yields maximum LCZ temperature.

Keywords: Salinity Gradient Solar Pond, Conduction heat loss, Ground heat loss, Maturation time.

1. Introduction

A need to explore solar energy technology arises from the concern of energy security and environmental issues. The various zones of salinity gradient solar pond (SGSP) are shown in Figure 1. In a non-convective zone (NCZ), a salinity gradient layer is established to prevent convection heat loss from the lower convective zone (LCZ). Thus, the thermal energy gets accumulated in LCZ and can be utilized for various applications (1). Recently, a number of applications of salinity gradient solar pond have been reported for desalination, power generation, and mineral processing (2–8).

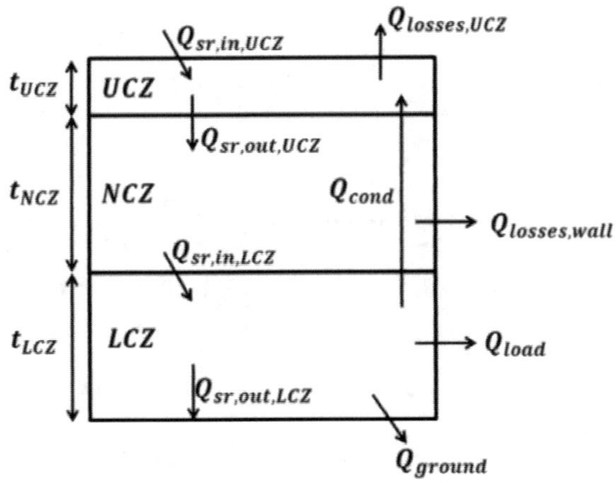

Figure 1. Schematic diagram exhibiting different zones of solar pond.

In the past few years, researchers have shown an increased interest in solar pond technology. Kumar et al. (5,9) determined the optimum thickness of zones for thermoelectric power generation application and also carried out investigations of peripheral heat conduction in solar pond. Verma and Das (10) studied the effect of wall profiles. Chakrabarty et al. (11) discussed the significance of ground conditions on the performance of salinity gradient solar pond. Ganguly et al. (12) examined the impact of bottom insulation thickness. The work carried out by the researcher shows the significance of studying the various operational parameters of a solar pond. Thus, the objective of the present work is to examine the impact of depths of different zone on the heat loss characteristics and temperature of salinity gradient solar pond.

2. Numerical model

A transient model described by [10] is used to assess the behaviour of the SGSP. The energy balance of UCZ and LCZ is expressed as Eq. (1) and Eq. (2) respectively.

$$\rho_{UCZ} C_{p,UCZ} t_{UCZ} A_{UCZ} \left(\frac{dT_{UCZ}}{d\tau}\right) = Q_{sr,UCZ} + Q_{cond} - Q_{losses,UCZ} \quad (1)$$

$$\rho_{LCZ} C_{p,LCZ} t_{LCZ} A_{LCZ} \left(\frac{dT_{LCZ}}{d\tau}\right) = Q_{sr,LCZ} - Q_{cond} - Q_{load} - Q_{ground} \quad (2)$$

Where, ρ is the fluid density in kg/m^3 and C_p is the fluid specific heat in J/kg-°C. The pond has an area A in m^2 with the zone thickness t in meters. The pond temperature T is in degree Celsius at time τ in hours. The amount of solar energy collected in the respective zone, Q_{sr} is the difference between the amount of radiation entering and leaving the zone. Thus, the solar radiation absorbed by UCZ and LCZ can be described as Eq. (3) and Eq. (4) respectively.

$$Q_{sr,UCZ} = Q_{sr,in,UCZ} - Q_{sr,out,UCZ} \quad (3)$$

$$Q_{sr,LCZ} = Q_{sr,in,LCZ} - Q_{sr,out,LCZ} \quad (4)$$

The heat loss that occurs due to conduction from LCZ to UCZ is denoted by Q_{cond}. The surface losses from UCZ are expressed as $Q_{losses,UCZ}$. Q_{load} is the amount of heat removed from LCZ in W/m^2 and Q_{ground} is heat loss to the ground. The wall losses are neglected in the present study. To assess the impact of various zone thicknesses on various operational parameters of solar pond, all the terms involved in Eq. (1) and Eq. (2) are determined according to the assumptions and procedure used by (13) under the metrological conditions of Nagpur city.

3. Result and Discussion

A numerical model is employed to solve Eq. (1) and Eq. (2) by using a finite difference method in MATLAB. Figure 2(a and b) shows the different meteorological input parameters of Nagpur city for the numerical model, which are obtained from "NASA," 2019 (14). The influence of NCZ and LCZ depth on conduction heat loss, ground heat loss, and temperature of LCZ are discussed in the subsequent sections.

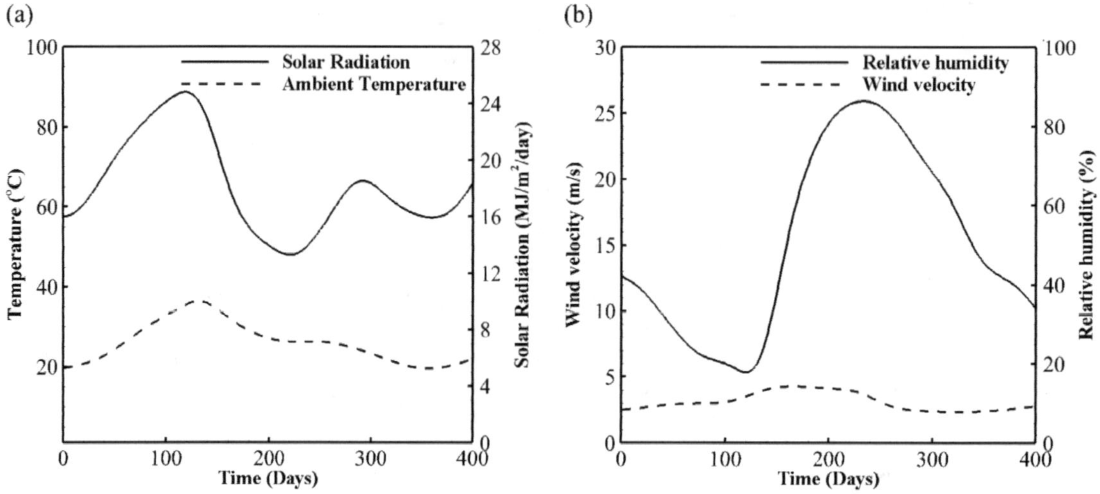

Figure 2. Meteorological conditions of Nagpur city, (a) ambient temperature and solar radiation and; (b) wind speed and relative humidity

3.1. Influence of LCZ and NCZ thickness on the maximum temperature of LCZ and maturation time

The impact of both NCZ thickness and LCZ thickness on maximum LCZ temperature and maturation time is shown in Figure 3(a and b). In the present analysis, NCZ thickness is considered from 0.5 m to 4.0 m and LCZ thickness is kept as 0.5 m, 1.0 m, and 1.5 m. Figure 3(a) exhibits the development of maximum LCZ temperature for variation in LCZ thickness. It can be seen that for a given LCZ thickness, the temperature of LCZ rises with an increase in the thickness of NCZ up to a certain limit. Furthermore, an increase in NCZ thickness results in the reduction of LCZ temperature. This phenomenon occurs due to a reduction in conduction loss and attenuation of solar radiation.

It can also be noticed that after the attainment of peak point, the drop in temperature is considerably large for the higher thickness of LCZ. The increase in LCZ thickness leads to a reduction in the peak temperature of LCZ. The peak temperatures are found to be 121.8 °C, 109.5 °C, and 98.5 °C for LCZ thicknesses of 0.5 m, 1.0 m, and 1.5 m respectively. The value of peak temperature reaches so high due to the absence of heat extraction process.

In Figure 3(a) the peak temperature occurs at NCZ thickness of 2.5 m, 2.0 m, and 2.0 m for LCZ thickness of 0.5 m, 1.0 m, and 1.5 m respectively, based on the given conditions. However, the variation in temperature between NCZ thicknesses of 1.5 to 2.0 m is found to be marginal for LCZ thickness of 1.0 m and 1.5 m. Thus, the selection of NCZ thickness of about 2.0 m will not provide appreciable output. Moreover, it will increase the financial investment. According to Hussain et al. (15) maturation period should also be considered for the selection of different zone thicknesses.

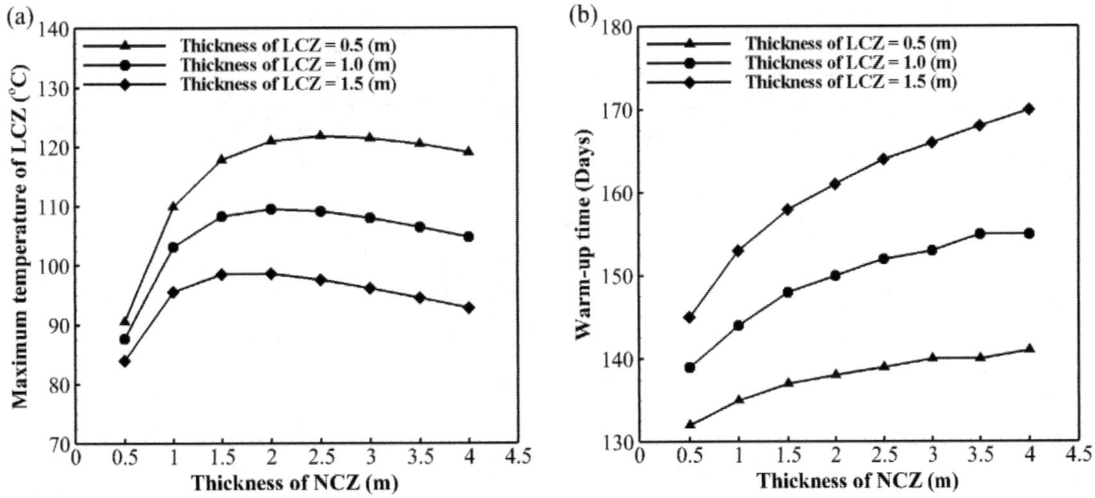

Figure 3. Effect of variation of LCZ thickness on (a) maximum LCZ temperature; (b) maturation time.

Figure 3(b) illustrates the maturation time for different thicknesses of LCZ. It can be seen that a longer maturation time is required with the increase in NCZ thickness owing to the attenuation of radiation intensity. The higher thickness of LCZ corresponds to a larger thermal mass of the solar pond, which also leads to a longer maturation period. A significant rise in maturation time is observed for higher NCZ and LCZ thickness. Thus, maximum temperature and maturation period should be taken into consideration to achieve the required parameters for the application.

3.2. Influence of NCZ thickness on conduction heat loss and ground heat loss

The performance of solar pond is significantly influenced by conduction and ground loss. To investigate the effect of NCZ thickness on conduction and ground loss from LCZ, a ratio of respective heat loss to solar radiation received in LCZ is employed as shown in Figure 4. The thickness of LCZ is considered as 1.0 m. In Figure 4 two peaks in conduction and ground heat loss profiles are observed due to the behaviour of solar radiation for the present location.

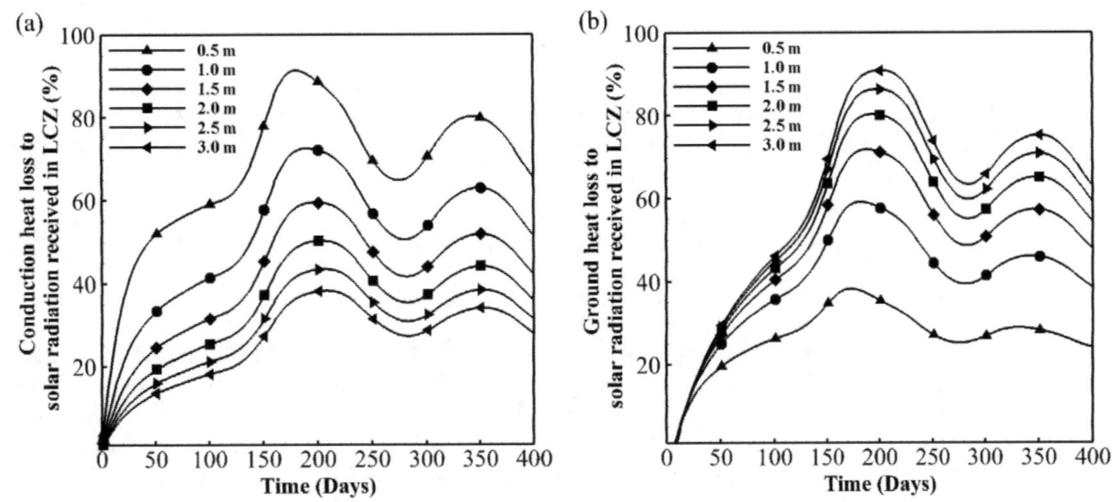

Figure 4. Impact of NCZ thickness on the heat loss (a) conduction; (b) ground.

For consecutive increase in NCZ thickness, it is found that the difference in the corresponding average percentage of conduction heat loss diminishes. A similar characteristic is observed for the ground heat loss profile. A reduction in conduction heat loss will enhance the accumulation of heat in LCZ which leads to an increase in ground heat loss. In a numerical study, Ganguly et al. (16) reported that the ground heat can be extracted and utilized to improve the output of the pond. Whereas, in Figure 4(b) it can be seen that ground heat loss is influenced by NCZ thickness. Hence, NCZ thickness should be taken into account while considering heat extraction from the ground.

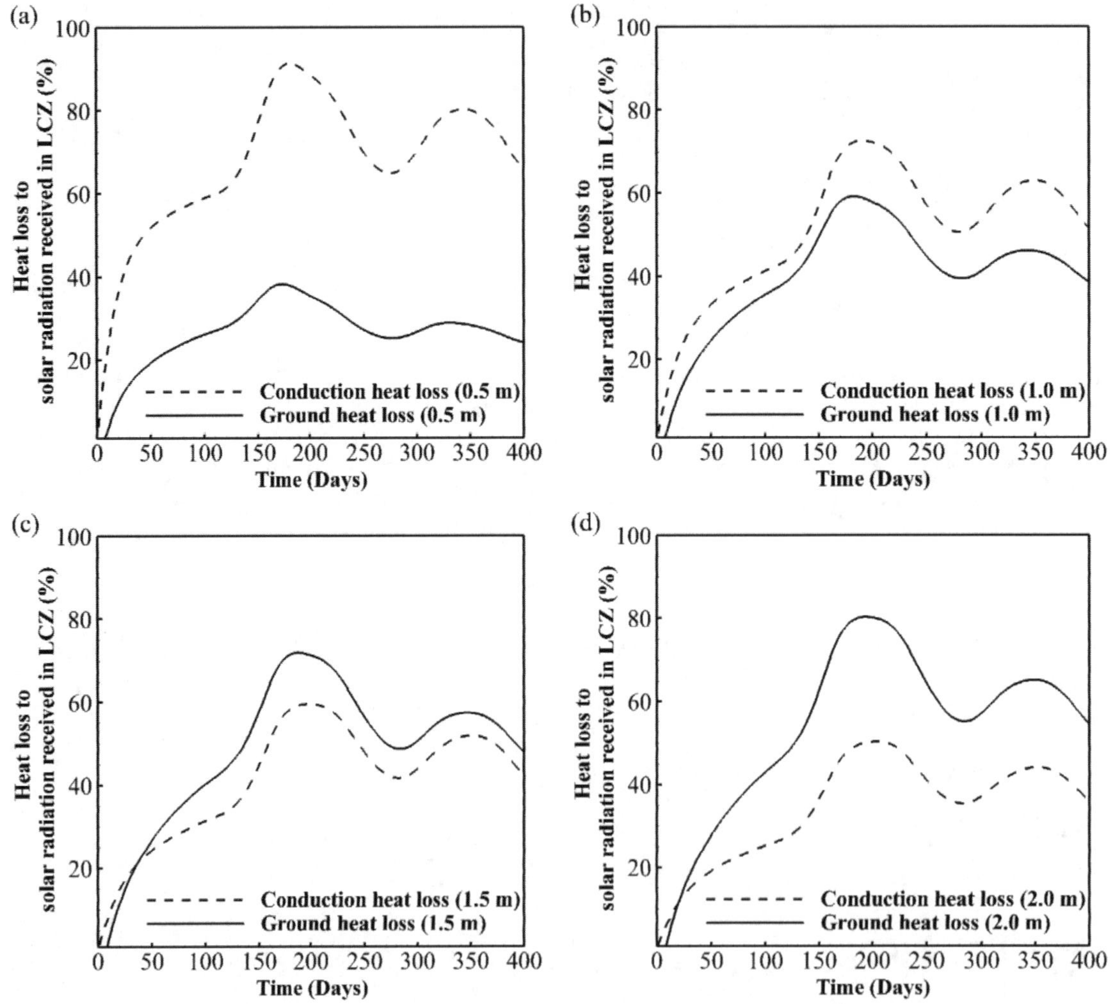

Figure 5. Comparison of conduction and ground heat loss under the influence of NCZ thickness (a) NCZ thickness = 0.5 m; (b) NCZ thickness = 1.0 m; (c) NCZ thickness = 1.5 m; (d) NCZ thickness = 2.0 m.

The comparison of conduction and ground loss for NCZ thickness of 0.5 m, 1.0 m, 1.5 m, and 2.0 m is presented in Figure 5. The LCZ thickness is kept constant as 1.0 m. The conduction loss is found to be higher than ground heat loss for the NCZ thickness of 0.5 m and 1.0 m. whereas, further increase in NCZ thickness results in lower conduction loss than ground loss under the given set of conditions. The difference in conduction loss and ground loss is found to be lower for NCZ thickness of 1.0 m and 1.5 m than 0.5 m and 2.0 m.

3.3. *Influence of LCZ and NCZ thickness on conduction heat loss and ground heat loss*

In order to understand the impact of both LCZ thickness and NCZ thickness on conduction and ground loss Figure 6 is presented. The LCZ thickness is considered as 0.5 m, 1.0 m, and 1.5 m. Figure 6(a, c, and e) shows the conduction heat loss for NCZ thickness of 0.5 m, 1.0 m, and 1.5 m respectively, and Figure 6(b, d, and f) exhibits the ground heat loss for NCZ thickness of 0.5 m, 1.0 m, and 1.5 m respectively.

It is found that for the higher thickness of NCZ, the conduction loss decreases, and ground loss increases. The conduction loss is noticed to be increased with the rise in LCZ thickness from 0.5 m, 1.0 m, and 1.5 m while considering NCZ thickness of 0.5 m. However, the impact of LCZ thickness on conduction loss is seen to be negligible for NCZ thicknesses of 1.0 m and 1.5 m after initial warm-up. The impact of LCZ thickness on ground heat loss is also found to be marginal for all the considered sizes of NCZ after initial warm-up.

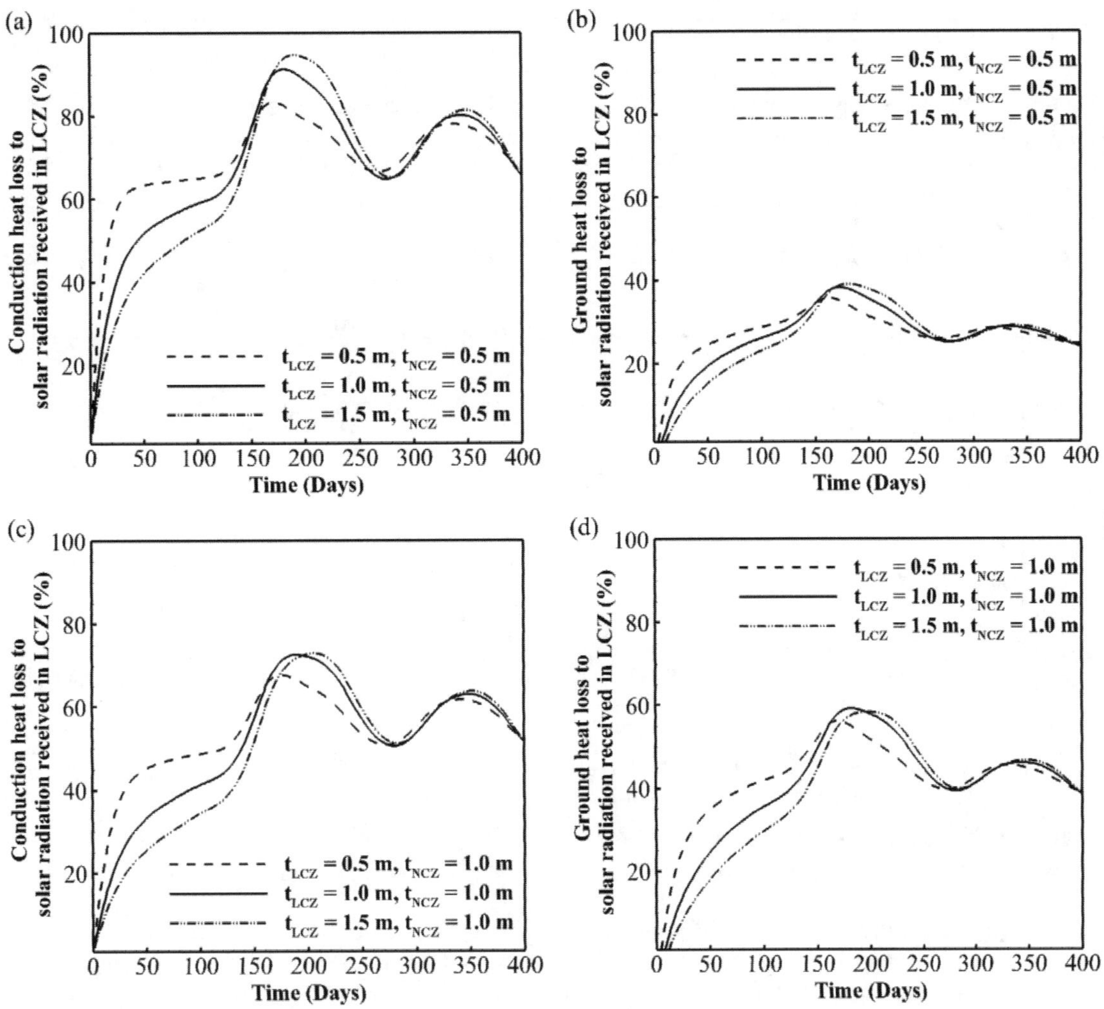

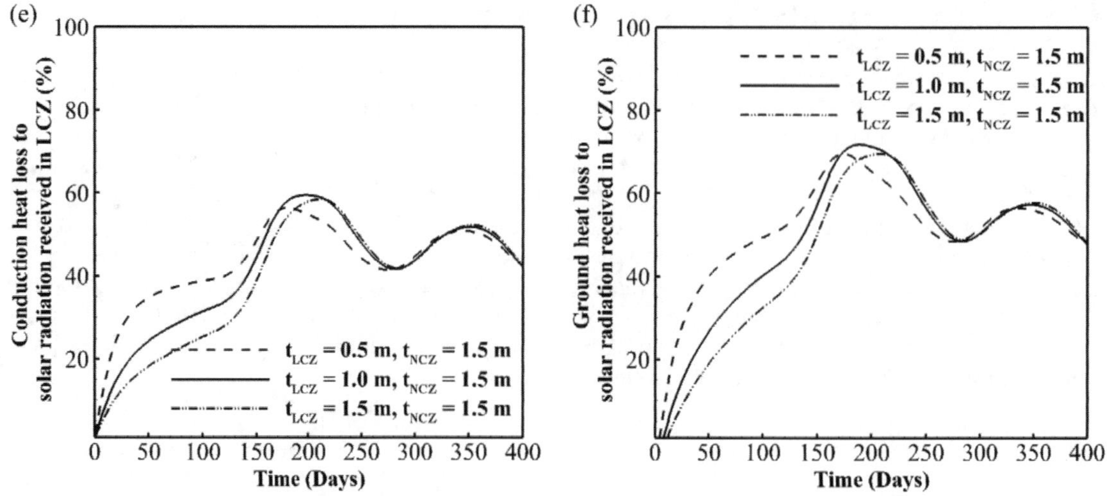

Figure 6. Variation of conduction and ground heat loss under the influence of LCZ and NCZ thickness, (a) Conduction heat loss (NCZ thickness = 0.5 m); (b) Ground heat loss (NCZ thickness = 0.5 m); (c) Conduction heat loss (NCZ thickness = 1.0 m); (d) Ground heat loss (NCZ thickness =1.0 m); (e) Conduction heat loss (NCZ thickness = 1.5 m); (f) Ground heat loss (NCZ thickness = 1.5 m).

3.4. Influence of LCZ and NCZ thickness on LCZ temperature

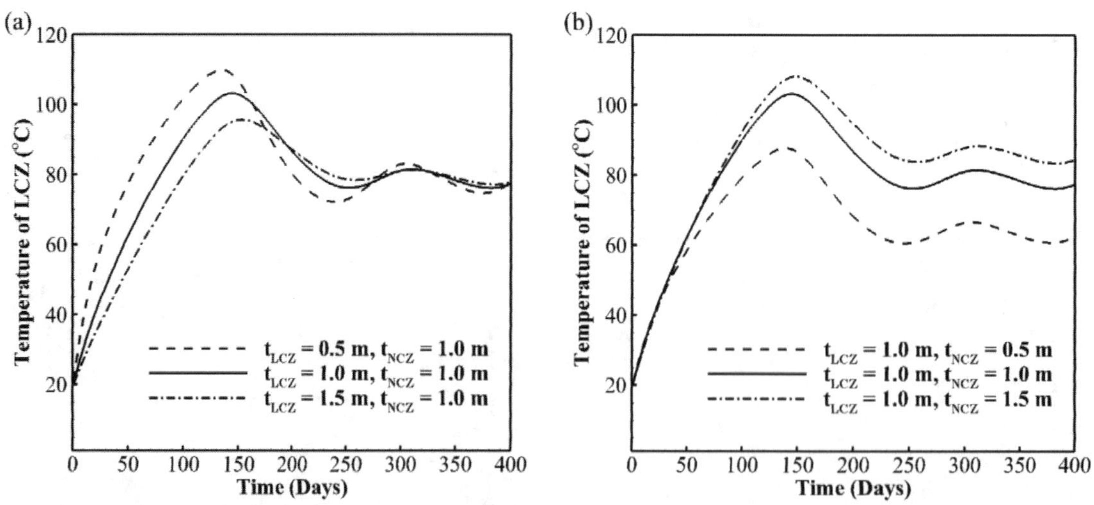

Figure 7. Variation of LCZ temperature under the influence of LCZ and NCZ thickness, (a) NCZ thickness = 1.0 m; (b) LCZ thickness = 1.0 m.

Figure 7(a) shows the variation of LCZ temperature by assuming constant NCZ thickness. It can be noticed that LCZ temperature decreases with the increase in LCZ thickness. However, the variation in temperature after the peak point is insignificant. It can be seen that a maximum temperature of about 109.9 °C can be obtained for LCZ thickness of 0.5 m. However, it also produces a higher fluctuation of temperature of about 37 °C. For LCZ depth of 1.5 m, the variation in temperature is observed to be depreciated by 20 °C than LCZ depth of 0.5 m. Such a variation between the highest and lowest temperature may inhibit the temperature required for the application. The impact of variation of NCZ thickness on the behaviour of LCZ temperature is shown in Figure 7(b) by considering the same thickness of LCZ. Enhancement in LCZ temperature is seen with the rise of NCZ thickness. Figure

7(b) exhibits a maximum LCZ temperature of 108.3 °C and 87.7 °C for NCZ thickness of 1.5 m and 0.5 m respectively. It can be seen that the configuration of a smaller thickness of LCZ and a higher thickness of NCZ yields maximum LCZ temperature. The LCZ temperature profiles are seen to be uniform for given NCZ thickness. Thus, NCZ thickness has a negligible effect on the variation of temperature.

4. Conclusion

In this present work, a transient investigation is carried out to examine the impact of thicknesses of the different zone on various operational parameters of solar pond. It is found that the maximum LCZ temperature and maturation time should be taken into the consideration for the selection of the suitable thickness of zones. For higher NCZ and LCZ thickness, a longer maturation time is required. For a higher thickness of NCZ, it is seen that the conduction loss decreases, and ground loss increases. The influence of LCZ thickness on conduction and ground heat loss is found to be marginal. The configuration of a smaller thickness of LCZ and a higher thickness of NCZ yields maximum LCZ temperature. For LCZ depth of 1.5 m, the variation in temperature is observed to be depreciated by 20 °C than LCZ depth of 0.5 m. The impact of LCZ thickness is found to be significant on the fluctuation of LCZ temperature; whereas, NCZ thickness has a negligible effect on the variation of temperature.

References

1. Date, A and Akbarzadeh A. Salinity gradient solar ponds. In: Akbarzadeh NE and A, editor. Solar Energy Sciences and Engineering Applications. CRC Press, Balkema, The Netherlands; 2014. p. 195–218.
2. Shekari Namin A, Rostamzadeh H, Nourani P. Thermodynamic and thermoeconomic analysis of three cascade power plants coupled with RO desalination unit, driven by a salinity-gradient solar pond. Thermal Science and Engineering Progress. 2020;18:100562.
3. Alcaraz A, Montalà M, Cortina JL, Akbarzadeh A, Aladjem C, Farran A, et al. Design, construction, and operation of the first industrial salinity-gradient solar pond in Europe: An efficiency analysis perspective. Solar Energy. 2018;164:316–26.
4. Montalà M, Cortina JL, Akbarzadeh A, Valderrama C. Stability analysis of an industrial salinity gradient solar pond. Solar Energy. 2019;180:216–25.
5. Kumar A, Singh K, Verma S, Das R. Inverse prediction and optimization analysis of a solar pond powering a thermoelectric generator. Solar Energy. 2018;169:658–72. Available from: https://www.sciencedirect.com/science/article/pii/S0038092X18304687
6. Ziapour B, Saadat M, Palideh V, Afzal S. Power generation enhancement in a salinity-gradient solar pond power plant using thermoelectric generator. Energy Conversion and Management. 2017;136:283–93.
7. Masoud Parsa S, Majidniya M, Alawee WH, Dhahad HA, Muhammad Ali H, Afrand M, et al. Thermodynamic, economic, and sensitivity analysis of salt gradient solar pond (SGSP) integrated with a low-temperature multi effect desalination (MED): Case study, Iran. Sustainable Energy Technologies and Assessments. 2021;47:101478. Available from: https://www.sciencedirect.com/science/article/pii/S2213138821004884
8. Mosaffa AH, Farshi LG. Thermodynamic feasibility evaluation of an innovative salinity gradient solar ponds-based ORC using a zeotropic mixture as working fluid and LNG cold energy. Applied Thermal Engineering. 2021;186:116488. Available from: https://www.sciencedirect.com/science/article/pii/S1359431120339636
9. Kumar A, Das R. Effect of peripheral heat conduction in salt-gradient solar ponds. Journal of Energy Storage. 2021;33:102084. Available from: https://www.sciencedirect.com/science/article/pii/S2352152X20319149
10. Verma S, Das R. Wall profile optimisation of a salt gradient solar pond using a generalized

model. Solar Energy. 2019;184:356–71.
11. Chakrabarty SG, Wankhede US, Walke P V., Gohil TB. Effect of Ground Condition on the Storage Zone Temperature of Salinity Gradient Solar Pond. In 2019.
12. Ganguly S, Date A, Akbarzadeh A. Effectiveness of Bottom Insulation of a Salinity Gradient Solar Pond. Journal of Solar Energy Engineering. 2018;140:1–5.
13. Chakrabarty SG, Wankhede US, Shelke RS, Gohil TB. Investigation of temperature development in salinity gradient solar pond using a transient model of heat transfer. Solar Energy. 2020;202.
14. NASA. Prediction of Worldwide renewable Energy Resources. 2019.
15. Husain M, Patil PS, Patil SR, Samdarshi SK. Optimum size of non-convective zone for improved thermal performance of salt gradient solar pond. Solar Energy. 2003;74(5):429–36. Available from: http://www.sciencedirect.com/science/article/pii/S0038092X03001579
16. Ganguly S, Date A, Akbarzadeh A. Heat recovery from ground below the solar pond. Solar Energy. 2017;155:1254–60. Available from: http://dx.doi.org/10.1016/j.solener.2017.07.068

Analysis on physical properties of micropolar nanofluid past a constantly moving porous plate

N Golden Stepha[1,*] and D Kavin Jacob[2]
[1]Department of Mathematics, R.M.K Engineering College, Kavaraipettai, Gummidipoondi, India.
[2]PG & Research, Department of Mathematics, Loyola College, Chennai, India
ngs.sh@rmkec.ac.in[1], kainjacob@gmail.com[2]
*Corresponding Author

Abstract. The computational analysis is presented for boundary layer heat and mass transfer flow of hydro magnetic micropolar nanofluid flow. In the flow model, viscosity of the fluid is taken as temperature-dependent and varies linearly and the other physical properties such as radiative heat flux, the magnetic field, the viscous dissipation, chemical reaction are additionally assumed in the energy equation and spices concentration equation respectively The PDEs representing the fluid flow have been changed into a framework of dimensionless ODEs and explained mathematically through the 4th order R-K and NS shooting technique. Temperature distribution, velocity distribution, micro rotation, and concentration distribution are explored graphically for a series of solid volume fraction ($0<\phi<2$) of nano-solid particles. All the findings for various flow parameters agreed perfectly with physical situation of the flow. It is observed that for increasing value of magnetic parameter, the concentration and temperature of the micropolar nano fluid near the boundary layer declines and increasing value of the volume fraction of nano-solid particle ϕ leads to decrease in velocity and micro rotation of the fluid within the boundary layer decreases.

Keywords. MHD Nanofluid, Boundary layer, Micropolar, heat transfer, and mass transfer

1. Introduction

In present technology, there is an important development in the utilization of colloids mass-production by nano-sized particles scattered in a base fluid. Heat controlling has been accepted as important tools in the recent techno-industrial world and it should be used effectively. Prior research shows that thermophysical properties can be enhanced by nanofluids than base fluids such as oil or water. Das, S. K et al. (2006), Liu M.S et al. (2005), and Choi et al. (2004). Yu W, France et al. (2008) provided an excellent review on the recent state of nanofluid innovation of heat transfer applications and area in which the data are presently questionable or disagreeing.

Due to the increasing production of metal and polymer sheets in engineering industries, the fluid dynamics of continuously moving plate plays an important role. The theory of continuously moving surfaces was introduced by Sakiadis (1961). Due to their cherished influences in various industrial and engineering processes such as plastic extrusion and polymer, the process of crystal development makes a micropolar fluid flow is a significant space of exploration. The concept of micropolar fluid was initiated by Eringen (1964). Later, many researchers have put forward their work on micropolar fluid past a constantly stretching plate and analyze the effect of energy transport on it [V.M. Soundalgekar, H.S. Takhar, (1983), A. Rapits, (1998).

Fluid viscosity is another important property which quantify the flow but in most of the previous studies on heat transfer in fluid flow it is supposed to be constant, however, if the temperature variation is large this assumption is not valid. So, in the present work viscosity is assumed as variable. Moreover, due to engineering applications, radiation energy plays a significant role in the energy transport flow of moving fluid. The problem of heat and mass transfer flow of micropolar fluid in the presence of radiation is analyzed by Loganathan and Golden Stepha (2012). Chemical reaction effects of the above problem were studied by Loganathan and Golden stepha (2012)

Abdul Rehman and Nadeem (2012) studied the mixed convective heat transfer effect of nanofluid over a cylinder with micropolar fluid is a base and the solution is obtained by HAM. Bourantas and Loukopoulos (2014) studied the natural convection of micropolar model in a square cavity and the analysis are performed for different nano particles. Kazimierz Rup and Konrad Nering (2014) studied unsteady natural convection of nanofluids using explicit finite difference scheme.

In literature, several investigations done on the MHD influences over the stretching sheet in numerous heat transfer-related flows such as MHD power producers, solar energy devices, and in biomedical field for hyperthermia cancer cure, brain tumor treatment. Due to their cherished influences in various industrial and engineering processes such as plastic sheets extrusion and polymer, the process of crystal growth makes a flow of the micropolar fluid in the boundary layer on a constantly moving surface as an important area of research. Heat transfer effect of Magnetohydrodynamic micropolar nanofluid past a permeable stretching/shrinking sheet with Newtonian heating is examined by Gangadhar et.al (2017). Recently Abdullah Dawar et al. (2020) studied Chemically reactive MHD micropolar nanofluid flow with velocity slips and variable heat source/sink and they solved the system of equation both analytically and numerically. The mhd micropolar-nanofluid flow in natural convection heat transfer over a radiative truncated cone was studied by Waqar A. Khan et al. (2020) and this study reveals an enhancement in the applicable parameters progresses the heat transmission rate.

Khuram Rafique et al. (2020) deliberated hydromagnetic flow of micropolar nanofluid and for investigation Buongiorno mathematical model of hydromagnetic micropolar nanofluid over a permeable inclined stretching sheet is considered and the system is solved by Keller- box scheme

The contemporary research aims to examine theoretically various physical properties of the micropolar model for Nano fluidic suspensions.

2. Problem Formulation

A Problem of two-dimensional, incompressible, steady, micropolar fluid flow on a flat absorbent plate which moves constantly with a continuous velocity in a water-based nanofluid containing different nanoparticles such as CuO, Al2O3, and TiO2, medium at rest is formulated mathematically. The rectangular coordinate system is taken to describe the problem and whose origin is located at the place where the plate is brought into the fluid medium as displayed in Figure 1.

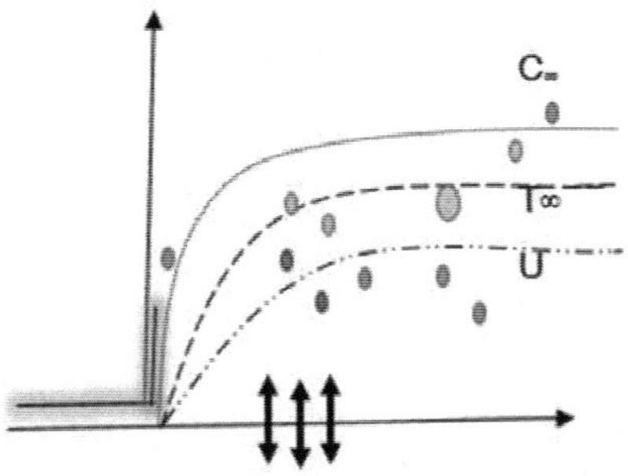

Figure 1. Flow Model

The surface is kept along the x-axis and the y-axis is vertical to it. The temperature and concentration on the surface are maintained at uniform value T_∞ and C_∞ respectively. The fluid is seen to be gray, radiating, and absorbing. Heat fluctuation in the y-direction is significant while comparing with the flux in the x-direction. The fluid viscosity is assumed as variable and it varies with respect to temperature linearly. The governing equations of the above flow model are

$$\frac{\partial u}{\partial x}+\frac{\partial v}{\partial y}=0 \tag{1}$$

$$u\frac{\partial u}{\partial x}+v\frac{\partial v}{\partial y}=\frac{1}{\rho_{nf}}\frac{\partial}{\partial y}\left(\mu_{nf}+k\right)\frac{\partial u}{\partial y}+\frac{k}{\rho_{nf}}\frac{\partial \sigma}{\partial y}-\frac{\alpha B^2 u}{\rho_{nf}} \tag{2}$$

$$\rho_{nf} j\left(u\frac{\partial \sigma}{\partial x}+v\frac{\partial \sigma}{\partial y}\right)=\gamma_{nf}\frac{\partial^2 \sigma}{\partial y^2}-k\left(2\sigma+\frac{\partial u}{\partial y}\right) \tag{3}$$

$$u\frac{\partial T}{\partial x}+v\frac{\partial T}{\partial y}=\frac{k_{nf}}{\left(\rho c_p\right)_{nf}}\frac{\partial^2 T}{\partial y^2}+\frac{\gamma_{nf}}{c_p}\left(\frac{\partial u}{\partial y}\right)^2-\frac{1}{\left(\rho c_p\right)_{nf}}\left(\frac{\partial q_r}{\partial y}\right)+\frac{\alpha B^2 u^2}{\left(\rho C_P\right)_{nf}} \tag{4}$$

$$u\frac{\partial C}{\partial x}+v\frac{\partial C}{\partial y}=D\frac{\partial^2 C}{\partial y^2}-R_C\left(C-C_\infty\right) \tag{5}$$

exposed to the following boundary condition

$$\begin{aligned} &At\ y=0,\quad u=U_0, v=V_W, \sigma=-n\frac{\partial u}{\partial y}, T=T_W, C=C_W \\ &At\ y=\infty,\ u=0, v=0, \sigma=0, T=T_\infty, C=C_\infty \end{aligned} \tag{6}$$

Where,

$\rho_{nf}=(1-\phi)\rho_f+\phi\rho_s$

ϕ - volume fraction of nano-solid particle

ρ_{nf} - density of nanofluid

ρ_s - the density of the solid particle

ρ_f - density of the base fluid

$$\left(\rho c_p\right)_{nf} = (1-\phi)\left(\rho c_p\right)_f + \phi\left(\rho c_p\right)_s$$

$$\mu_{nf} = \frac{\mu_f}{(1-\phi)^{2.5}}$$

Now considered the following dimensionless similarity variables

$$\eta = y\sqrt{\frac{U_0}{2\gamma_f x}},\ \psi = \sqrt{2\gamma_f U_0 x} f(\eta)\quad \sigma = \sqrt{\frac{U_0^3}{2\gamma_f x}} g(\eta),\quad \theta = \frac{T-T_\infty}{T_W - T_\infty},\quad \lambda = \frac{C-C_\infty}{C_W - C_\infty} \tag{7}$$

Where, $f(\eta)$ and $g(\eta)$ are non-dimensional stream functions, with respect to the equation (7), equations (2)-(5) are changed to the system of ODE given below

$$\frac{\left(\frac{\mu_{nf}}{\mu_f}+k\right)}{\left((1-\phi)+\phi\frac{\rho_s}{\rho_f}\right)} f''' + ff'' + \frac{\gamma_r}{(1-\phi)^{2.5}\left((1-\phi)+\phi\frac{\rho_s}{\rho_f}\right)} f''\theta' + \frac{k}{\left((1-\phi)+\phi\frac{\rho_s}{\rho_f}\right)} g' - M\left(f'\right)^2 = 0 \tag{8}$$

$$\frac{\left(\frac{\mu_{nf}}{\mu_f}+\frac{k}{2}\right)}{\left((1-\phi)+\phi\frac{\rho_s}{\rho_f}\right)} g'' - gf' + fg' + \frac{2k}{\left((1-\phi)+\phi\frac{\rho_s}{\rho_f}\right)}\left(2g+f''\right) = 0 \tag{9}$$

$$(3N+4)\theta'' + \frac{3N\,\mathrm{Pr}\,Ec}{(1-\phi)^{2.5}\left((1-\phi)+\phi\frac{\rho_s}{\rho_f}\right)}\left(f''\right)^2 + 3N\,\mathrm{Pr}\,f\theta' + 3M\,Ec\,\mathrm{Pr}\,N\left(f'\right)^2 = 0 \tag{10}$$

$$\frac{1}{(1-\phi)^{2.5}\left((1-\phi)+\phi\frac{\rho_s}{\rho_f}\right)}\lambda'' + Sc\lambda' f + ScR\lambda = 0 \tag{11}$$

Where,

$$\mathrm{Pr} = \frac{\gamma_f \rho c_p}{k_{nf}},\quad N = \frac{k' k_{nf}}{4\sigma_1 T_\infty^3},\quad Ec = \frac{U_0^2}{c_p(T_W - T_\infty)},$$

$$F_W = -V_W\sqrt{\frac{2x}{\gamma_f U_0}}\quad R = -R_C\left(\frac{2x}{U_0}\right), M = \frac{\alpha B^2}{\rho_{nf} U_0}$$

The initial and boundary conditions corresponding to dimensionless quantities are given by

$$f(0) = F_W, f'(0) = 1, g(0) = -nf'', \theta(0) = 1, \lambda(0) = 1\ \ as\ \ \eta = 0 \tag{12}$$

$$f'(\infty) = 0,\ g(\infty) = 0,\ \theta(\infty) = 0, \lambda(\infty) = 0\ \ as\ \ \eta = \infty$$

3.Numerical Solution

The set of non-linear PDEs (2) - (5) with the equation (6) are converted into ODEs by similarity transformation technic. The system of non-linear ODEs (7) - (10) with the equation (11) are solved by using the 4th order R-K method along with NS procedure (Adams and Rogers (1973)) for the proposed parameter ϕ, Fw, N, Pr, K, N, M and Sc. The system of ODE (7) – (10) which is subjected to the condition (11) solved by implementing a computer program. A step size of $\Delta\eta = 0.01$ was designated to fulfill the convergence criterion of 10-4 in all cases.

4. Results and Discussion

The mathematical solution of the above system is given in terms of graphs.

Figure 2 expresses the influence of hydro magnetic parameter on fluid velocity near the plate. Figure 2 displays that the velocity of the fluid decrease as the hydro magnetic parameter increases.

Figure 3 displays the microrotation of the micropolar nanofluid close to the boundary layer for different values of hydro magnetic parameter. It observes that the angular velocity of the micropolar nanofluid decreases as M increases.

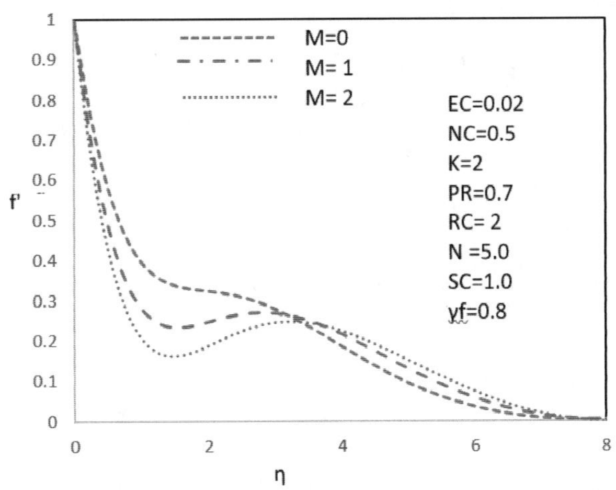

Figure 2 Impact of hydro magnetic parameter on velocity of the fluid.

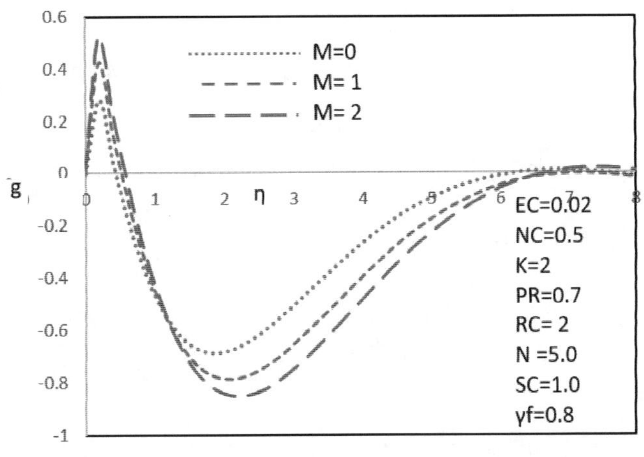

Figure 3. Impact of hydro magnetic parameter on Micro rotation of the fluid

IOP Conf. Series: Materials Science and Engineering 1206 (2021) 012004 doi:10.1088/1757-899X/1206/1/012004

Figure 4 exhibits the concentration distribution of the micropolar nanofluid and it is observed that fluid concentration close to the plate decrease as the hydro magnetic parameter increases. Figure 5 exhibits the temperature distribution of the micropolar nanofluid and it states a slight variation in the temperature of the fluid when hydro magnetic parameter changes.

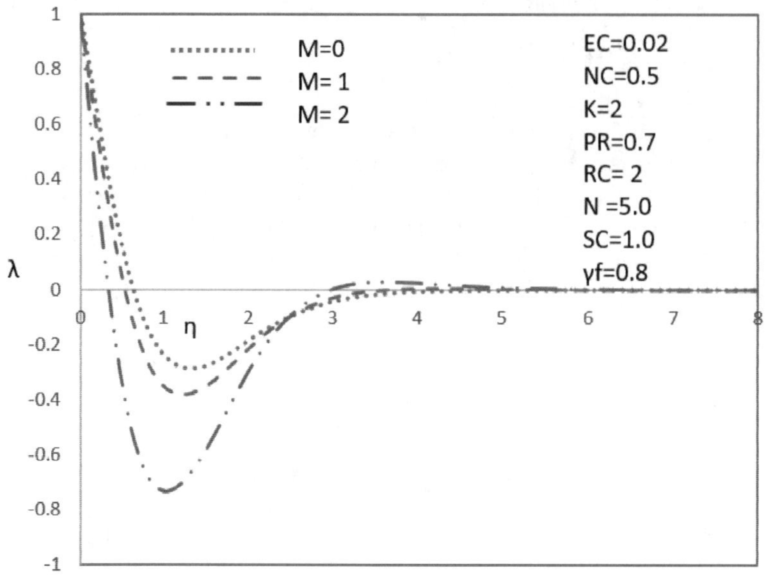

Figure 4 Influence of hydro magnetic parameter on Concentration.

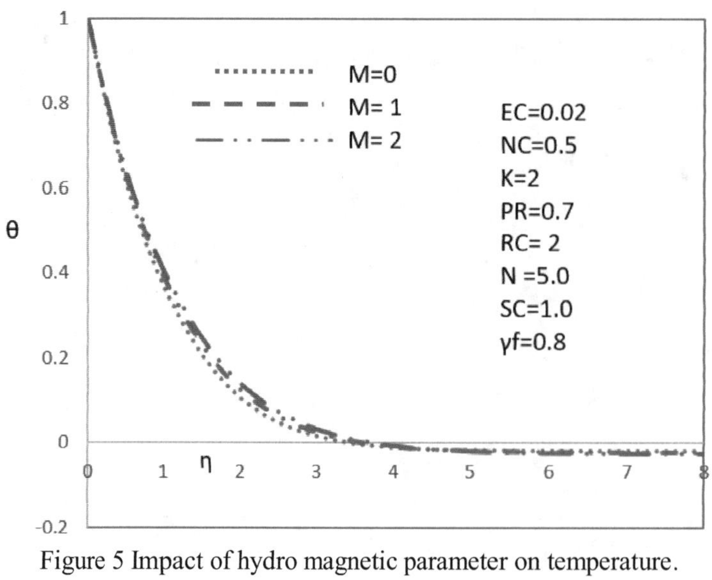

Figure 5 Impact of hydro magnetic parameter on temperature.

Figure 6 exhibits the influence of coupling constant on Velocity of the fluid with in the boundary layer and it is found that for increasing K velocity of the fluid with in the boundary layer increases.
Figure 7 exhibits the Influence of coupling constant on angular Velocity of the fluid with in the boundary layer and it reveals that for increasing K angular velocity of the fluid with in the boundary layer decreases.
Effect of coupling constant on concentration of the fluid with in the boundary layer of the flow is shown in figure 8 and it reveals that for increasing K concentration of the fluid increases with in the boundary layer.

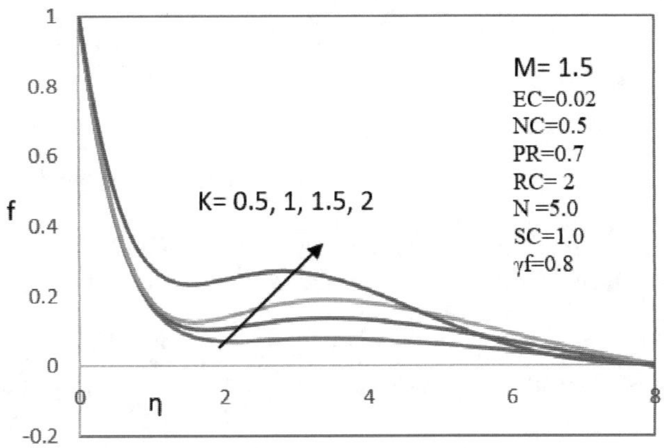

Figure 6. Influence of different K on velocity of the fluid.

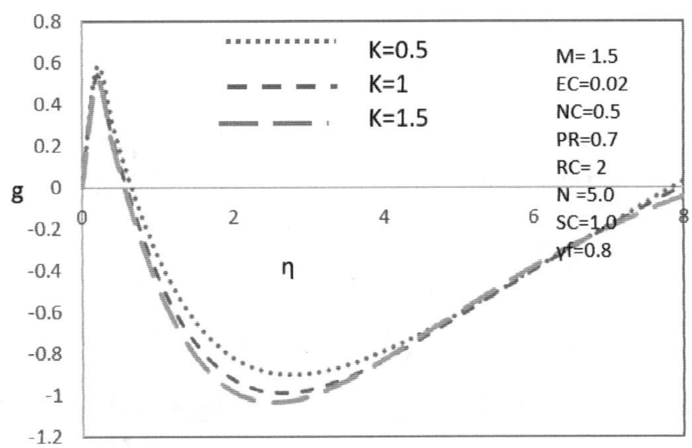

Figure 7. Impact of different K on angular Velocity of the fluid.

η

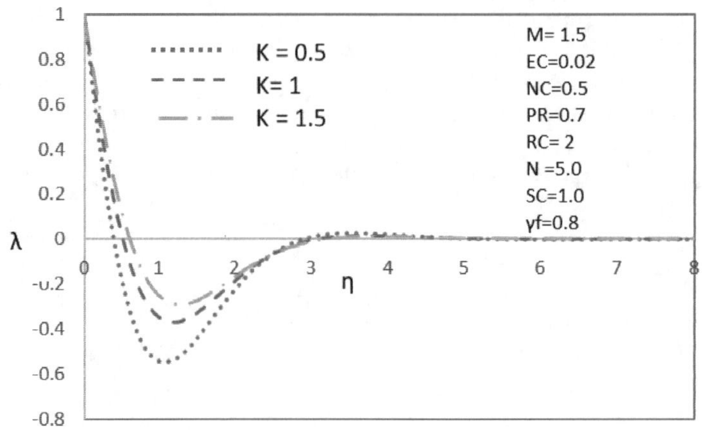

Figure 8. Impact of different K on Concentration of the fluid.

Figure 9 exposes the velocity of the micropolar nanofluid in the boundary layer of the flow for various values of volume fraction of nano-solid particles. Figure 9 shows that the velocity of the fluid decrease as the volume fraction of nano-solid particles increases.

Figure 10 shows the micro rotation of the micropolar nanofluid near the boundary layer for different values of volume fraction of nano-solid particles. It observes that the angular velocity of the micropolar nanofluid decreases as ϕ rises.

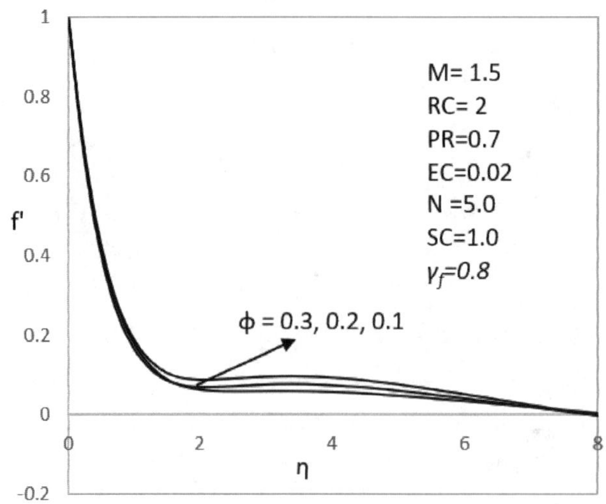

Figure 9. Velocity profile for different ϕ.

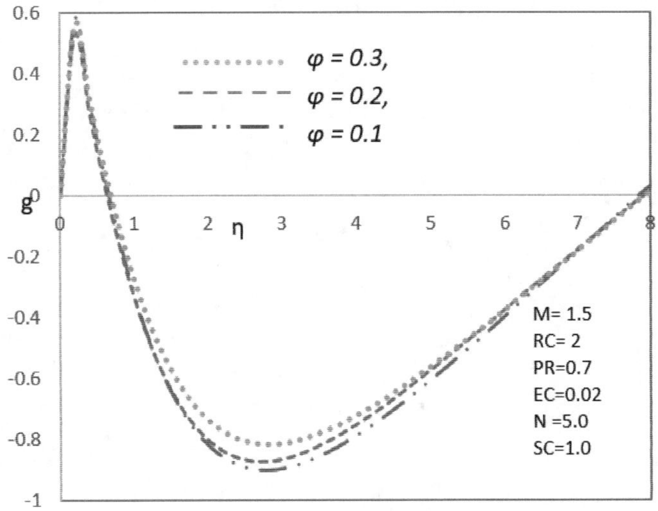

Figure 10. Micro rotation profile for different φ.

Figure 11 exhibits the temperature distribution of the micropolar nanofluid and it states a slight variation in the temperature of the fluid when the volume fraction of nano-solid particle changes.
Figure 12 exhibits the concentration distribution of the micropolar nanofluid and it shows that the concentration of the fluid falls when the volume fraction of nano-solid particle increases changes.

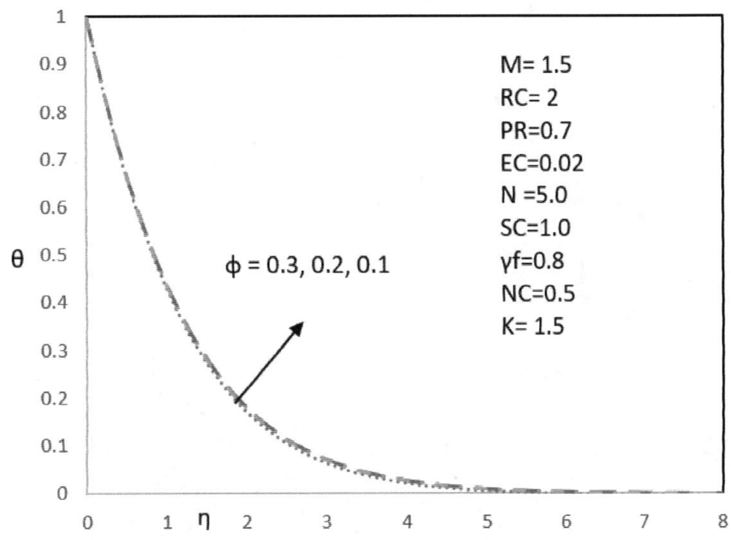

Figure 11. Temperature profile for different φ.

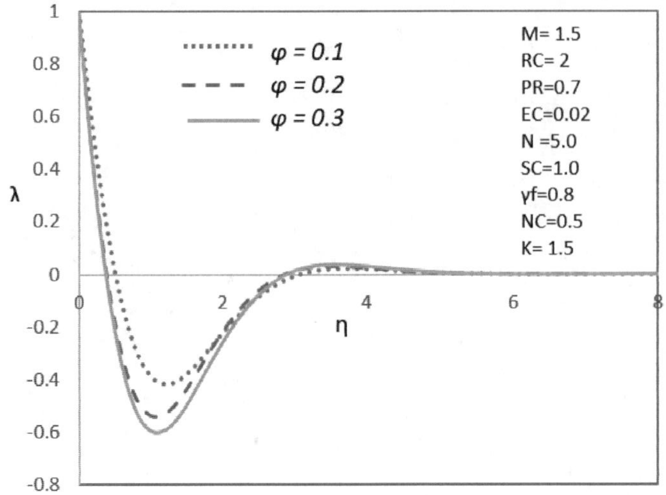

Figure 12. Concentration profile for different φ.

Figure (13)-(15) indicate the temperature, concentration, velocity, for various values of Pr. It shows that the velocity and temperature of the fluid close to the boundary layer decreases as Pr increases, but concentration of the fluid inside the boundary layer declines as Pr decreases.

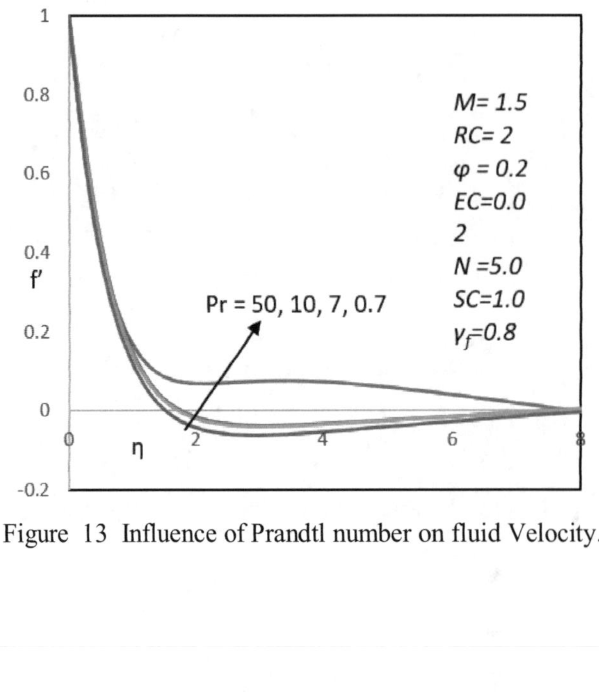

Figure 13 Influence of Prandtl number on fluid Velocity.

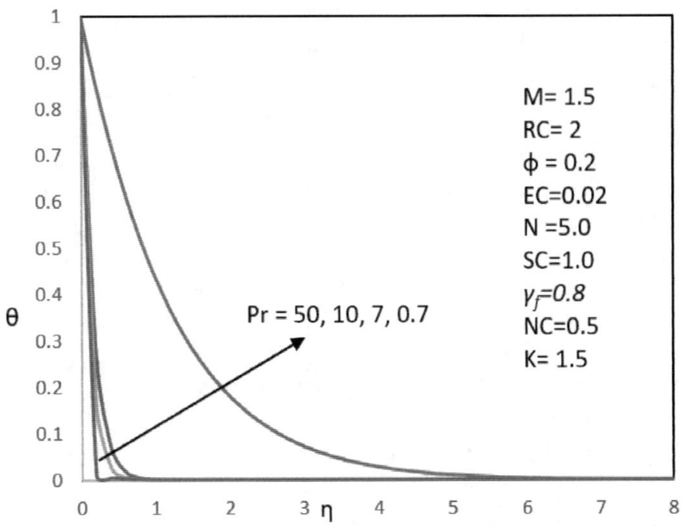

Figure 14 Influence of Prandtl number on fluid temperature.

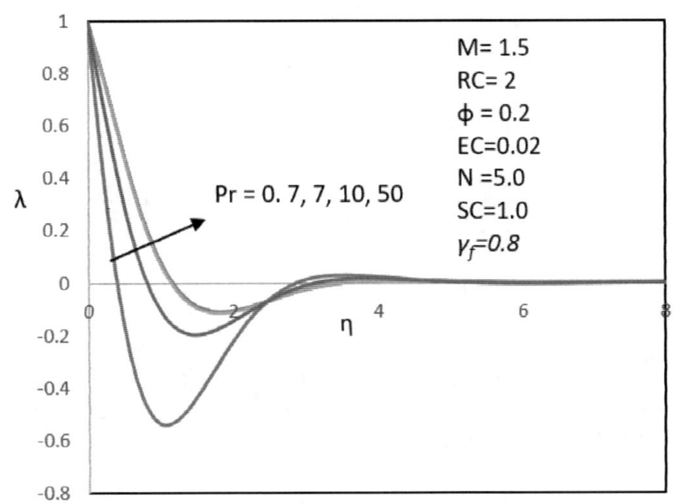

Figure 15. Influence of Prandtl number on fluid concentration.

5. Conclusion

The mathematical exploration has been performed on the influence of microrotation, velocity, concentration and temperature of the fluid flow inside the boundary layer.

Execution is done for the parameters such as hydro magnetic parameter M, volume fraction of nano-solid particle ϕ, coupling constant K, and Prandtl number. Moreover, the effect of the above-mentioned parameters on the fluid properties in the boundary layer are presented graphically.

Final conclusion are as follows;

- The concentration and temperature of the fluid in the boundary layer declines when magnetic parameter increase.
- The velocity and microrotation of the fluid close to the boundary layer drops when magnetic parameter increases.
- For increasing K, the fluid velocity close to the boundary layer increases, but angular velocity of the fluid decreases near the boundary layer.
- For increasing value of the volume fraction of nano-solid particle ϕ, the velocity and micro rotation of the fluid within the boundary layer decreases.
- An upsurge in ϕ reduces the concentration and temperature of the fluid flow significantly while it raises the velocity of the fluid flow.
- The temperature and velocity declines due to an rise in Prandtl number and concentration declines as the Prandtl number decreases.

References

[1]. Choi, S. U. S. (2009). Nanofluids: from vision to reality through research. Journal of Heat Transfer, 131(3): 1–9.

[2]. Yu, W., France, D. M., Routbort, J. L., and Choi, S. U. S. (2008). Review and comparison of nanofluid thermal conductivity and heat transfer enhancements. Heat Transfer Engineering, 29(5): 432–460.

[3]. Tyler, T., Shenderova, O., Cunningham, G., Walsh, J., Drobnik, J., and McGuire, G. (2006). Thermal transport properties of diamond-based nanofluids and nanocomposites. Diamond and Related Materials, 15(11–12): 2078–2081.

[4]. Das, S. K., Choi, S. U. S., and Patel, H. E. (2006) Heat transfer in nanofluids—a review. Heat Transfer Engineering, 27(10): 3–19.

[5]. Liu, M.S., Lin, M. C.C., Huang, I.T., and Wang, C.C. (2005). Enhancement of thermal conductivity with carbon nanotube for nanofluids. International Communications in Heat and Mass Transfer, 32(9): 1202–1210.

[6]. Choi, S. U. S., Zhang, Z. G., and Keblinski, P., Nalwa, H. S. (2004). "Nanofluids," Encyclopedia of Nanoscience and Nanotechnology, 6, 757–737, American Scientific, Los Angeles, Calif, USA.

[7]. Eringen A.C. (1964), Simple Microfluids. Int. J. Eng. Sci., 2: 205 - 207.

[8]. Soundalgekar V.M, Takhar H.S. (1983) Flow of a micropolar fluid on a continuous moving plate. Int. J.Eng. Sci., 21: 961-965

[9]. Rapits A, (1998) Flow of a micropolar fluid past continuously moving plate by the presence of radiation. Int. J. Heat Mass Transfer 4, 2865-2866.

[10]. Hassan A. M. El. Arabawy, (2003) Effect of suction/injection on the flow of a micropolar fluid past a continuously moving plate in the presence of radiation. *Int. J. Heat and Mass Transfer.* 46 1471 – 1477.

[11]. Loganathan P, Golden Stepha N. (2012) Radiation and mass transfer effects on flow of micropolar fluid past a continuously moving plate with suction/injection. *Int. Review of Mechanical Engineering*, 6(1): 22-26.

[12]. Loganathan P, Golden Stepha N. (2012). Chemical Reaction Effect on Forced Convective Heat and Mass Transfer Flow of Micropolar Fluid Past a Continuously Moving Porous Plate in the Presence of Radiation. *International Journal of Heat and Technology*, 30 (2): 107-144.

[13] Abdul Rehman, Nadeem. S. (2012) Mixed Convection Heat Transfer in Micropolar Nanofluid over a Vertical Slender Cylinder, *Chin. Phys. Lett*. Vol. 29, No. 12 124701

[14] Bourantas G.C, Loukopoulos V.C, (2014) Modeling the natural convective flow of micropolar nanofluids, *International Journal of Heat and Mass Transfer*, 68, 35–41

[15] Kazimierz Rup, Konrad Nering, (2014), Unsteady natural convection in micropolar nanofluids, *archives of thermodynamics,* Vol. 35, No. 3, 155–170.

[16] Gangadhar K, Kannan T, Jayalakshmi P,(2017) Magnetohydrodynamic micropolar nanofluid past a permeable stretching/shrinking sheet with Newtonian heating, J Braz. Soc. Mech. Sci. Eng.

[17]. Abdullah Dawar, Zahir Shah Poom Kumam , Hussam Alrabaiah ,Waris Khan , Saeed Islam, Nusrat Shaheen, (2020) Chemically reactive MHD micropolar nanofluid flow with velocity slips and variable heat source/sink, Scientific Reports volume 10, Article number: 20926

[18]. Waqar A. Khan , Rashad A.M , EL-Kabeir S.M.M, EL-Hakiem A.M.A, (2020) Framing the MHD Micropolar-Nanofluid Flow in Natural Convection Heat Transfer over a Radiative Truncated Cone, Processes, 8, 379.

[19]. Khuram Rafique , Muhammad Imran Anwar, Masnita Misiran , Ilyas Khan, Dumitru Baleanu, Kottakkaran Sooppy Nisar , El-Sayed M. Sherif, Asiful H. Seikh, Hydromagnetic Flow of Micropolar Nanofluid, (2020) Symmetry, 12, 251.

[20]. J.A. Adams, D.F. Rogers (1973). Computer-Aided Heat Transfer Analysis. McGraw-Hill.

Green hydrogen economy and opportunities for India

Ujwal Sontakke[1,*], Santosh Jaju[1]

[1]Department of Mechanical Engineering, G H Raisoni College of Engineering, Nagpur, India.
sontakke_ujwal.mtechcc@ghrce.raisoni.net[1], santosh.jaju@raisoni.net[2]

*Corresponding Author

Abstract. With increasing numbers of vehicles on roads, India is facing the issue of large vehicular emissions. The burning of crude oil is the major issue behind these emissions. India doesn't have enough resources to fulfill all the energy demands of vehicles and hence, imports crude oil from oil-rich countries. To tackle the issues associated with oil imports and vehicular emissions, there is a need to search for carbon-free alternate fuel that is available locally in sufficient quantity to meet India's energy demands. The green economy is a new concept evolving and gaining attention worldwide, the concept focuses on sustainable and environmentally friendly solutions. Hydrogen is such a carbon-free fuel that can help to achieve the targets of the green economy and the best means to store energy for a long time. Hydrogen is a high energy content fuel and has about zero greenhouse gas emissions when used in fuel cells. Hydrogen is not directly available in free form, but it can be produced using electrolyzers and various other techniques. India's continuously growing renewable power generation capacity gives the advantage to produce hydrogen from green sources like solar, and wind at the time of lower demand. The present review work focuses on the opportunities for India in green hydrogen production as the adaption of green hydrogen offers many benefits to India including energy security, and decarbonizing the transport sector.

Keywords. Green Hydrogen, Green Economy, Fuel Cell, Energy Security, Alternate fuels, Renewable Energy.

Nomenclature

PEMFC	Proton Exchange Membrane Fuel Cells
IC Engine	Internal Combustion Engine
NO_x	Nitrogen Oxides
CNG	Compressed Natural Gas
GHG	Greenhouse Gases
FCEVs	Fuel Cell Electric Vehicles
PHEV	Plug-In Hybrid Electric Vehicles

1. Introduction

Global warming is an issue; the whole world is trying to resolve. India and China both represent around 36% of the total world population. India is also one of the largest emitters of greenhouse gases in the world. Considering, the population and pollution, it is the responsibility of India to take important steps to limit global warming. The burning of crude oil is one of the major reasons behind the large emission of greenhouse gases in India. To tackle the issues associated with global warming, many countries are now adopting the concept of a green economy. Hence, to effectively contribute to controlling global warming, along with other countries, India also needs to shift its focus on a green economy [1].

Globally, the automobile sector alone is responsible for 24% of CO_2 emissions while in India it contributes to 13.5% of total CO_2 emissions. General environmental management by reducing the net carbon emissions from the transport sector is an important consideration in the green economy. A large proportion of India's energy demand depends upon fossil fuels as shown in Figure 1 [1,2].

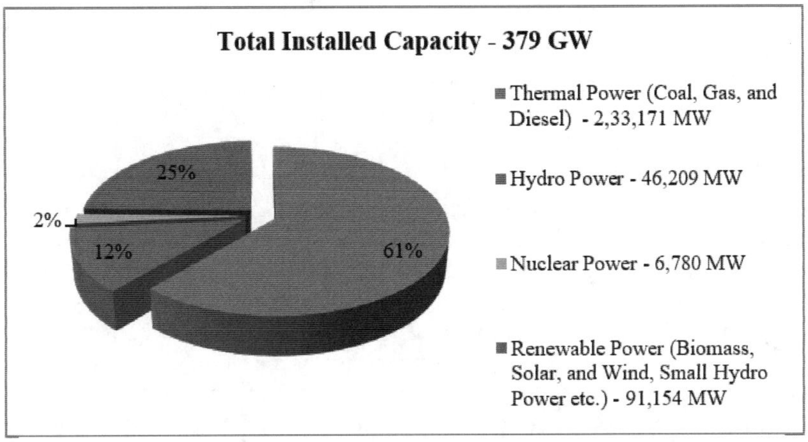

Figure 1. India's total installed power capacity from various renewable and non-renewable sources [2].

To achieve a target of zero-emission, a hydrogen-based economy can make a great impact. Despite many advantages, the hydrogen-based green economy is still facing many challenges including hydrogen storage, commercialized acceptance in the Indian market, standards, regulating policies, safety, and cost. The low volumetric density of the hydrogen is the basic issue with its storage and limits the storage amount of hydrogen on board. The cost of fuel cell stack is a major concern, recently due to the developments and research, the cost of fuel cells has slightly reduced and is expected to be reduced further in future time. The mass production and cost-effective non-platinum catalysts can help to bring the cost down [3,4].

Considering the future scope of hydrogen in energy transition the current review work presents opportunities for India in the energy sector. Section 2 of the context represents the hydrogen as a future fuel option and focuses on hydrogen production, its storage and PEMFCs. Section 3 of the review is focused on using green hydrogen energy for Indian automobiles sector, its importance, and benefits. Section 4 of the review describes the opportunities for India in hydrogen energy and challenges related to them.

2. Hydrogen as a Future Fuel

The properties of hydrogen and its availability makes it better fuel for the future than other fuels. Hydrogen is a clean fuel with high energy density and can be used with both IC engines and fuel cells. With Fuel cells, water in the form of steam is only emittent whereas, in the case of IC engines, NOx is only emittent. Hydrogen is a light gas and non-toxic [5]. Table 1 of the context shows the properties of hydrogen and some other common automobile fuels. Clearly, hydrogen has the highest energy content and octane number when compared to other fuels. The only major issue associated with hydrogen is its low volumetric density.

Table 1. Properties of Automobile fuels [6].

Properties	Gasoline	Diesel	CNG	Hydrogen
Molecular Weight	100-105	200	16.04	2.02
Specific Gravity	0.72-0.78	0.81-0.89	0.424	0.07
Density (kg/m^3)	715-780	800-890	0.7-0.9	0.08375
Energy Content	9.3 kWh/lit	10 kWh/lit	13.1 kWh/kg	33.3 kWh/kg
Octane Number	86-94	-	120+	130+
Cetane Number	5-20	40-55	-	-

2.1. Hydrogen Production

There are various methods to produce hydrogen commercially. These production methods can use both renewable and non-renewable energy sources. But still, conventional fossil fuel dominates hydrogen production with the cost of 1.34 - 2.27$/kg (~₹100 – 200/kg). Hydrocarbon reforming and pyrolysis are the two most widely used and well-developed methods of hydrogen production. Hydrocarbon reforming converts hydrocarbon fuel through reforming techniques to produce hydrogen. Steam reforming involves a catalytic conversion of hydrocarbon whereas partial oxidation is the conversion of steam and hydrocarbon into hydrogen and CO_2. Auto thermal reforming involves exothermic partial oxidation which can increase the hydrogen production rate. Pyrolysis is another important technique in which hydrocarbons go under thermal decomposition [7,8]. Pyrolysis and hydrocarbon reforming both methods consume hydrocarbons but considering fossil fuel depletion and global warming, there is a need to switch on renewable sources for hydrogen production. Grid-based electrolytic hydrogen production is more cost-effective as it increases the utilization of electrolyzers when compared to only renewable power. The cost of hydrogen production from electrolyzers can fall up to the range of 12.6 $/kg (~₹900/kg), as it depends on electricity tariffs, transportation, and storage cost, etc. [9].

Table 2. Types of Hydrogen [10–12].

	Grey Hydrogen	Blue Hydrogen	Green Hydrogen
Derived From	Natural Gas	Natural Gas	Water Splitting and Renewable Power
Method	Steam Methane Reforming	Advanced Gas Reforming	Electrolysis
GHG Emissions	High	Low	Zero
Carbon Capture	CO_2 emitted in the atmosphere	CO_2 captured and stored	No CO_2 emissions
Cost of Production	₹ 100 – 200 /Kg	₹ 300 / Kg	₹ 350 – 400 /Kg

Table 2 shows the classification of hydrogen based on color codings and also clarifies the other information that includes the method of production, GHG emissions, carbon capture, and cost of production. The estimated cost of Grey hydrogen in India is around ₹150 – 200/kg and with the cheapest renewable power; the cost of green hydrogen is around ₹350/kg. Hydrogen production from green energy sources can successfully help to reduce the overall carbon emissions from vehicular emissions. Hydrogen from water electrolyzers is pure enough to be used in fuel cell electric vehicles. Whereas hydrogen from gas transport pipelines requires purification before end-use in FCEVs to remove odor, dust, etc. [7,8]. Figure 2 of the context represents the approximate cost of hydrogen production by different methods.

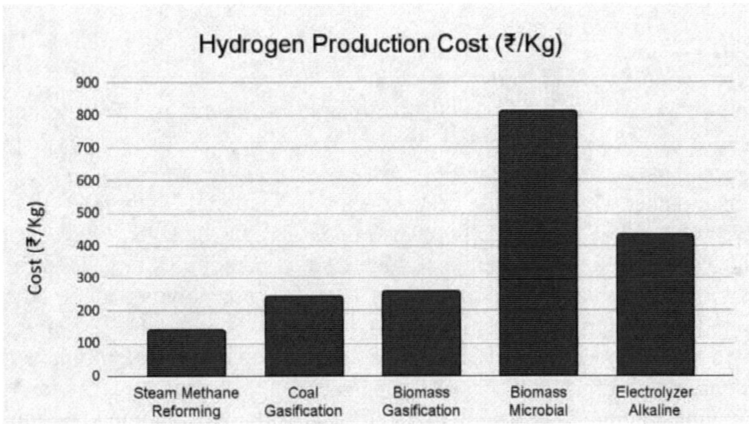

Figure 2. Hydrogen Production Cost by different methods [18].

2.2. *Hydrogen Storage*

The main issue with hydrogen is its storage and transportation as hydrogen is a highly combustible gas. The very low volumetric density makes it difficult to store hydrogen even in small quantities. Liquid and compressed storage are widely used to store hydrogen on a large scale [13]. The hydrogen consumes a large volume even after compressing it at very high pressure. Commercially available fuel cell vehicles opt for 700 bar storage pressure as hydrogen occupies a large space at low pressure. Similarly, high-pressure tanks for decentralized storage of hydrogen especially for transport applications are necessary. However, tanks capable of holding such high pressure are generally made up of carbon fiber which is a very expensive material. As the pressure requirement increases, the quantity of carbon fiber required for the tank rises along with the up-gradation of this compression system specification which can increase the initial cost of storage [14]. Hence many researchers are now focusing on the hydrogen production methods, transportation of hydrogen, and its storage.

Liquefaction of hydrogen requires a significant energy input as the boiling point of hydrogen is very low (-253 °C) but liquid hydrogen provides comparatively a high storage density. Liquefaction consumes about 30% of hydrogen energy. The high volumetric density is the main advantage of liquid hydrogen storage. Another means of hydrogen storage is adsorption which exhibits van der Waals bonding between hydrogen molecules and materials that store hydrogen in the solid phase. Metal hydrides and chemical hydrides exhibit these reactions and operate at low pressure. All three storage options have their respective limitations and hence currently there is no perfect solution for hydrogen storage. The cost involved for hydrogen storage for each of these options is given in Table 3. Many researchers are continuously working in this field to provide a better solution for hydrogen storage and with development, it is improving day by day [15,16].

Table 3. Hydrogen Storage Options [14].

Hydrogen Storage Methods	Compressed Gas Storage	Liquid Hydrogen Storage	Metal Hydride Storage
Hydrogen Transportation by Trailers/ Trucks	63 – 460 Kg	360 – 4300 Kg	Up to 500 Kg
Cost of H_2 Storage	17.9 ₹/Kg H_2	48 ₹/Kg H_2	101.1 ₹/Kg H_2
Cost of H_2 Transportation	35.88 ₹/Kg H_2	2.16 ₹/Kg H_2	31.44 ₹/Kg H_2
Cost of H_2 Fueling Station	30.03 ₹/Kg H_2	37.95 ₹/Kg H_2	[-]

2.3. Proton Exchange Membrane Fuel Cells

PEMFC's are the oldest and well-tested fuel cells that mainly use hydrogen, methanol, and ethanol as fuel. PEMFCs are evolving as a better alternative for conventional fossil fuel vehicles [17]. Automobile companies like Honda, Toyota, and Hyundai have already started the production of commercialized PEM fuel cell vehicles; many other companies are now working in this field. Membrane Electrode Assembly is the most important component of the fuel cells consisting of the Catalyst layer, Microporous layer, and Gas diffusion layer. The Platinum catalysts are mostly used which contributes largely to the overall cost and electrochemical reaction takes place at these catalyst layers. These polymer membranes act as a charge carrier for protons and also separates reactant gases hydrogen and oxygen [18]. The only byproducts that come from the hydrogen fuel cells are heat and water. Zero greenhouse gas emissions, higher efficiency, short refilling time and high power density have placed fuel cell technologies at the top level and opened a large scope for the applications that include powering commercial vehicles, industrial trucks, boats, airplanes, portable appliances, etc. [4]. From the year 2002 to 2017 significant rise in shipments of PEMFC was observed. The transport sector is a major buyer of the PEMFC which contributes about 90% of total shipments. Major companies like Toyota, Honda, and Hyundai have already started the research and development on fuel cell vehicles, their testing and are now focusing on their commercialization [19].

3. Green Hydrogen For Indian Automobile Sector

A total of 11 % of CO_2 emissions in India comes from road transport. This air pollution from vehicular emission can cause various hazardous effects like nausea, breathing difficulties, skin disease, and cancer. Vehicular pollution is also the reason due to which many Indian cities have not succeeded to maintain their air quality index with the safety guidelines of the World Health Organization (WHO) [20,21]. In India, fossil fuel-based vehicles are the major contributor to air pollution followed by industries and thermal power plants. According to the Government of India from the year 1951 to 2012, the number of vehicles has increased from 0.3 million to 159.5 million [22]. Still, the number of vehicles on Indian roads is continuously increasing, Figure 3 of the context shows the data related to total number of vehicles sold in 2019-20.

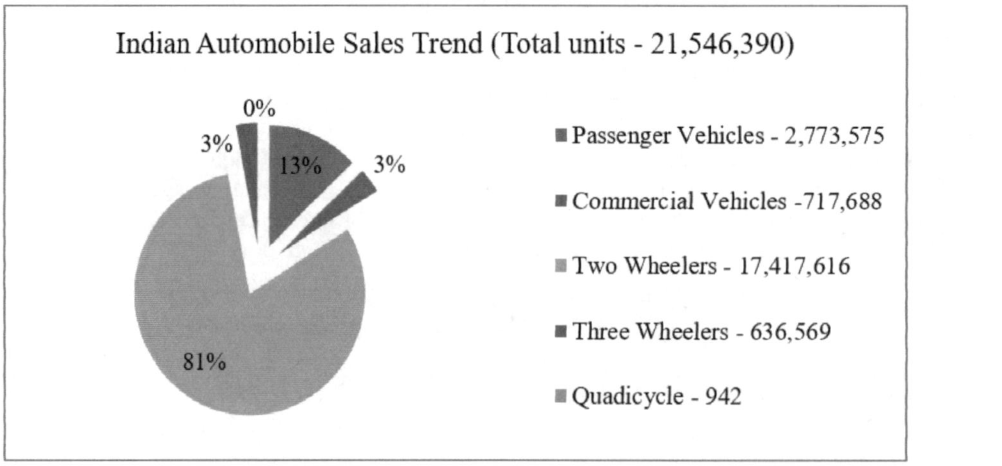

Figure 3. Indian Automobile Sales Trend in India in 2019-20 [22].

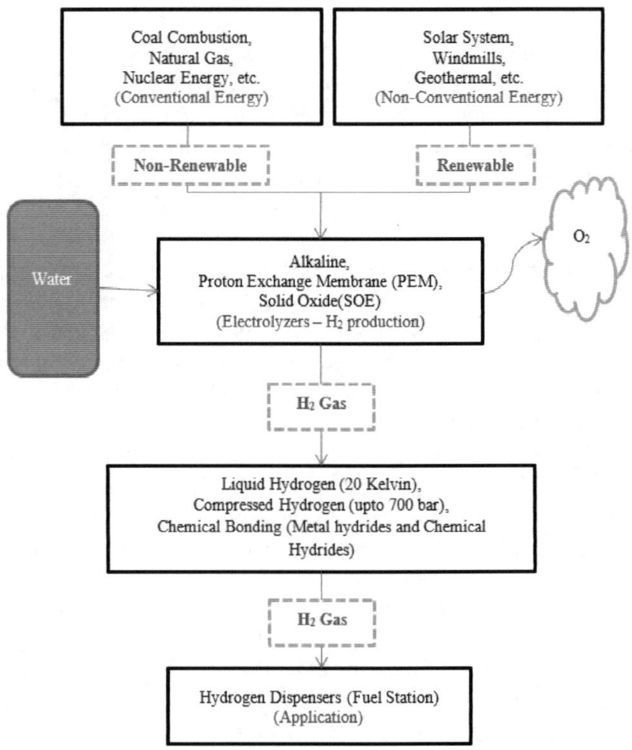

Figure 4. Hydrogen Energy Flow.

The rise in population and economic growth results in the large consumption of fossil fuel-based resources which contribute to the high level of emissions of harmful gases and particulates. Many Indian cities are now witnessing higher Air Quality Index value especially metropolitan areas like Delhi, Mumbai, Kolkata, etc. A major fraction of this air pollution comes from vehicular emissions and the reason behind this is unplanned urbanization, uncontrolled vehicular density, and low turnover of old vehicles [22,23]. Emissions from automobiles depend upon the type of fuel, vehicle types, and maintenance. The automobile sector is responsible for 60% of greenhouse gas emissions. One of the major reasons for large vehicular emissions is the type of fuel and the age of the vehicle. In India, a large number of passenger vehicles run on a diesel engine, which contributes heavily to vehicular emissions as compared to petrol and electrically powered vehicles [23]. Hence, there is a need to find better alternate fuel, which is emission-free and available in abundant in the local market.

3.1. Hydrogen as an Alternative Fuel for Indian Vehicles

Since 1970, Over 90% increase in CO_2 emissions has been recorded as per the United States Environmental Protection Agency [24]. Many nations are now focusing on shifting to electric vehicles by replacing their fleets of ICE vehicles. The reason behind this is very low emissions from electric vehicles. Also, overall emissions can be reduced if renewable electricity sources are available to power electric vehicles [25]. With the increase in demand and positive governmental policies, interest in research and development in this field has been increased. Many improved battery systems are now available in the market, which is the outcome of new research. Li-ion batteries are now the first choice for many automobile companies with their low weight, high energy storage capacity and continuously decreasing price [26]. Two types of electric vehicles are more popular in the market, plug-in hybrid electric vehicles (PHEV) and battery-powered electric vehicles. PHEV is powered by both internal combustion engines and batteries whereas BEVs are fully electric-powered and take power from batteries alone [27]. These are considered potential solutions to the current environmental challenges. Battery recharge time, less availability of charging infrastructure, and low travel range are the main reasons because of which BEVs are still not popular in India. However, with research and development in this field, electric vehicles are now gaining the attention of the public. To overcome all these issues associated with PHEV and BEV, Fuel cell electric vehicles are now becoming a more popular option.

Fuel Cell Electric Vehicles are equipped with fuel cells that consume hydrogen to supply the necessary power to the vehicle's motor. These fuel cells are more efficient and also more environmentally friendly as compared to many other conventional energy conversion systems. Fuel cells available in the market are of various types and can be differentiated based on load requirement, type of electrolyte used, operating conditions, fuel used, and the type of application [17]. The proton exchange membrane fuel cell phosphoric acid fuel cells, alkaline fuel cells, and solid oxide fuel cells are the common types of fuel cells. Among all this, Hydrogen based PEMFC's are more popular in the automobile sector. Some automobile manufacturers have already started working on the manufacturing of commercial fuel cell-based vehicles. These vehicles have great potential considering the range and short refueling time to replace diesel and gasoline-based vehicles. The availability of hydrogen is not an issue as it is abundant in nature, but it is not readily available. Hydrogen can be a better solution for an energy transition. Hydrogen has a huge potential to provide an economical and environmental solution to greenhouse gas emissions.

3.2. Indian Renewable Power Generation Capacity and Green Hydrogen

A large quantity of hydrogen produced worldwide comes from steam reforming of natural gas, coal gasification which contributes heavily to greenhouse gas emissions. To adopt a green hydrogen economy, hydrogen production from renewable sources must be the priority. But a major hurdle in the adoption is its efficiency and cost. Table 1 of the context represents the features, uses, opportunities, and challenges for the future adoption of hydrogen. Coupling solar PV modules and other green energy sources with electrolyzers can greatly benefit the future of green hydrogen production and will help to replace the fossil fuel-based economy with a green hydrogen economy [28]. Ibrahim Dincer in his article [28] has presented many alternatives to produce green hydrogen and some case studies on the same.

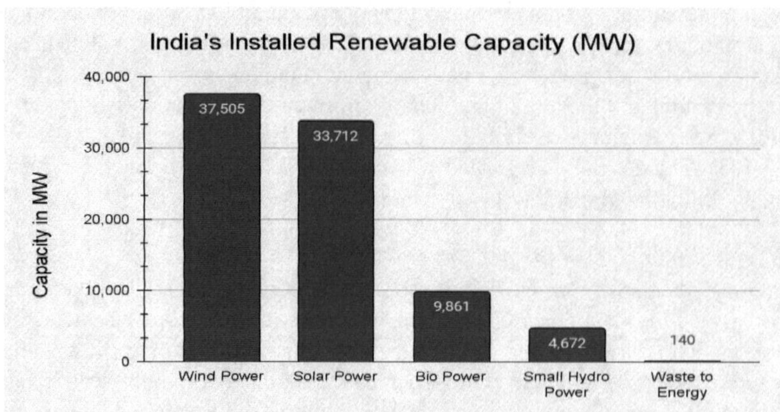

Figure 5. India's Total Installed Renewable Energy Capacity 2019 [29].

From the last decade, a significant fall in the cost of renewable energy has been noticed. Much ongoing research has successfully contributed to the cost reduction of renewable energy. These declining costs will help to produce hydrogen economically where abundant solar and wind energy are available.

Currently in India, the cost of renewable power is near ₹4/KWh and it needs to fall up to ₹1.5/KWh to successfully compete with the grey hydrogen [30]. The government of India is continuously focusing on the increase in renewable source installations. The main intention behind this is to reduce the dependency on fossil fuel consumption and to reduce emissions.

The wind and solar installation in the region can produce sufficient electricity. At the time of lower energy demand, this excess energy can be significantly utilized to produce hydrogen on-site and then can be transported for end-use. Figure 5 of the context gives a detailed idea of India's renewable power generation capacity. India's huge potential in renewable energy gives many advantages to produce green hydrogen which can be then utilized to decarbonize the industries and transport sector significantly. The next section of the context reviews the status of hydrogen and fuel cell technology in India.

3.3. The Blending of Hydrogen in CNG

Natural gas is currently available in various forms such as LNG, CNG, LPG, and also LCNG. Similarly, a new form of natural gas i.e. HCNG (Hydrogen and CNG) is now becoming popular. Hydrogen as a better alternative fuel can be used with CNG. Many kinds of research have shown that Hydrogen blended with CNG (i.e. HCNG) improves the engine performance and significantly reduces emission [31]. Hydrogen can be blended with compressed natural gas in various proportions ranging from 8 to 50% of hydrogen. The combustion of hydrogen only emits NOx and there is no emission of major pollutants like CO, HC, SOx. These benefits from hydrogen blending are gaining the attention of the international community. The blending of hydrogen in CNG even in fewer amounts (5 to 30%) improves the calorific value and flame characteristics of the fuel [19]. With 18% Hydrogen blending in CNG, Sehgal et al. tested the SI engine and found that there was a 5% reduction in fuel consumption, 20% reduction in HC emissions, and 10-20% increase in NOx with better performance characteristics. NOx emissions depend on combustion temperature. The exhaust gas circulation significantly reduces the reaction temperature and hence NOx emission also reduces. Existing vehicles which use CNG as fuel can also run on hydrogen blended CNG with modifications. Similarly, Existing CNG refueling stations can also easily be made compatible with HCNG with some modifications [11,12].

4. Opportunities for India in Green Hydrogen Economy and Challenges

4.1. Opportunities for India

Based on India's current progress in the renewable energy sector, it is clear that green hydrogen will make a greater impact on India's overall energy sector. Green hydrogen will help to provide a sustainable solution for the Indian transport sector. The Energy and Research Institute (TERI) of India has predicted that the demand for hydrogen will increase from 6 Mt to 28 Mt by 2050 and the cost of hydrogen from renewables will fall by 50% by 2030. TERI claimed that about 80% of hydrogen in India will be produced from renewables by 2050 [32]. India has fewer reserves of natural gas and green hydrogen production from renewables can make a difference in this scenario. Under the 'Make in India' program, India has the opportunity to start the production of electrolyzers and fuel cells which will allow capturing a large share in this market worldwide. As compared to other parts of the world, India has a low cost of electricity from the solar photovoltaic systems, this generated power in the future will be helpful to scale up green hydrogen production. Water consumption by electrolyzers is another issue that needs to be discussed. Electrolyzers consume about 9 liters of water to produce 1 kg of hydrogen. In this scenario, seawater electrolysis may be helpful that requires further development and research work [32].

The existing hydrogen infrastructure is not enough to promote the larger acceptance of fuel cell vehicles. For further developments, hydrogen refueling stations will be playing a very important role. At the end of 2019, there were 470 operating hydrogen refueling stations in the world. Countries like Japan, Germany, the United States, and China have a large number of hydrogen stations. Currently, India has only two established hydrogen refueling stations one at the Indian Oil R&D Center, Faridabad, and the second at the National Institute of Solar Energy, Gurugram. To give a boost to the public acceptance of fuel cell vehicles in India, the number of hydrogen fuel stations need to be increased. Figure 6 shows the number of fueling stations in some of the major countries where it can be seen that India has only 2 fueling stations that are used majorly for research purposes [29,33].

For a country like India with its growing economy, controlling pollution and energy security is an important issue. In this scenario, the great potential of hydrogen is far away from gaining popularity in India. India is exploring renewables especially solar, wind and hydropower but hydrogen application is still not in focus. Many Indian institutions including the Indian Space Research Organization (ISRO), Indian Oil Corporation (IOCL), Bharat Heavy Electricals Limited (BHEL), and Tata Motors are actively working on hydrogen and fuel cell technologies. This ongoing research will help India to face challenges regarding fuel cell acceptance, regulations, and standards formulations, etc. [3]. Most of the sustainable energy technologies are developed and manufactured outside India. However, the sizable renewable power generation capacity of the country provides the opportunity to produce green hydrogen and maximize the benefits from the energy transition. In the future, India will see huge growth in the heavy transport sector. Features of fuel cell electric trucks like zero carbon emissions with long-range travel and fast refueling time make these trucks better alternatives for diesel trucks for the Indian heavy transport sector. The increase in the use of locally produced green hydrogen can significantly reduce the dependency on petroleum imports in India. In the same scenario, Table 4 of the context shows some important benefits, opportunities, and challenges of shifting to hydrogen energy.

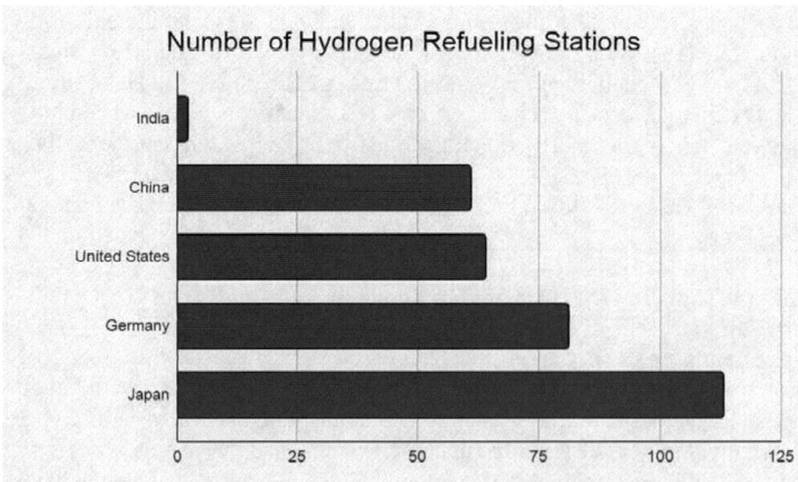

Figure 6. Hydrogen Fuel Stations by Countries [33][29].

Following are some benefits of using green hydrogen as a fuel in India (Source: Department of Science and Technology, New Delhi).

- Reduced Petroleum Imports and Energy Security,
- Decarbonization of Transport sectors,
- Integration of Renewables with Hydrogen production,
- Addresses Climate Change Issues.

The flexibility of fossil fuels to use wherever we want and their high density is not available with renewables. However, hydrogen production integrated with renewables can bring this flexibility. Various Governmental and public agencies in India are investing in the research and development of hydrogen technology, particularly in the development of new materials, processes, systems, etc. India is one of the participants in Mission Innovation (MI) and working on MI's eight Innovation Challenges (IC8) that are focused on renewable and clean hydrogen. Many Indian Institutions and Researchers are constantly working on improvement in hydrogen production, its storage, and transport which will in the future help to increase the utilization of hydrogen [12,29]. Similarly, many organizations and researchers are working on developing more economical and efficient fuel cells to promote their commercial use for the transport sector. Indian organizations are also promoting many demonstration projects. The ongoing developments by the Indian government and public organizations created a lot of opportunities to become a major player in the energy sector.

Table 4. Hydrogen Energy - Features, Uses, Opportunities, and Challenges [34].

Features	Potential - Clean, secure, and affordable future energy source. Versatile – It can be produced from many resources such as renewable energy, nuclear energy, coal, natural gas, etc. Also, hydrogen is the leading option for energy storage. Application – Hydrogen can be used in wide applications such as transport, industry, building, and power generation, etc.
Uses	Industry - Metalworking, Oil refining, Flat glass production, Generator Cooling, etc. Transport – Hydrogen fuel cell-based Passenger cars, Aircraft, Material handling equipment, Buses, Trains, etc. Building – Hydrogen blending with natural gas in commercial buildings, Heating and Cooling purposes in buildings, etc. Power Generation – Gas turbines, Backup power, Stationary fuel cells for remote applications, etc.
Opportunities	Industrial ports, logistic parks – Make industrial ports, logistic parks, etc. hydrogen compatible and encouraging them to shift to a green hydrogen economy. Building a hydrogen infrastructure – To adopt a hydrogen economy and to take advantage of the future market scenario. In transport – Using hydrogen in transport for freights, fleets, and corridors. In trade – International hydrogen trade needs to start soon which can make an impact on the global energy market.
Challenges	Hydrogen production from renewable energy – The cost of hydrogen production from green energy is not efficient and costly. Mass manufacturing of fuel cells, electrolyzers, refueling equipment is not still implemented which contributes heavily to their price. Slow development of hydrogen infrastructure – Less number of refueling stations and other hydrogen infrastructure slows down the wide adoption of hydrogen. The purity of produced hydrogen for fuel cell applications. Governmental regulations - for safety in storage and transportation for large volumes of hydrogen.

4.2. Challenges

The hydrogen economy is a broad concept from the production of hydrogen to its distribution. The successful implementation of which requires great effort. The hydrogen production from renewables is very costly as compared to fossil fuel-based production. Similarly, the mass acceptance of hydrogen requires further improvements in both fuel cell technology and electric drivetrains. Hybrid vehicles bring the benefits of both conventional and fuel cell vehicles, hence their demand will increase in the future. These developments will heavily affect the crude oil and natural gas market worldwide, which can be a major challenge for the global supply chain [35].

In the current scenario, the cost of fuel cells, electrolyzers, and refueling equipment is very high. Mass production of this equipment locally in India can decrease their cost significantly. India is an evolving manufacturing hub but has no major manufacturer of this equipment in the country, which is an issue that needs to be solved. Another challenge for India is to increase the hydrogen production capacity by simultaneously reducing the cost of hydrogen by utilizing its renewable power generation capability [36].

The important challenge associated with the hydrogen economy is the development of hydrogen storage facilities and delivery networks. The low volumetric energy density makes the handling of the hydrogen difficult and more expensive. Safety and financial security is also an important concern need to be addressed with the implementation of new improved technologies and research. Government policies and regulations will play a major role in the whole development. Under the FAME India scheme, the Government of India provides attractive subsidies on electric vehicles to increase awareness and acceptance in its population. Such kind subsidies, if given will encourage the building of a market for hydrogen fuel cell vehicles and hydrogen infrastructures [36,37].

5. Conclusion

India has a large growing population and economy, but comparatively has limited availability of fossil fuels to fulfill its energy demands. The consumption of fossil fuels is contributing to the heavy emission of greenhouse gases. A large number of vehicles that are based on petroleum are the major reason for increased petroleum imports in India. To reduce the environmental pollution and petroleum imports in India, there is a need to look for an alternate fuel for the transport sector. Considering the need to search for an alternate fuel, the context focuses on the opportunities offered by Green Hydrogen Economy and related challenges.

The concept of Green hydrogen economy brings many opportunities for India to become energy independent. For the last decade, India is constantly focusing on growing its renewable energy capacity by taking advantage of its geography. Integrating hydrogen production with these renewables can scale up the green hydrogen production in India. India can take the advantage of its renewable energy scenario and can scale up its hydrogen production facilities. The mass production offers India an opportunity to export green hydrogen to other nations. Green Hydrogen when used with fuel cells can help India significantly reduce its petroleum imports and environmental pollution. Renewable energy in India provides the opportunity to produce green hydrogen and to develop hydrogen infrastructure but for adoption, many challenges still need to be solved. These challenges include hydrogen production cost, storage, transportation, policies, regulations, public awareness, etc.

The world is slowly moving towards the adoption of a Hydrogen economy and India is also taking important initiatives. Indian organizations which include both government and public are investing in the research of hydrogen technologies. Many Ongoing research and demonstration projects are very important to develop hydrogen and fuel cell technology economically. The progress in this development will play a key role in the commercialization of the technology. Well-developed fuel cell technology and locally produced green hydrogen will be key players to decarbonize the Indian transport sector by replacing the current petroleum-based vehicle engines.

6. Reference

[1] Warner KJ, Jones GA. The 21st century coal question: China, India, development, and climate change. Atmosphere (Basel) 2019;10:1–17. https://doi.org/10.3390/atmos10080476.
[2] Ministry of Power G of I. Power Sector at a Glance ALL INDIA 2021. https://powermin.gov.in/en/content/power-sector-glance-all-india.
[3] Medisetty VM, Kumar R, Ahmadi MH, Vo DVN, Ochoa AAV, Solanki R. Overview on the

Current Status of Hydrogen Energy Research and Development in India. Chem Eng Technol 2020;43:613–24. https://doi.org/10.1002/ceat.201900496.
[4] Rath R, Kumar P, Mohanty S, Nayak SK. Recent advances, unsolved deficiencies, and future perspectives of hydrogen fuel cells in transportation and portable sectors. Int J Energy Res 2019;43:8931–55. https://doi.org/10.1002/er.4795.
[5] Ansari A, Hapani B, Kathrotia D, Gokani R, Ajudiya CD. A Review Paper on Hydrogen Gas as Alternate Fuel for Four Stroke IC Engine. IRE Journals 2017;1:11–5.
[6] DOE US. Alternative Fuels Data Center - Fuel Comparison Chart 2021;100:1–4.
[7] Staffell I, Scamman D, Abad V, Balcombe P, Dodds PE, Ekins P, et al. Environmental Science The role of hydrogen and fuel cells in the global energy system 2019:463–91. https://doi.org/10.1039/c8ee01157e.
[8] Nikolaidis P, Poullikkas A. A comparative overview of hydrogen production processes. Renew Sustain Energy Rev 2017;67:597–611. https://doi.org/10.1016/j.rser.2016.09.044.
[9] Mallapragada DS, Gençer E, Insinger P, Keith DW, O'Sullivan FM. Can Industrial-Scale Solar Hydrogen Supplied from Commodity Technologies Be Cost Competitive by 2030? Cell Reports Phys Sci 2020;1:100174. https://doi.org/10.1016/j.xcrp.2020.100174.
[10] Woodside Energy. Hydrogen - Woodside Energy. WoodsideComAu n.d. https://www.woodside.com.au/what-we-do/hydrogen (accessed March 23, 2021).
[11] matthey.com. Hydrogen for a cleaner future. MattheyCom n.d. https://matthey.com/en/markets/energy-generation-and-storage/hydrogen (accessed March 23, 2021).
[12] Department of Science and Technology ND. India Country Status report on Hydrogen and fuel cells 2020.
[13] Acar C, Dincer I. Review and evaluation of hydrogen production options for better environment. J Clean Prod 2019;218:835–49. https://doi.org/10.1016/j.jclepro.2019.02.046.
[14] Deloitte China, Ballard. Fueling the Future of Mobility Hydrogen and fuel cell solutions for transportation. Financ Advis 2019;1:Volume 1.
[15] Yadav K, Sircar A. Hydrogen Compressed Natural Gas and Liquefied Compressed Natural Gas: Fuels for Future. J Energy Manag 2017;2:29–33.
[16] Anderson J, Gronskvist S. Large-scale storage of hydrogen | Elsevier Enhanced Reader 2019.
[17] Abdelkareem MA, Elsaid K, Wilberforce T, Kamil M, Sayed ET, Olabi A. Environmental aspects of fuel cells: A review. Sci Total Environ 2021;752. https://doi.org/10.1016/j.scitotenv.2020.141803.
[18] Peighambardoust SJ, Rowshanzamir S, Amjadi M. Review of the proton exchange membranes for fuel cell applications. vol. 35. 2010. https://doi.org/10.1016/j.ijhydene.2010.05.017.
[19] Wang Y, Ruiz Diaz DF, Chen KS, Wang Z, Adroher XC. Materials, technological status, and fundamentals of PEM fuel cells – A review. Mater Today 2020;32:178–203. https://doi.org/10.1016/j.mattod.2019.06.005.
[20] Yadav SK, Sahay M. A Study on Automobile Industry Growth in India and Its Impact on Air Pollution 1991:1–7.
[21] Khandar C, Kosankar S. A review of vehicular pollution in urban INDIA and its effects on human health. J Adv Lab Res Biol 2014;V:54–61.
[22] Gurjar BR, Ravindra K, Nagpure AS. Air pollution trends over Indian megacities and their local-to-global implications. Atmos Environ 2016;142:475–95. https://doi.org/10.1016/j.atmosenv.2016.06.030.
[23] Arunkumar M, Dhanakumar S. Vehicular pollution and its implications. Poll Res 2019;38:634–47.
[24] United States Environmental Protection Agency. Global Greenhouse Gas Emissions Data | Greenhouse Gas (GHG) Emissions | US EPA n.d. https://www.epa.gov/ghgemissions/global-greenhouse-gas-emissions-data (accessed June 10, 2021).

[25] Towoju OA, Ishola FA. A case for the internal combusion engine powered vehicle.pdf 2020:315–21.
[26] Iclodean C, Varga B, Burnete N, Cimerdean D, Jurchiş B. Comparison of Different Battery Types for Electric Vehicles. IOP Conf Ser Mater Sci Eng Pap 2017. https://doi.org/10.1088/1757-899X/252/1/012058.
[27] Shafiq S, Irshad U Bin, Al-Muhaini M, Djokic SZ, Akram U. Reliability Evaluation of Composite Power Systems: Evaluating the Impact of Full and Plug-in Hybrid Electric Vehicles. IEEE Access 2020;8:114305–14. https://doi.org/10.1109/ACCESS.2020.3003369.
[28] Dincer I. Green methods for hydrogen production 2011;7. https://doi.org/10.1016/j.ijhydene.2011.03.173.
[29] Ministry of New and Renewable Energy G. Hydrogen Energy | Ministry of New and Renewable Energy, Government of India. MnreGovIn n.d. https://mnre.gov.in/new-technologies/hydrogen-energy (accessed March 23, 2021).
[30] MEC Intelligence. Abridged India Hydrogen and Fuel Cells Show 2020 | MEC + 2021. https://mecintelligence.com/mec-perspectives/articles/abridged-india-hydrogen-fuel-cells-2020/ (accessed June 10, 2021).
[31] Yadav S. Review On Effectiveness Of Hydrogen Enrichment In Cng Fuelled International Conference on Efficient Engineering Systems 2020. Int Conf Effic Eng Syst 2020 Conf Proceeding Vol 01 *Corresponding 2021.
[32] Hall W, Spencer T, Renjith G, Dayal S. The Potential Role of Hydrogen in India. Energy Resour Inst 2020:1–24.
[33] Bermudez J, Hasegawa T, Bennett S. Hydrogen – Analysis - IEA. Track Report, IeaOrg 2020. https://www.iea.org/reports/hydrogen (accessed March 23, 2021).
[34] The Future of Hydrogen – Analysis - IEA. IEA 2019. https://www.iea.org/reports/the-future-of-hydrogen (accessed March 23, 2021).
[35] Tseng P, Lee J, Friley P. Hydrogen Economy: Opportunities and Challenges * n.d.
[36] Pudukudy M, Yaakob Z, Mohammad M, Narayanan B, Sopian K. Renewable hydrogen economy in Asia - Opportunities and challenges: An overview. Renew Sustain Energy Rev 2014;30:743–57. https://doi.org/10.1016/j.rser.2013.11.015.
[37] Ministry of Heavy Industries and Public Enterprises G of I. National Automotive Board (NAB) 2020. https://fame2.heavyindustry.gov.in/ (accessed June 13, 2021).

Mechanical Characterization of TiO_2 nanoparticles based on Glass Fiber Reinforced Polymer Composite

Abhishek Singh[1], S. C. Jayswal[2]

Madan Mohan Malaviya University of Technology, Gorakhpur, Uttar Pradesh
[1]aksinghgida@gmail.com, [2]scjme@mmmut.ac.in

Abstract. Nanotechnology has become the best truly developing innovation in the field of engineering science. Numerous examinations have been completed by different exploration researchers in the prior many years. In my examination work research, the impact of cross breed E-glass built up fiber with epoxy Nano composite. The Nano composite covers overlays were set up by hand layup procedures by shifting layers of Titanium Dioxide (TiO_2) nanoparticles of 0.6% individually. The nano added substances are utilized to improve the strength from destroy opposition, hardness of the polymer composite and high strength to weight ratio. The Nano composite laminates this prepared are characterized by the compression and flexural test. The flexural properties of the glass fiber built up plastic improved with expansion of nanoTiO_2 filler particles. At 0.6 wt% of TiO_2 and having 12 layers the force at yield is 327.99N and bending stiffness 63.11 N/mm and in 9 layers force at yield is 149.06 and bending stiffness 36.22 N/mm. True interfacial bonding b/w the fiber and epoxy turned into the primary motive for reaching higher flexural properties.

Keywords. Nanoparticles, flexural strength, compression test, E-glass fiber, hand lay-up

1. Introduction

A composite material is made of two materials with different physical and chemical properties. When the two materials are got mixed, they create a material. Which is specialized for a specific job. It becomes stronger, lighter or resistant to electricity. It can improve the strength & stiffness. This new era there is a large growth in the material of composite. The composite material is a real venture for the fibers. The composite is made of epoxy base material. Epoxy base composite matrix has brilliant potential to supernumerary the conventional steel substances. Polymer matrix is the alternate procedures to growth new elegance of polymer structural substances. This modification can be completed with the aid of addition of various ceramic powders of varied sizes to reap the desired mechanical property [1]. In recent years, the incorporation of nanoparticles into polymers has been attracting great interest. Numerous studies deal with the outcome of nanoparticles on the mechanical properties of composite materials. Composites are moreover known as Fiber-Reinforced Polymer (FRP) composites. It is made-up of polymer matrix. This is bolstered with an engineering material. The man-made fiber (like glass, carbon or aramid) or particular reinforcing fiber. The matrix protects the fibers from environmental or outer damage and transmissions of the load on numerous fibers. The fibers, in turn, offer electricity and stiffness to enhance the matrix—and assist it resist cracks and fracture. Amit Chatterjee et. al. [2] investigated the property of the matrix of nanoparticles size fillers into epoxy resin. The ultrasonic mixing process via sonic cavitation was employed to disperse the

particle into resin system. The nanoparticles are spread evenly throughout the entire volume of resin. The nano composite stronger the storage modulus, Tg, tensile modulus, and flexural modulus from neat epoxy resin. The mechanical performance and thermal stability of the nano composites epoxy are based on the dispersion of the $TiO2$ within the epoxy matrix. Pallabi Bhuyan et.al. [3] investigated the effect of Cu powder and nanostructured Cu powder on the property of Cu-E-glass fiber composites changed into examined.
The file show pinnacle interplay among the matrix and the reinforcement and there can be comparable distribution of the reinforcement within the matrix. The Cu-E-glass fiber composites hardness turn out to be determined to boom from 0.8GPa to 2.7GPa. As the volume weight percentage of fiber glass is increased for unmilled case. As in milled case it also increases. Chandra Shekar et. al. [4] the use of filament winding technique there are few polymers matrix composite was manufactured. The glass fiber reinforced epoxy, carbon fiber reinforced epoxy matrix polymer composites and hybrid fiber reinforced polymer composites are manufactured successfully. And it is found that fracture toughness value for carbon reinforced polymer composite is higher as compared to E-glass fiber reinforced polymer composite. Mahavir Choudhary et. al. [5] build marble dust in various weight percentage by the used of vacuum-assisted resin transfer molding (VARTM) technique. It filled with glass fiber with epoxy resin composite materials. The thermo mechanical and erosive wear property of glass fiber composite was determined. The tensile strength and modulus of the not filled polymer composite is around 430.76 MPa and 3.85 GPa. Mawarnie Ismail et. al. [6] investigated the composite material which is made by reinforcing the bidirectional glass fibre and RH short fiber with epoxy resin. The best combinations result of hybrid composite is having 5 wt% of RH fiber and 25 wt% of glass fiber gives best mechanical properties like tensile test, flexural test and impact test. M. Sreenivasan et. al. [7] concluded the analysis of a coil spring which is composed of a polymer matrix composite material and performing analysis of the coil spring on Ansys 16.0 software he proposed that the equivalent stress, strain, deformation on applying a load of 5000 N on coil spring there is lesser change as compared to the carbon fibre and Kevlar fibre on applying the same load on it. S. Nallusamy et. al. [8] studied the effect of nanoparticles of titanium oxide as glass fibre is reinforced with it. The outcome of hybrid E-glass bolstered fiber with epoxy nanocomposite. The nanocomposite laminate is prepared with the aid of hand lay-up technique with the aid of various the composition of titanium oxide. It is observed that at 3 wt % of $TiO2$ the most flexural power become attained because of true interfacial bonding b/w the fibre and epoxy. Raut et.al [9] The analytical and experimental investigation on Automotive outdoor door deal with. The production of Handle the use of composite material with Glass fiber as reinforcement and Polyamide as a matrix turned into successful. We examine the experimental test as well as theoretical test and there may be infrequently any difference in load sporting capability. Hence, we can conclude that most reliable amount of fabric is used to fabricate the Handle no any extra material is used.

1.1 Classification of composites:

Composite are classified in two main terms as they are mentioned below:
1. **Matrix phase**: The primary segment, having a non-stop character, is referred to as matrix. Usually, high ductile and lots much less hard segment is the nature of matrix. It holds the scattered portion and offers a heap with it.

2. Dispersed (reinforcing) phase: The second phase is embedded within the matrix in a discontinuous shape. This secondary segment is known as moreover as dispersed section.

1.2 Classification of composites based totally on matrix material:

a). **Metal Matrix Composites (MMC):** MMC are group of materials incorporated with various reinforcing phases such as whiskers or continuous fiber or particulate fibers.
b). **Ceramic Matrix Composites (CMC):** CMC are special type of composite material in which both the reinforcement and matrix material are ceramics.
c). Polymer Matrix Composites (PMC): PMC is the material consist of a polymer (resin) matrix combined with a fibrous reinforcing dispersed phase.

1.3 Polymer Matrix Composite (PMC): Polymer Matrix Composite (PMC) is the material consisting of polymer (resin) matrix combine with a fibrous reinforcing dispersed phase Polymer matrix composite are very popular due to their low cost & simple fabrication method. PMCs are currently the most widely utilized composite of the available composites. Ceramic fibres are used to reinforce the matrix in PMCs because they have a higher strength than the matrix material. The matrix, reinforcement, process parameters, microstructure, composition, and interphase all have an impact on the capabilities of PMCs. PMCs are well-known for their low cost and simple manufacturing process. PMC producers can use a variety of manufacturing strategies to generate cost-effective items. Each manufacturing process contains structures that define the type of product that will be produced. The manufacturer is able to provide this knowledge as the best option for the customer. Matrix of polymer the thermoplastic or thermosetting plastic serves because the matrix, with one or greater reinforcements including carbon, glass, metal, and herbal fibres. Polymers produce correct components because they can be treated effectively. Lightweight is the most significant attribute of polymers. PMCs (polymer matrix composites) have a wide range of characteristics. High strength, super impact, compression, and fatigue characteristics; value-effective production and tooling strategies; first rate chemical and corrosion resistance; chemical inertness and superior mechanical features are only a few of them. PMCs are used in rockets, planes, and other vehicles, safety equipment and helmets.

2. Materials and Methodology

2.1 Materials

a) **E-glass fiber:** In present paper work E-glass fiber was chosen for the matrix material. This fiber is 300 gsm it has a good strength than the other glass fibers as it shows in below figure 3. It has a large application in automobile, aerospace, structural, paper industry, defence, food processing, chemical industry and cooling towers, fireproof suits and due to its fabulous properties, such as high strength to weight ratio, light weight and high tensile strength. It is cheaper than the other polymer matrix composite material. From the diverse varieties of glass fiber. This fiber is drastically used due to maximum regularly used fibers due to its immoderate electricity, low weight and has nicely warmth stiffness, cold resistance, moisture and rust resistance [10].

Table 1: Properties of E-glass fibre

Property	Minimum value	Maximum value	Units
Density	2.55	2.6	Kg/m^3
Young modulus	72	85	GPa
Compressive Strength	4000	5000	MPa
Hardness	3000	6000	MPa
Tensile strength	1950	2050	MPa

b) **Titanium dioxide nanoparticle (TiO2):** Titanium dioxide is an inorganic chemical with a white tint that has been utilized in a extensive variety of merchandise for over a century. It is valued for its non-poisonous, non-reactive, and luminous houses, which effectively lighten the whiteness and brightness of a massive range of substances. It's the brightest and whitest. It has a numeral of particular dispositions which make it perfectly suited to various unique applications. It has a virtually immoderate melting problem of 1843°C and boiling trouble of 972°C, so appear virtually as a robust. It is far un-dissolvable in water. It works as an insulator. Another critical asset of titanium dioxide is that it may show photocatalytic interest beneath UV mild. Which makes it most operative for environmental purification, for specific styles of protecting coating layers, sterilisation and anti-fogging surfaces, or even in maximum cancers therapy for maximum cancers affected character.

c) **Copper oxide (CuO):** Copper nanoparticles are generally proper away available in most volumes. Researchers have furthermore endorsed the use of silver and copper ions as superior disinfectants chemical factors for wastewater generated from hospitals containing infectious micro-organism. The antibacterial property of copper nanoparticles, which show that copper nanoparticles have a extraordinary capacity as bacterial agent. Copper non-public immoderate ductility and actually excessive thermal and electric conductivity. Copper particles are in brown to black powder.

Fig 1 Copper Oxide

Fig 2 Titanium dioxide nanoparticles (TiO2)

Fig 3 E-glass fiber of 300 gsm

2.2 Bonding Material: The resin epoxy it truly is used have become Araldite (LY-556) epoxy resin and chemically belongs to 'epoxide' circle of relatives and hardener (HY -951), aliphatic 951 are provided through Ashwani Composite GIDA, Uttar Pradesh, India. Araldite, which, is used to bind the glass fiber layers collectively. Hardener is picked to decorate the strong point and grip to the composite. Epoxy resin is highly viscous and is mixed with a chemical substance to make curing substance which is called as hardner. Araldite LY556 is notable property like mechanical, dynamic, thermal & chemical resistance at temperatures under 75°C. Nature of hardener is colourless. Its aroma like ammonia with its pH value 13 [9]. The individual layers of composite the mild steel roller is used to roll on the layers of composite to remove the air which is present inside layer during coating of epoxy and preserve uniform thickness of epoxy inside the layers of fiber. Protects are cured at room temperature for as a minimum 48 hours earlier than characterization. Bonding material figure is shown as in fig 4.

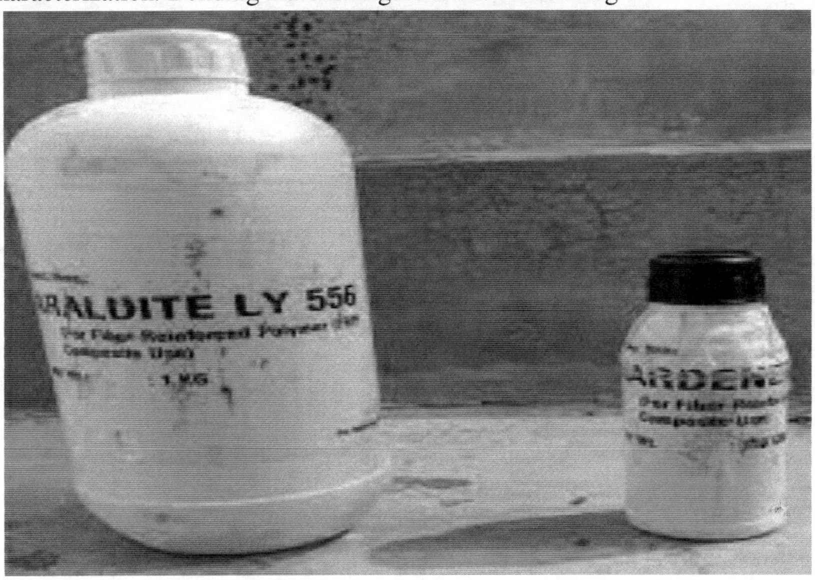

Fig 4 Araldite LY556 and Harder 951

2.3 Fabrications of composite material

The work material was fabricated by hand lay-up method. This method is the moulding method. In this method the fibers are fabricated from reinforcement and covered with the matrix epoxy resin by using laying down. The coating is applied on the layer of each fabric and according to our required thickness the layers are used. It is more time-consuming method and it required more labour for productions. The important disadvantage of this method is undesirable shaped products are made. Such advantages are utilized in low performance composite including fiber-glass boat and bath tub manufacturing. The ultrasonication of TiO_2 with acetone is performed at 36 KHz frequency for 60 minutes. Afterward the TiO_2 nanoparticles is keep to cool for one hours. The proper mixing of titanium dioxide (TiO_2) nanoparticles with epoxy LY 551 by the stirrer at 800 rpm for 1 hours. And then the mixture of TiO_2 with epoxy is ready to apply the paste on the glass fiber. The titanium dioxide nanoparticles are present in 0.6 wt % and glass fiber with epoxy resin and coating of glass fiber with copper oxide. The steps involved in Hand lay-up method.

Hand –Lay-Up method is also called as manual method. The following are the steps involved in the preparation of glass fiber (FRC laminates).

Step 1: Cut the glass fiber polymer of 300gsm into 210 mm X 297 mm size.

Step 2: Preparation of the sample is done by cutting the glass fiber, placing the glass fibre of the size of A4 sheet paper.

Step 3: Then, prepare the matrix by mixing of titanium dioxide nanoparticles (TiO_2) with epoxy LY556 and hardener HY 951 in the ratio of 10:3.

Step 4: The A4 sheet is placed on the wooden plate by placing the weight of 40 kg.

Step 5: After placing the OHB sheet in the mould, the matrix is placed above OHB sheet in the mould and spread evenly.

Step 6: Once the Epoxy is placed, the glass fibre is placed above the OHB sheet with Epoxy.

Step 7: After placing the cotton fibre in the mould, the epoxy should be poured on the cotton fibre in the mould and the fibre should be immersed in the epoxy resin.

Step 8: OHB sheet is placed over the immersed glass fibre and it should be free of voids otherwise the desired strength should not be achieved.

Step 9: The laminate is allowed for curing in atmospheric condition for 48 hours.

Step 10: The specimen should be taken out from the mould and after that the OHB sheet is also removed after the time period is over that is 2 days.

Fig 5 Glass fiber with epoxy titanium dioxide

3. Result and Discussion

3.1 Compression Test

Compression test are conducted by applying loads on the specimen. The specimen is kept in b/w the two plates and then the suitable load is applied to know the effect occurred on the specimen. Specimens are prepared according to ASTM D 638 Standards. The dimension is 260 mm x 24 mm x 5 mm. This composite sample is made of having 0.6% of titanium dioxide nanoparticles with glass fibers with epoxy resin and coating with copper oxide. This composite specimen is made of having 9 layers. The composite specimen A is compressed as it is shown in fig-6 and it is found the composite is break when the force is applied of 6.87 KN and compressive displacement at break is 0.50 mm. This composite specimen C is made of 12 layers but the titanium percentage is fixed and during the testing it is found that the composite is break when force is applied of 10.50 KN and compressive displacement at break is 0.49 mm. This composite sample B having 9 layers specimen is compressed and it is found the composite is break when the force is applied of 2.31 KN and compressive displacement at break is 0.50 mm. The composite specimen D is made of 12 layers but the titanium percentage is fixed and during the testing it is found that the composite is break when force is applied of 3.72 KN and compressive displacement at break is 0.50 mm as shown in fig-7. Compression test result of all the sample is as mentioned below in tabulated form:

Table 2 Compression results of all four specimen

S.No.	Sample	Fiber layers	With filler/without filler	Force at break [KN]	Compressive displacement at break [mm]	Coefficient of correlation at Modulus
1.	A	9L	With filler	6.87	.50	1.00
2.	B	9L	Without filler	2.31	.50	1.00
3.	C	12L	With filler	10.50	.49	1.00
4.	D	12L	Without filler	3.72	.60	1.00

3.2 Flexural strength

A flexural test is easier than the tensile test. Specimens are prepared according to ASTM D 638 Standards. The dimension is 150 mm x 15 mm x 3 mm. The sample is made of having 0.6% of titanium dioxide nanoparticles with glass fibers with epoxy resin and coating with copper oxide. This composite specimen is made of having 9 layers of glass fiber. The composite specimen is flexural and it is found that composite is bend when the force at yield of 149.09 and flexural displacement at 4.51 mm in sample A as shown in fig 8. The sample C is made of 12 layers the composite is flexible when force at yield point is 327.29 N and flexural displacement at 5.70 mm. The composite sample B is made of glass fibers with epoxy resin and coating with copper oxide as shown in fig 9. The specimen is made of having 9 layers. The composite specimen is flexural and it is found the composite is bend when the force at yield of 112.11 N and flexural displacement at yield is 4.96 mm and bending stiffness 25.29 N/mm. In sample D specimen is made of 12 layers and during the testing it is found that the composite is bend when force at yield 130.03 N and displacement at yield is 3.28 mm.

Flexural test of all the four sample is shown in tabulated form:

IOP Conf. Series: Materials Science and Engineering 1206 (2021) 012006 doi:10.1088/1757-899X/1206/1/012006

Table 3 flexural test result of all four samples.

Sr. No.	Samples	Fiber layers	With filler/without filler	Force at yield (offset 0.2%) [N]	Displacement at yield (offset0.2%) [mm]
1.	A	9 L	With filler	149.06	4.51
2.	B	9L	Without filler	112.11	4.96
3.	C	12L	With filler	327.99	5.70
4.	D	12L	Without filler	130.03	3.28

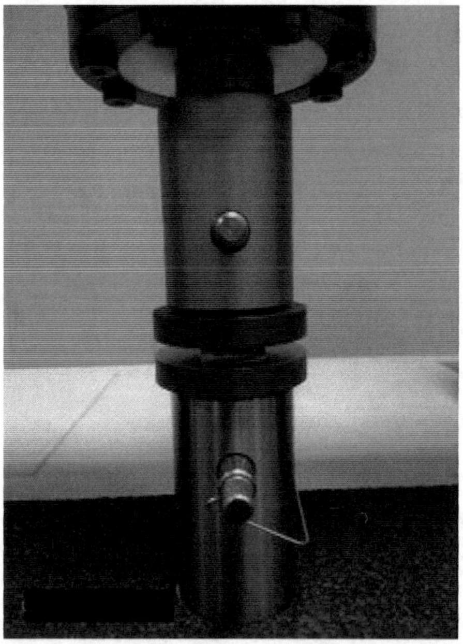
Fig-6 Compression sample A

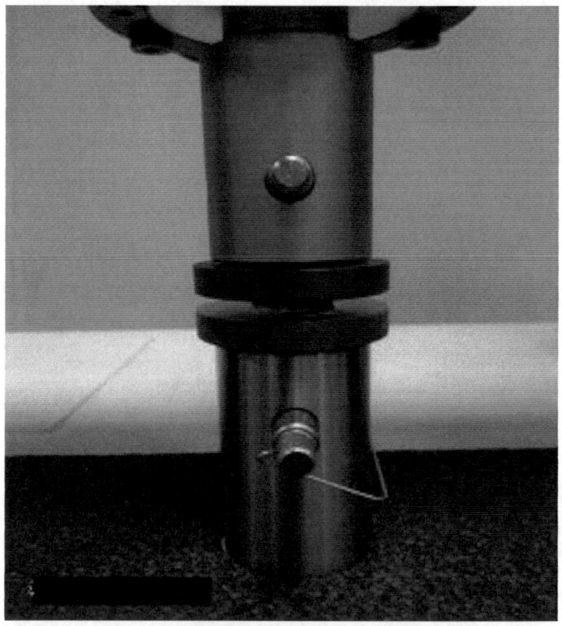
Fig-7 Compression sample D

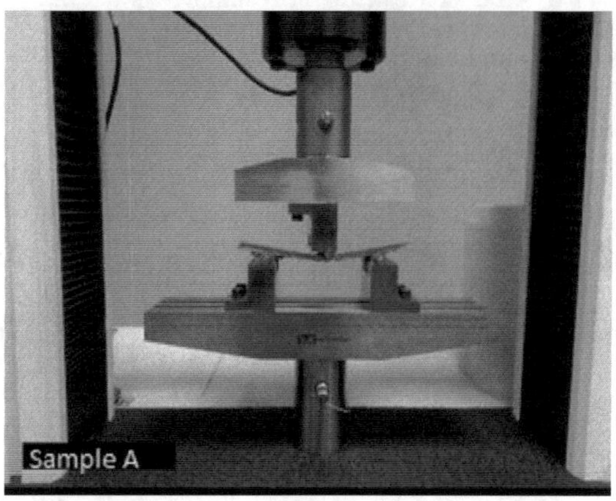

Fig-8 flexural sample A

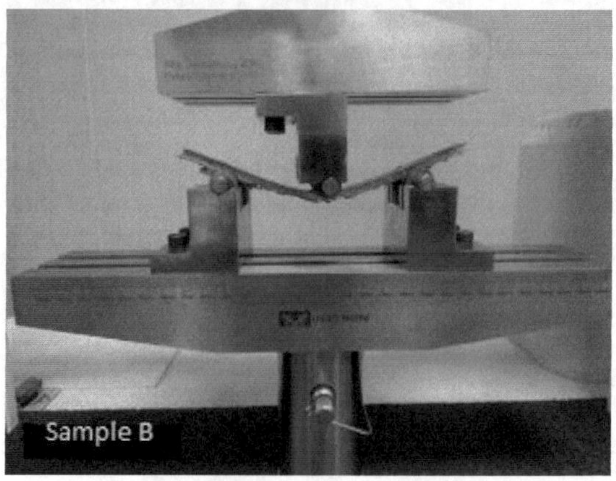

Fig-9 flexural sample B

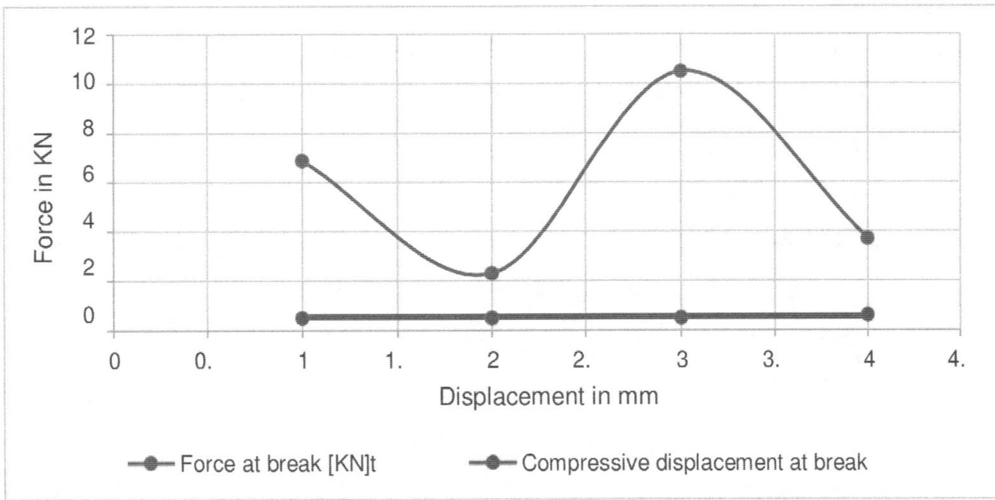

Fig 10 Compression test of composite sample

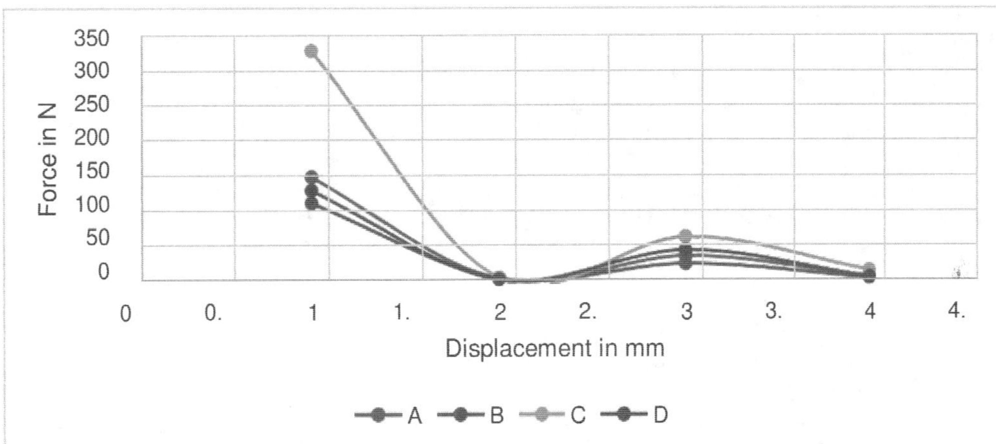

Fig 11 Flexural test of composite samples

4. Conclusion

In this paper work the glass fiber epoxy titanium dioxide nanoparticles with copper coating consist of 0.6 weight percent of nanoparticles of TiO_2 with different no. of fiber layers specimen is prepared. The different composite specimen shows tremendous mechanical property.

There is two testing were performed for the prepared composite material such as flexural test (three point bending) and compression test.On the comparision of the compression test sample having 9 layers and 12 layers respectively with filler TiO2 nanocomposite. The strength to weight ratio of this composite is higher. The result of compression test of the sample the force at break in 9 and 12 layers are 6.87 KN and 10.50 KN respectively.As we found that in 12layer composite is more stronger than the 9layer composite sample. As the layer is increases the strength but filler material is remains constant.But during the neat (without filler material) composite the compression test of the 9 layers and 12 layers. It is found that the force at break is 2.31 KN and 3.27 KN respectively. There is little increase in force at break.In the flexural testing of this composite material with filler material it shows an excellent flexural strength.The composite which is made of glass fiber with reinforcement TiO2 epoxy composite. The composite with 9 layer and 12 layers the force at yield are 149.06 N and 327.99 N respectively. The composite sample with neat (without TiO2) the testing of sample flexural test. The force at yield in 9 layers and 12 layers are 112.11 N and 130.03 N respectively.

Acknowledgments

It gives me great pleasure to express my thanks and indebtedness to my supervisor, Prof. S.C. Jayswal, for his counsel and direction from the beginning of this study. I am also grateful to Vibronics Medical Pvt Ltd Kundli, Sonipat Haryana, for allowing me to test my composite samples there.

Reference

1. Rankin, Susan M., Mikayla K. Moody, Amit K. Naskar, and Christopher C. Bowland. "Enhancing functionalities in carbon fiber composites by titanium dioxide nanoparticles." *Composites Science and Technology* 201 (2021): 108491.
2. Chatterjee, Amit, and Muhammad S. Islam. "Fabrication and characterization of TiO2–epoxy nanocomposite." *Materials Science and Engineering: A* 487, no. 1-2 (2008): 574-585.
3. Sharma, Devesh, Shailesh Yadav, Suraj Chand, Umesh Kumar, Bhaskar Chandra Kandpal, Ashok Kumar, and D. K. Gupta. "Fabrication and characterization of Glass Fiber/SiC reinforced polymer composite." *International Journal of Applied Engineering Research* 13, no. 6 (2018): 175-178.
4. Shekar, K. Chandra, B. Singaravel, S. Deva Prasad, N. Venkateshwarlu, and B. Srikanth. "Mode-I fracture toughness of glass/carbon fiber reinforced epoxy matrix polymer composite." *Materials Today: Proceedings* 41 (2021): 833-837.
5. Choudhary, Mahavir, Tej Singh, Ankush Sharma, Maheshwar Dwivedi, and Amar Patnaik. "Evaluation of some mechanical characterization and optimization of waste marble dust filled glass fiber reinforced polymer composite." *Materials Research Express* 6, no. 10 (2019): 105702.
6. Ismail, Mawarnie, M. R. M. Rejab, J. P. Siregar, Zalinawati Mohamad, M. Quanjin, and A. A. Mohammed. "Mechanical properties of hybrid glass fiber/rice husk reinforced polymer composite." *Materials Today: Proceedings* 27 (2020): 1749-1755.
7. Sreenivasan, M., M. Dinesh Kumar, Ram Krishna, T. Mohanraj, G. Suresh, D. Hemanth Kumar, and A. Sai Charan. "Finite element analysis of coil spring of a motorcycle suspension system using different fibre materials." *Materials Today: Proceedings* 33 (2020): 275-279.
8. Nallusamy, S. "Characterization of epoxy composites with TiO2 additives and E-glass fibers as reinforcement agent." In *Journal of Nano Research*, vol. 40, pp. 99-104. Trans Tech Publications Ltd, 2016
9. Raut, Girish Naresh, and V. P. Mali. "Optimizations of outer door handle in automobiles using composite material." (2017).
10. VijayKiran, B., K. Mallikarjuna Rao, and D. LingaRaju. "An investigation on mechanical properties of e-glass fiber reinforced polymer nanocomposites." *Materials Today: Proceedings* 18 (2019): 5454-5463.

Comparative assessment of hard turning under dry and minimum quantity lubrication

Avez Shaikh[1], Ajinkya Shinde[1,*], Satish Chinchanikar[2], Guruprasad Zagade[1] and Sonia Pardeshi[1]

[1]Department of Mechanical Engineering, Vishwakarma Institute of Information Technology, Pune, India.
[2]Professor, Department of Mechanical Engineering, Vishwakarma Institute of Information Technology, Pune, India.

avez.21820048@viit.ac.in, ajinkyashinde.work@gmail.com[*]

Abstract. Hard turning with CBN and ceramic inserts is commonly regarded as a cost-effective alternative to grinding. However, there have been few studies comparing hard turning with low-cost carbide tools to high-cost CBN and ceramic cutting tools. However, when it comes to the usage of cutting coolant during severe turning, there are mixed outcomes. In this study, a PVD-coated TiSiN-TiAlN carbide tool was used to hard turn AISI 52100 steel in a dry and MQL environment. Through multi-objective optimization, a comparative assessment in terms of surface roughness, cutting force, and tool life under various cutting settings is provided. In terms of three components of cutting force, surface roughness, and tool life, mathematical models were constructed to forecast and improve machining performance. Under both dry and MQL conditions, the study discovered an optimal cutting speed of 108 m/min, a feed value of 0.09 mm/rev, and a depth of cut of 0.16 mm. Under MQL, hard turning produced optimal surface roughness and tool life of 0.88 m and 64 minutes, respectively. In comparison to hard turning under dry cutting, the optimal surface roughness was 1.07 m and the tool life was 49 minutes. Under MQL, tool life increased by over 31%, according to the findings of the experiments. Under dry and MQL conditions, however, no significant differences in cutting forces and surface roughness were identified.

Keywords. Machining performance, hard turning, dry and MQL quantity lubrication.

1. Introduction

Manufacturing is evolving fast nowadays as a result of environmental replenishment and new technology [1]. The use of CBN and ceramic inserts in hard turning is generally recognised as a low-cost alternative to the more expensive grinding procedure. Hard turning using a low-cost carbide tool vs high-cost ceramic and CBN cutting tools has been the subject of few research. However, when it comes to the usage of cutting coolant during severe turning, there are mixed outcomes. Cutting fluids constitute a severe concern for environmental preservation and human health, hence demand for dry or machining with a minimum quantity lubrication (MQL) is growing [2].

The majority of the researchers have put forth a lot of work to compare dry cutting, wet cutting, and MQL cutting. Shokrani et al. [3] found a variety of cooling strategies that can be used in conjunction to reduce or eliminate the need of traditional cutting fluids. During turning, Amrita et al. [4] found that MQL nano-graphite augmented cutting fluid performed better. However, it has been found that the concentration of nanoparticles in the base fluid has a significant impact on torque and power consumption during the micro-drilling process under MQL [5].

During turning, a group of researchers [6-8] found that MQL performed better than dry and flood cooling. Yldrm et al. [9] found that turning Ni-based Inconel 625 with hBN nanofluid under the MQL reduced surface roughness and increased tool life compared to turning under dry circumstances. Bruni et al. [10], on the other hand, found no benefit in using the MQL to reduce tool wear. Furthermore, moist cutting resulted in increased surface roughness, according to their findings. MQL enhanced tool life, according to Liao and Lin [11], which they attribute to the formation of a protective oxide layer between the tool-chip contacts that acts as a diffusion barrier. Duc et al. [12] observed a decrease in cutting forces, friction coefficient, and tool wear during their experimental work on hard milling utilising aluminium oxide nanofluid under MQL. MQL increased tool life, according to Chinchanikar and Choudhury [13], due to improved lubrication and cooling. However, the efficiency of cooling is highly dependent on the choice of cutting speed, feed, and cut depth [14].

MQL reduced tool wear and surface roughness, according to Dhar et al. [15-16]. This could be due to a shift in tool-work and tool-chip interactions caused by a drop in cutting temperature. Hegab et al. [17-18] used multi-walled carbon nanotubes suspended in vegetable oil to test the MQL turning performance of Titanium alloy. The bonding between the workpiece and tool interface improved surface quality, tool life, and power consumption in their experiments using the nanofluid MQL. Chinchanikar et al. [19] found that while dry hard turning generated a superior surface finish, coconut oil as a cutting fluid produced a better surface finish at higher cutting speeds. MQL, according to Khan et al. [20], aids in reducing tool wear caused by abrasion, adhesion, and diffusion type wear. Under MQL conditions, Kilickap [21] noticed superior surface qualities than compressed air and dry drilling. MQL provides certain challenges for lubricants in reaching the exact cutting zone, according to Attanasio et al. [22].

According to the literature review, there are conflicting results on the usage of cutting fluids during severe turning. Furthermore, parametric optimization is frequently documented with ceramic and CBN tools during hard turning. However, just a few studies on multi-objective optimization of hard turning with a carbide tool have been published. In light of this, cutting parameters were optimised under dry and MQL conditions in order to discuss the use of coolants during hard turning in this study. Experimental observations were used to construct correlations between cutting parameters and performance indicators, specifically three components of cutting force, tool life, and surface roughness.

2. Experimental details

Experiments were carried out utilising a TiSiN-TiAlN coated carbide tool on AISI 52100 steel (62 HRC). The insert was mounted using a right-hand style tool holder defined by ISO as PCLNR 2525M12. Experiments were planned using a central composite rotatable design matrix [23]. Experiments were carried out with a mist made up of a minimum of 60 ml/hr of water-based cutting oil and 5 bar of compressed air. By combining compressed air and oil immediately beyond the nozzle, a minimum amount of fluid flow was obtained. The nozzle contains two inlets, one for the cutting fluid and the other for the air, which are 90 degrees apart.

Table 1. Experimental conditions.

Parameters	Levels				
	1	2	3	4	5
Cutting speed (*V*) (m/min)	100	108	125	142	150
Feed (*f*) (mm/rev)	0.081	0.088	0.113	0.142	0.15
Depth of cut (*d*) (mm)	0.1	0.2	0.3	0.4	0.5

3. Results and discussion

A strain-gauge type dynamometer was used to measure the average values of the cutting force components. A Qualitest surface roughness tester was used to measure the roughness of the surface. Dino-Lite, a USB-powered portable digital microscope, was utilised to track flank wear and growth at regular intervals along the length of the cut. Hard turning studies on AISI were carried out on a CNC lathe with PVD-coated nano laminated TiSiN-TiAlN carbide inserts, changing feed, depth of cut, and speed. ISO has assigned the CNMG 120408 geometry to all of the insets (80o diamond shape with 0.8 mm nose radius). The inserts were mounted using a right-hand style tool holder defined by ISO as PCLNR 2525M12.

Twenty hard turning experiments were carried out, each with different cutting parameters under dry and MQL cutting circumstances. Experiments were carried out with a fluid flow rate of 60 ml/hr and a compressed air pressure of 5 bar. Table 1 shows the experimental matrix for dry and MQL cutting used in this work. Cutting parameters were chosen based on the manufacturer's instructions and a review of the literature. Based on the experimental results, mathematical models are created to estimate three components of cutting force, surface roughness, and tool life under MQL and dry circumstances, taking into account the effect of feed (f), cutting speed (V), and depth of cut (d),

Cutting Force = f (V, f, and d) and can be written as:

Cutting force = $K\ V^a f^b d^c$

where, the constant K and exponents a, b, and c are to be determined by substituting the experimental cutting force values at the given speed, feed, and depth of cut. The similar type of work can be referred from authors earlier published work [13,19, 25].

Table 2. Coefficients of force model under dry cutting.

Performance index	Cutting environment	Experimental-based mathematical models	R^2 value	Equation no.
Tangential cutting force (Fc)	Dry	$Fc = 2974.1\ V^{-0.2669} f^{0.7359} d^{0.0644}$	0.9	(1)
	MQL	$Fc = 4191.03\ V^{-0.3376} f^{0.682} d^{0.1176}$	0.92	(2)
Feed cutting force (Ff)	Dry	$Ff = 4049.45\ V^{-0.4531} f^{0.5305} d^{0.3347}$	0.97	(3)
	MQL	$Ff = 4525.48\ V^{-0.5232} f^{0.3897} d^{0.3187}$	0.95	(4)
Radial cutting force (Fr)	Dry	$Fr = 512.613\ V^{-0.0314} f^{0.2775} d^{0.2281}$	0.94	(5)
	MQL	$Fr = 543.34\ V^{-0.0368} f^{0.2873} d^{0.1928}$	0.93	(6)
Surface roughness (Ra)	Dry	$Ra = 33.508\ V^{-0.3502} f^{0.6886} d^{0.0766}$	0.87	(7)
	MQL	$Ra = 12.445\ V^{-0.0673} f^{0.845} d^{0.1547}$	0.95	(8)
Tool life (T)	Dry	$T = 16839513\ V^{-3.1217} f^{-0.643} d^{-0.38}$	0.92	(9)
	MQL	$T = 4456.21\ V^{-0.3369} f^{0.8202} d^{0.0884}$	0.91	(10)

A strain-gauge type three-component dynamometer was used to measure the tangential or primary cutting force (Fc), feed force (Ff), and radial cutting force (Fr) in this study. The least square error method was used to examine the experimental results. Data-Fit software was used to determine the unknown coefficients (Version 8.1).

The three components of cutting force, the tangential or main cutting force (Fc), feed force (Ff), and radial cutting force (Fr), tool life, and surface roughness, are generated using experimental data. The constant and exponent values for developed models are shown in Table 2. The R-Squared values of the developed models; equations (1) to (10) are in the range of 0.85 to 0.95, indicating that the developed models are valid and could be used to predict the investigated performance during AISI 52100 steel hard turning with PVD-applied nanocrystalline TiSiN-TiAlN coated carbide tool within the domain of the process parameters selected in this study.

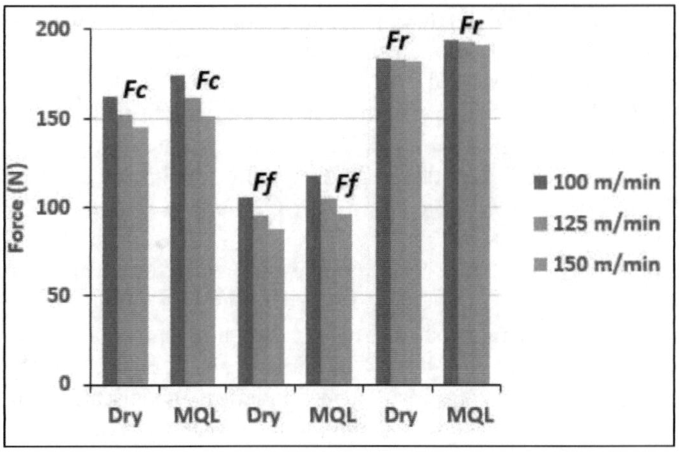

(a)

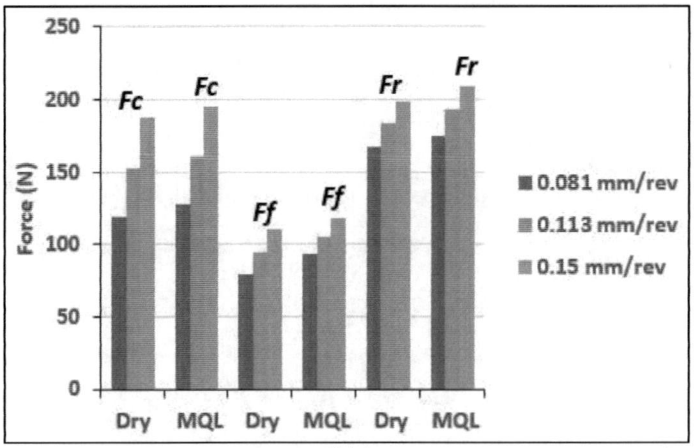

(b)

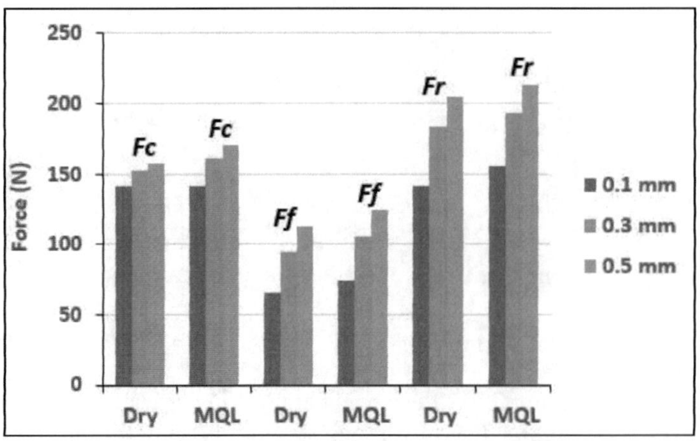

(c)

Figure 1. Variation of cutting force components with (a) Cutting speed, (b) Feed, and (c) Depth of cut.

The standard RSM technique was used to analyse the experimental findings in order to study the effect of cutting parameters on performance measures and to identify the region of interest where performance measures attained their optimum or near optimum value. Using the derived equations, the cutting force varies with process parameters.

Figure 1(a) displays the cutting force components as a function of cutting speed and depth of cut of 0.113 mm/rev and 0.3 mm, respectively. Figure 1(b) displays the cutting force as it varies with feed, whereas Figure 1(c) depicts the cutting force as it varies with depth of cut. Figure 1(b) depicts the results obtained at V = 125 m/min and d = 0.3 mm. Figure 1(c) shows the cutting force components measured at V = 125 m/min and f = 0.2 mm/rev.

Figure (1) shows that cutting forces are larger under MQL and decrease with increasing cutting speed while increasing with increasing d and f. It is also clear that the radial component of force has a higher value than the tangential and feed forces. Cutting forces were seen to be greatly reduced at higher values of V when paired with higher values of feed, which may be attributed to a shear strength reduction of the workpiece due to greater temperature generation, particularly during dry cutting conditions [24-25].

Figures (2) and (3) show the variance in surface roughness and tool life under various cutting conditions. Table 3 shows the cutting condition matrix for figures 2 and 3.

Table 3. Cutting condition matrix

Cutting condition	Cutting speed (m/min)	Feed (mm/rev)	Depth of cut (mm)
1	100	0.113	0.3
2	125	0.113	0.3
3	150	0.113	0.3
4	100	0.081	0.3
5	150	0.15	0.3
6	100	0.113	0.1
7	150	0.113	0.5

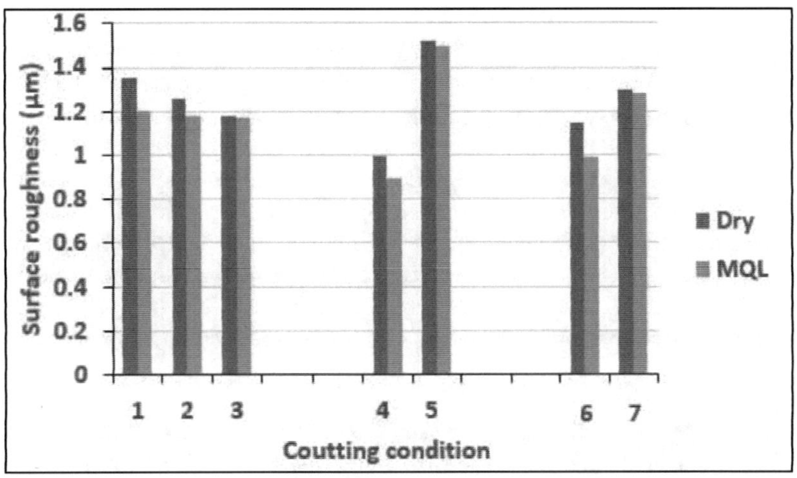

Figure 2. Surface roughness at different cutting conditions.

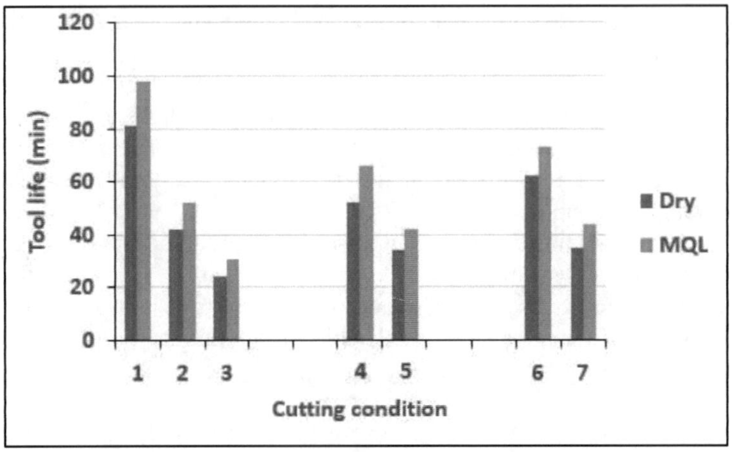

Figure 3. Tool life at different cutting conditions.

From figure (2), it can be seen that superior surface polish is obtained under MQL cutting and at higher V, lower f, and d values. However, with higher cutting parameters, this result is not as noticeable. The improved surface finish under MQL circumstances could be due to improved chip breaking and chip control over dry cutting. However, the improvement in surface smoothness under MQL was not as considerable since the greater cutting pressures used in MQL caused more vibrations and thus chatter marks on the completed workpiece. In conclusion, there was no substantial benefit to employing MQL during hard turning in terms of improving surface smoothness and lowering cutting force.

Figure 3 demonstrates increased tool life when MQL cutting is used and process parameters are kept to a minimum. However, at higher cutting parameters, there is no discernible improvement in tool life. The values of the exponents and constants ((9) and (10) show that V is the most influential parameter on tool life. However, this outcome is more noticeable while dry machining. The low tool life found under dry conditions was due to crater wear causing a rapid fracture of the tool. However, at greater V, this effect was severe.

4. Multi-objective optimization

Furthermore, for various cooling settings, the optimum process parameters that result in greater tool life, surface finish, and minimum cutting force were established. To determine the best cutting parameters, the Design-Expert software was employed. The optimal option is the one with the maximum desirability. Under both the dry and MQL circumstances, V = 108 m/min, d = 0.16 mm, and f = 0.09 mm/rev are shown to be optimal. Hard turning with MQL produced the best tool life and surface roughness of 64 min and 0.88 m, respectively. In comparison to dry cutting, we achieved an optimal surface roughness of 1.07 m and a tool life of 49 minutes. Under MQL, experimental results reveal a considerable increase in tool life, nearly by 31%. However, there was no discernible change in surface smoothness or force between chilling strategies. The increased tool life under MQL can be attributed to the generation of a minimal cutting temperature and, as a result, a minimum tool wear due to adhesion wear.

5. Conclusion

A comparative study of machining performance during hard turning with MQL and dry conditions is addressed in this paper. Experiments on AISI 52100 steel (6-62 HRC) were carried out with a coated TiSiN-TiAlN carbide tool. For installing the insert, an ISO PCLNR 2525M12 right-hand style tool holder was employed. The trials were conducted with a MQL of 60 ml/hr at 5 bar air pressure. When compared to dry cutting, the MQL condition produced a considerable increase in tool life of over 31 percent. However, in MQL conditions, no substantial benefit in terms of obtaining a superior surface finish or lowering the force has been noted. This study found that a cutting speed of 108 m/min, a depth of cut of 0.16 mm, and a feed rate of 0.09 mm/rev are the best parameters for both dry and MQL conditions. Hard turning with MQL produced the best tool life and surface roughness of 64 min and 0.88 m, respectively. In comparison to dry cutting, an optimum surface roughness of 1.07 m and tool life of 49 minutes were recorded. As dominant wear forms, nose wear, chipping off at the nose, and clearance face were noticed.

References

[1] Grzesik W 2008 *Advanced machining processes of metallic materials: theory, modelling and applications* (Elsevier).

[2] Diniz A E and Ferreira J R 2003 Influence of refrigeration/lubrication condition on SAE 52100 hardened steel turning at several cutting speeds *Int. J. Mach. Tools Manuf.* **43**(3) pp 317–26.

[3] Shokrani A, Dhokia V and Newman S T 2012 Environmentally conscious machining of difficult-to-machine materials with regard to cutting fluids *Int. J. Mach. Tools Manuf.* **57** pp 83–101.

[4] Amrita M, Srikant R R and Sitaramaraju A V 2014 Performance evaluation of nanographite-based cutting fluid in machining process *Mater. Manuf. Processes* **29**(5) pp 600–05.

[5] Garg A, Sarma S, Panda B N, Zhang J and Gao L 2016 Study of effect of nanofluid concentration on response characteristics of machining process for cleaner production *J. Cleaner Prod.* **135** pp 476–89.

[6] Kumar C R V and Ramamoorthy B 2007 Performance of coated tools during hard turning under minimum fluid application *J. Mater. Process. Technol.* **185**(1-3) pp 210–16.

[7] Kang M C, Kim K H, Shin S H, Jang S H, Park J H and Kim C 2008 Effect of the minimum quantity lubrication in high-speed end-milling of AISI D2 cold-worked die steel (62 HRC) by coated carbide tools *Surf. Coat. Technol.* **202**(22-23) pp 5621–24.

[8] Varadarajan A S, Philip P K and Ramamoorthy B 2002 Investigations on hard turning with minimal cutting fluid application (HTMF) and its comparison with dry and wet turning. *Int. J. Mach. Tools Manuf.* **42**(2) pp 193–200.

[9] Yıldırım Ç V, Sarıkaya M, Kıvak T and Şirin Ş 2019 The effect of addition of hBN nanoparticles to nanofluid-MQL on tool wear patterns, tool life, roughness and temperature in turning of Ni-based Inconel 625 *Tribol. Int.* **134** pp 443–56.

[10] Bruni C, Forcellese A, Gabrielli F and Simoncini M 2006 Effect of the lubrication-cooling technique, insert technology and machine bed material on the workpart surface finish and tool wear in finish turning of AISI 420B *Int. J. Mach. Tools Manuf.* **46**(12-13) pp 1547–54.

[11] Liao Y S and Lin H M 2007 Mechanism of minimum quantity lubrication in high-speed milling of hardened steel *Int. J. Mach. Tools Manuf.* **47**(11) pp 1660–66.

[12] Minh D T, The L T and Bao N T 2017 Performance of Al_2O_3 nanofluids in minimum quantity lubrication in hard milling of 60Si2Mn steel using cemented carbide tools *Adv. Mech. Eng.* **9**(7) 1687814017710618.

[13] Chinchanikar S and Choudhury S K 2014 Hard turning using HiPIMS-coated carbide tools: Wear behavior under dry and minimum quantity lubrication (MQL) *Measurement* **55** pp 536–48.

[14] Leppert T 2011 Effect of cooling and lubrication conditions on surface topography and turning process of C45 steel *Int. J. Mach. Tools Manuf.* **51**(2) pp 120–26.

[15] Dhar N R, Kamruzzaman M and Ahmed M 2006 Effect of minimum quantity lubrication (MQL) on tool wear and surface roughness in turning AISI-4340 steel *J. Mater. Process. Technol.* **172**(2) pp 299–304.

[16] Dhar N R, Ahmed M T and Islam S 2007 An experimental investigation on effect of minimum quantity lubrication in machining AISI 1040 steel *Int. J. Mach. Tools Manuf.* **47**(5) pp 748–53.

[17] Hegab H, Umer U, Deiab I and Kishawy H 2018 Performance evaluation of Ti-6Al-4V machining using nano-cutting fluids under minimum quantity lubrication Int. J. Adv. Manuf. Technol. 95.

[18] Hegab H, Kishawy H A, Gadallah M H, Umer U and Deiab I 2018 On machining of Ti-6Al-4V using multi-walled carbon nanotubes-based nano-fluid under minimum quantity lubrication *Int. J. Adv. Manuf. Technol.* **97**.

[19] Chinchanikar S, Salve A V, Netake P, More A, Kendre S and Kumar R 2014 Comparative evaluations of surface roughness during hard turning under dry and with water-based and vegetable oil-based cutting fluids *Procedia Mater. Sci.* **5** pp 1966–75.
[20] Khan M M A, Mithu M A H and Dhar N R 2009 Effects of minimum quantity lubrication on turning AISI 9310 alloy steel using vegetable oil-based cutting fluid *J. Mater. Process. Technol.* **209**(15-16) pp 5573–83.
[21] Kilickap E, Huseyinoglu M and Ozel C 2011 Empirical study regarding the effects of minimum quantity lubricant utilization on performance characteristics in the drilling of Al 7075 *J. Braz. Soc. Mech. Sci. Eng.* **33** pp 52–57.
[22] Attanasio A, Gelfi M, Giardini C and Remino C A R L O 2006 Minimal quantity lubrication in turning: Effect on tool wear *Wear* **260**(3) pp 333–38.
[23] Cochran W G 1957 *Experimental designs 2nd ed.* (John wiley & sons).
[24] Chinchanikar S, Choudhury S K and Kulkarni A P 2013 Investigation of chip-tool interface temperature during turning of hardened AISI 4340 alloy steel using multi-layer coated carbide inserts *Advanced Materials Research* **701** pp 354–58.
[25] Chinchanikar S and Choudhury S K 2013 Effect of work material hardness and cutting parameters on performance of coated carbide tool when turning hardened steel: An optimization approach. *Measurement* **46**(4) pp 1572–84.

Simulation-based Study of Graphene-water Nanofluid flow through Microchannel Heatsink

Ritwik Bhattacharya[1,3] **and Pranab Samanta**[2]

[1] Undergraduate Student, Department of Mechanical Engineering, IIEST Shibpur, West Bengal 711 103, India
[2] Principal Scientist, Department of Surface Engineering and Tribology, CSIR-Central Mechanical Engineering Research Institute, Durgapur, West Bengal 713 209, India

Abstract. This study presents a CFD analysis of the laminar flow of graphene-water nanofluid through a Silicon microchannel heatsink using commercial software ANSYS FLUENT. The microchannel has a rectangular cross-section of given dimensions, and the base of the heatsink is subjected to a constant heat flux. Simulations of the coolant flow are performed at different fluid inlet velocities for nanoparticle concentrations of 0%, 3% and 6% in the base fluid- Water. Results for temperature and pressure distributions in the microchannel heatsink are presented. The cooling performance of the MCHS improves significantly by increasing the flow velocity and enhancing the nanoparticle concentration in the coolant.

Keywords. microchannel, nanoparticle, base fluid, thermal conductivity, cooling.

1. Introduction

The electronics industry has witnessed some phenomenal advancements in the past few decades. However, this progress has been made possible only because of continuous improvements in thermal management technologies that have ensured that the systems work steadily- by removing the high heat generated by the devices.

The path-breaking work of Tuckerman and Pease [1] has given rise to extensive research and development of the microchannel heat sink (MCHS) as a viable device for electronic cooling [2–4]. MCHS comes in handy in micro-level MEMS devices, micro-fluidics and sensitive applications such as micro drug delivery- where the extremely small area of heat dissipation creates significant challenges for the structural and thermal stability of the device and hampers its longevity. For nearly two decades now, research in this domain has focussed on enhancing heat transfer in an MCHS for improvement in cooling performances. Chai et al. conducted a numerical investigation on heat transfer enhancement in a silicon MCHS having offset ribs on sidewalls under laminar flow conditions and inferred that the proposed MCHS model displays better cooling performance than a smooth MCHS [5]. Besides modification of microchannel geometry, heat transfer can also be enhanced by employing nanofluids as coolants in the microchannel. Nanofluids are produced when nanoparticles are suspended in

[3] Author to whom any correspondence should be addressed.
E-mail ID: ritwikbhattacharya99@gmail.com

conventional heat transfer fluids, and they exhibit better thermal conductivity than the base fluids [6]. Lee and Mudawar experimentally investigated the performance of a microchannel heatsink using Al_2O_3-water nanofluid and demonstrated an enhancement in the single-phase heat transfer coefficient, particularly in the laminar flow domain [7]. Al-Baghdadi et al. performed the CFD analysis of some metal-oxide containing water-based nanofluids flowing through a Silicon MCHS in the range of 100<Re<900 and made a thorough investigation on physical parameters such as pressure drop across the microchannel, temperature distributions in the computational domain, etc. for each nanofluid [8].

Graphene is a carbon allotrope with a special 2-D crystal structure that provides it with excellent electrical and thermal properties [9]. The use of graphene as nanoparticles has found much attention in recent times [10]. Several experimental investigations have confirmed the suitability of graphene-water nanofluid as an effective coolant in various systems [11,12]. In this paper, the cooling performance of graphene-water nanofluid in a Silicon MCHS is studied using commercial CFD software ANSYS FLUENT by simulating the laminar flow of the coolant through the microchannel. Various performance measures such as temperature distributions and pressure drops across the microchannel are investigated for different nanoparticle concentrations in the base fluid.

2. Model Description

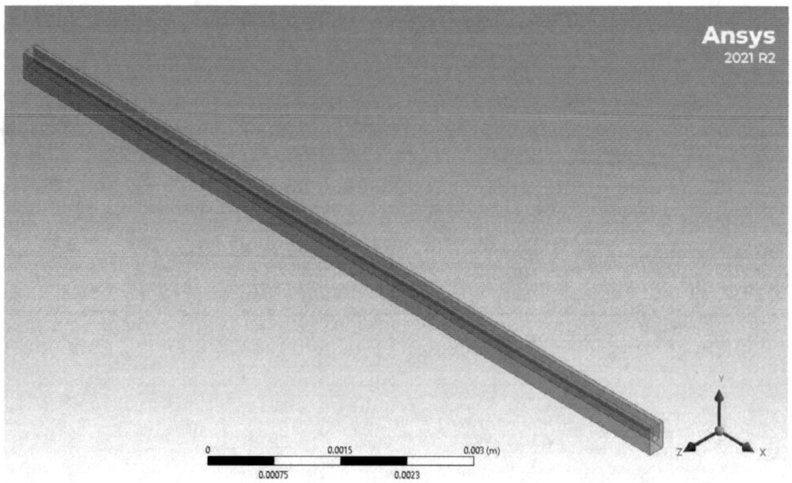

Figure 1. 3D Model of Microchannel Heatsink.

Table 1. Dimensions of Microchannel Heatsink.

Length	Dimension (m)
Length of the MCHS (L)	0.01
Height of the MCHS (H_T)	0.00035
Width of the MCHS (W_T)	0.0002
Height of the Microchannel (H)	0.0002
Width of the Microchannel (W)	0.0001

Figure 1 shows a diagram of the rectangular microchannel heatsink along with the orientation, and Table 1 presents the dimensions. The MCHS is made of Silicon and the same MCHS model has been used by

Al-Baghdadi et al. in their study as well [8]. For easy computation, we consider only a single microchannel for fluid flow. A 3D model of the microchannel heatsink is created in Autodesk Fusion 360 and then exported to the CFD software, where the simulations are carried out.

3. Numerical Modelling

3.1. Conservation Equations

The numerical model presented in this study assumes a laminar, single-phase flow of incompressible fluid across a microchannel and negligible effect of gravity. Thus, the steady-state Navier-Stokes equations can be used to obtain solution [8].

For the fluid "f", the equations for mass and momentum conservations respectively are [8]:

$$\nabla.(\rho_f.u)=0 \tag{1}$$

$$(u.\nabla).\rho_f= -\nabla p+ \mu_f.\nabla^2 u \tag{2}$$

where p refers to the pressure (Pa), ρ_f refers to the fluid density (kg/m^3), u refers to the flow velocity (m/s) , and μ_f refers to the dynamic viscosity (kg/m s) of the fluid.

The energy equation for the fluid is [8]:

$$u.\nabla T= \frac{k_f}{\rho_f.\ C_p}\nabla^2 T \tag{3}$$

where k_f denotes the fluid's thermal conductivity (W/m K), T denotes the temperature(K) and C_p refers to the specific heat capacity (J/kg K) of the fluid.

For the solid "s", energy equation is [8]:

$$k_s.\nabla^2 T=0 \tag{4}$$

where k_s denotes the thermal conductivity of the solid (W/m K).

3.2. Thermophysical Properties

Table 2. Thermophysical properties of Water and Graphene.

Fluid	Density ρ (kg/m^3)	Specific Heat C_p (J/kg K)	Thermal Conductivity k (W/m K)	Dynamic Viscosity μ [kg /m s]
Water*	996.7	4180	0.608	0.000855
Graphene	2267	75	4000	

* Fluid properties at 300 K.

Table 3. Thermophysical properties of Silicon.

Material	Density (kg/m^3)	Thermal conductivity (W/m K)	Specific Heat (J/kg K)	Young's modulus (Pa)	Thermal Expansion (1/K)	Poisson's Ratio
Silicon	2329	130	700	$170e^9$	$2.6e^{-6}$	0.28

The temperature-correlated equation of density (kg/m^3) for water is obtained from the Best-Fit Curve, extrapolating from Density variation with Temperature data [13].

$$\rho_{bf}= 999.9+0.0556(T\text{-}273.15) -(8*10^{-3})(T\text{-}273.15)^2+(6*10^{-5})(T\text{-}273.15)^3\text{-}(2*10^{-7})(T\text{-}273.15)^4 \tag{5}$$

The equations for specific heat capacity (J/kg K), thermal conductivity (W/m K) and dynamic viscosity (kg/m s) respectively for water are as follows [8]:

$$C_{p_{bf}} = 8959.9 - 40.535(T-273.15) + 0.11243\,(T-273.15)^2 - 1.014*10^{-4}*(T-273.15)^3 \quad (6)$$

$$k_{bf} = -0.58166 + 6.3556*10^{-3}T - 7.964*10^{-6}T^2 \quad (7)$$

$$\mu_{bf} = 2.414*10^{-5}*10^{\frac{247.8}{T-140}} \quad (8)$$

The thermophysical properties of the base fluid are considered at the inlet temperature of 300K to resemble real-life conditions closely.

The thermophysical properties of Water at 300 K and graphene are shown in Table 2 [14].

For the nanofluid "nf", the equations of its thermophysical properties are as follows [15]:

$$\rho_{nf} = (1-\varphi).\rho_{bf} + \varphi.\,\rho_{np} \quad (9)$$

$$C_{p_{nf}} = \frac{(1-\varphi).\rho_{bf}C_{p_{bf}} + \varphi.\,\rho_{np}C_{p_{np}}}{\rho_{nf}} \quad (10)$$

$$k_{nf} = k_{bf}*\frac{k_{np}+2k_{bf}-2\varphi.(k_{bf}-k_{np})}{k_{np}+2k_{bf}+\varphi.(k_{bf}-k_{np})} \quad (11)$$

$$\mu_{nf} = \frac{\mu_{bf}}{(1-\varphi)^{2.5}} \quad (12)$$

where φ denotes the volume concentration of nanoparticles "np" in the coolant.

The current study involves Silicon as heatsink material. Table 3 states the thermophysical properties of Silicon.

3.3. Procedure and Boundary Conditions

Table 4. Inlet and Boundary Conditions.

Coolant	Graphene-water
Inlet Fluid Temperature $T_{f,in}$ [K]	300
Outlet gauge pressure [Pa]	0
Inlet velocity of coolant u_m [m/s]	1.3; 2.5
Concentration of nanoparticles φ	0, 3, 6 vol%
Heat Flux on MCHS Bottom wall q [W/m^2]	10^6

The Finite Volume Method is implemented in the present study to discretize the governing equations- which are solved using ANSYS FLUENT 2021 R2. The important inlet and boundary conditions employed are summarized in Table 4. The two different inlet velocities fall within 100<Re<500 for both the base fluid and the nanofluids and hence constitute laminar flow through the microchannel. For the same inlet velocity and boundary conditions, performances of graphene-water nanofluid at various nanoparticle concentrations are analyzed and compared.

The Reynolds number of the fluid flow is defined as:

$$Re=\frac{\rho_f\, u_m D_h}{\mu_f} \tag{13}$$

where D_h refers to the microchannel's hydraulic diameter (m) and u_m refers to the fluid inlet velocity (m/s).

Here, the hydraulic diameter for the rectangular microchannel is:

$$D_h=\frac{2WH}{W+H} \tag{14}$$

where W denotes the microchannel's width (m) and H indicates the microchannel's height (m).

Table 5. Grid Independence Test.

	Elements	Nodes	Relative error
MESH-1	800000	912912	
MESH-2	900000	1018017	0.004421726
MESH-3	1000000	1123122	0.002119347
MESH-4	1100000	1228227	0.001222145
MESH-5	1200000	1333332	0.000781618

The 3D Model is linearly meshed. The grid independence of the solution is established by executing several simulation trials using base fluid water for a fluid inlet velocity of 1.3 m/s. Table 5 lists the relative error $[(\Delta P_{mesh(n)} - \Delta P_{mesh(n-1)})/\Delta P_{mesh(n)}]$, calculated for 5 different meshes. MESH-4 is chosen-which contains 1100000 elements and 1228277 nodes. It is found to provide good resolution, yet it is economical in terms of processing time. SIMPLE Algorithm is selected for implementing Pressure-Velocity coupling. The computational domain is initialized with the inlet boundary conditions before commencing the iterative process. For spatial discretization of energy and momentum equations, a second-order upwind scheme is chosen. For achieving solution convergence, a residual of 1.0×10^{-6} is selected for the continuity, velocity and energy equations.

3.4. Model Validation

For validating the model, the Average Nusselt Numbers of the simulated model using water at distinct flow Reynolds Numbers within the laminar domain are compared with the results experimentally obtained by Chai et al. [5].

The average Heat Transfer Coefficient (W/m^2 K) and Nusselt Number respectively are [5]:

$$h=\frac{q.A_{bot}}{A_{con}.(T_{w,\,area}-T_{f,\,m})} \tag{15}$$

$$Nu=\frac{h.D_h}{k_{f,m}} \quad (16)$$

where the applied heat flux upon the bottom wall of the MCHS (W/m^2) is denoted by q, A_{bot} denotes the silicon bottom wall area (m^2), A_{con} denotes the inner wall-fluid contact area, $k_{f,m}$ indicates the mass-average thermal conductivity of the fluid, $T_{w,area}$ denotes the area-weighted temperature of the MCHS bottom wall (K) and $T_{f,m}$ indicates the mass-average temperature of the base fluid domain.

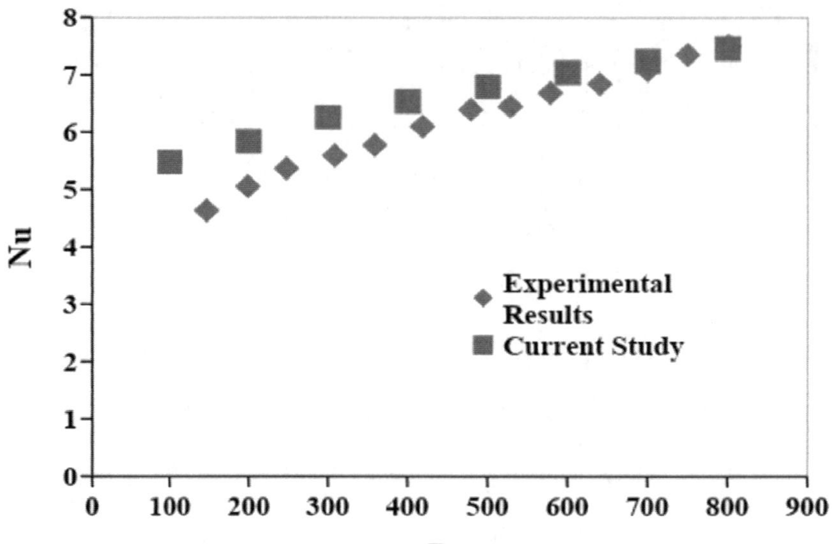

Figure 2. Comparison of Average Nusselt Numbers: Current results using Water as fluid through the microchannel versus experimental results.

Figure 2 clearly indicates that the simulation results are in reasonable agreement with the experimentally obtained results.

4. Results and Discussion

Table 6. Numerical Results.

Inlet velocity, u_m [m/s]	Fluid	Volume Fraction φ (in %)	Pressure Drop ΔP [Pa]	Fluid outlet temperature, $T_{f,out}$ [K]	Maximum wall temperararature, $T_{w,max}$ [K]
	Water	0	20501.651	318.463	337.566
1.3	Water + Graphene	3	22075.858	319.008	336.701
	Water + Graphene	6	23826.222	319.587	335.988

	Water	0	41244.996	309.605	328.379
2.5	Water + Graphene	3	44342.870	309.885	327.464
	Water + Graphene	6	47778.207	310.186	326.655

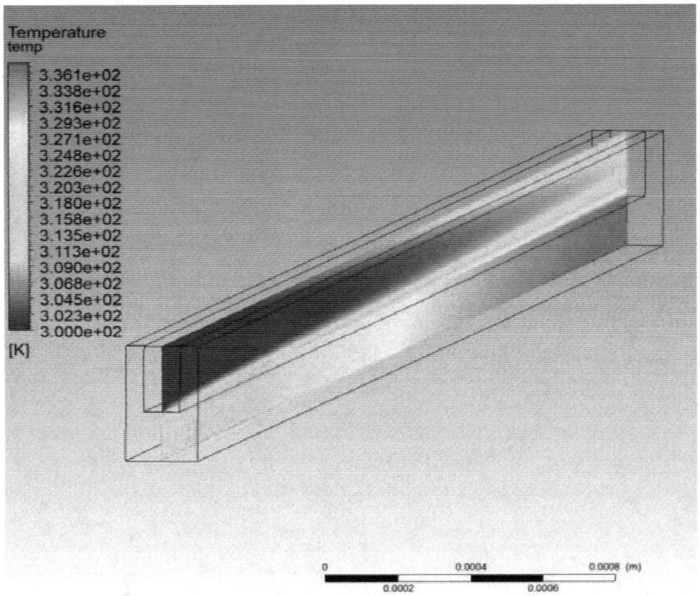

Figure 3. Temperature Distribution at inlet velocity 1.3 m/s using Water.

4.1. Influence of Nanoparticle concentration on Pressure Drop
Table 6 clearly suggests that nanofluids display greater pressure drop (Pa) across the microchannel than the base fluid at the same coolant inlet velocity. When the nanoparticle concentration in the coolant is increased, the pressure drop also increases.

Pumping power (W) is required to steadily drive the flow of coolant in the microchannel and is given by:

$$P_{pump} = u_m.\Delta P.A_{cs} \tag{17}$$

where A_{cs} denotes the area of cross-section of the rectangular microchannel (m^2), ΔP denotes the pressure drop across the microchannel (Pa). Therefore, the pumping power increases on increasing the concentration of nanoparticles in the coolant.

4.2. Influence of Nanoparticle Concentration on Wall Temperature
For the MCHS, the cooling performance is determined by evaluating the overall thermal resistance (K/W), which is defined as:

$$R_T = \frac{T_{w,\,max} - T_{f,\,in}}{q} \tag{18}$$

where $T_{w,max}$ denotes the maximum wall temperature of the MCHS (K), $T_{f,in}$ represents the fluid inlet temperature. Table 6 suggests that the maximum wall temperature decreases on increasing the nanoparticle concentration in the coolant and consequently, the overall thermal resistance decreases.

4.3. Influence of Nanoparticle Concentration on Coolant Exit Temperature

For a particular fluid inlet velocity, the coolant temperature at the microchannel outlet increases on increasing the nanoparticle concentration.

4.4. Effect of Flow Velocity on Wall Temperature

It is evident from Table 6 that for a given coolant, a higher flow velocity through the microchannel leads to lower wall temperature of the heatsink and hence, the overall thermal resistance decreases.

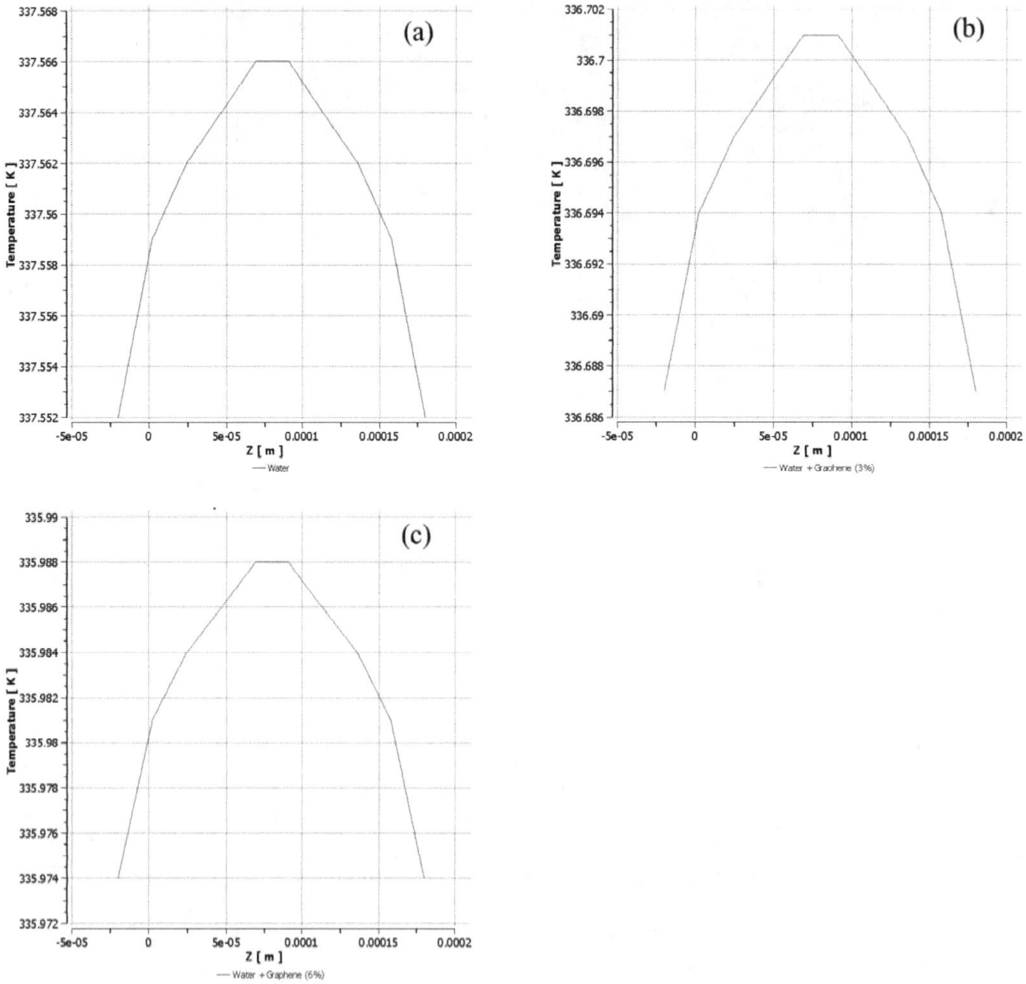

Figure 4. Temperature Gradient of MCHS Bottomwall edge in the outlet plane (i.e. X= L, Y= 0) due to inlet velocity 1.3 m/s for (a) Water (b) Graphene(3%)+Water (c) Graphene(6%)+Water.

4.5. Effect of Flow Velocity on Pressure Drop

An increase in flow velocity enhances pressure drop. It is evident from Table 6 that the same nanofluid will induce a more significant pressure drop at a higher fluid velocity.

4.6. Overall Variation in Temperature Distribution

Figure 3 displays the temperature distribution across the axial plane of the MCHS for the flow of base fluid water at an inlet velocity of 1.3 m/s. For both the solid and fluid zones, maximum temperature occurs at the outlet. Similar trends are displayed when nanofluids flow through the microchannel.

Figures 4 and 5 depict the temperature variation in the Silicon bottom wall edge in the outlet plane. Temperatures are minimum at either end of the MCHS (Z=0 and Z= W_T), increase as the centre-line of the bottom wall (Z= $W_T/2$) approaches, where it finally attains the maximum wall temperature. As visible, the wall temperature decreases with an increase in nanoparticle concentration in the coolant.

Figures 6 and 7 show the temperature variation of the Silicon bottom wall along the length of fluid flow in the X-direction. At the same inlet velocity, the axial bottom wall temperature of the heatsink decreases on increasing the concentration of nanoparticles in water along the length of fluid flow.

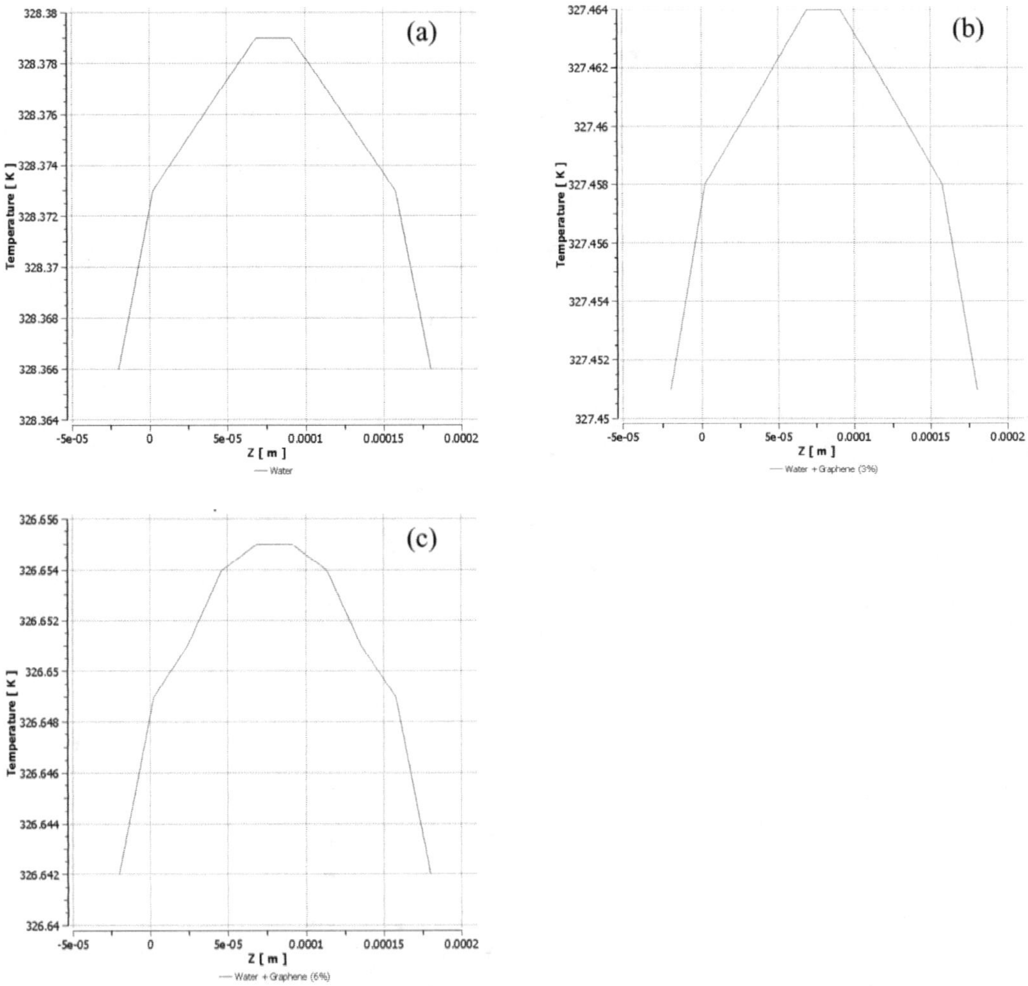

Figure 5. Temperature Gradient of MCHS Bottomwall edge in the outlet plane (i.e. X= L, Y= 0) due to inlet velocity 2.5 m/s for (a) Water (b) Graphene(3%)+Water (c) Graphene(6%)+Water.

5. Conclusion

This study analyses the enhanced heat transfer performance of nanofluids compared to the base fluid in the Silicon MCHS. The enhancement in heat transfer is attributed to a significant improvement in the coolant's thermal conductivity- obtained by adding very small concentrations of nanoparticles in the coolant. The thermal conductivity of the nanofluid containing 6% graphene by volume concentration exceeds the thermal conductivity of water by almost 9.5%. Using nanofluids in the cooling device reduces its thermal resistance and wall temperature- compared to water. Therefore, it is established that the cooling performance of the MCHS gets better on increasing the graphene nanoparticle concentration in the coolant. Furthermore, the findings on temperature distribution in the MCHS in this study agree with the results obtained by Farsad et al. [16] using other nanofluids- which reinforces the consistency of this study.

The nanofluid-cooled MCHS presents an efficient device for cooling electronic components. However, it is accompanied by a considerable rise in pressure drop across the microchannel, which increases pumping power requirements. Hence, the future scope of research lies in addressing this limitation and suggesting ways to balance optimum nanoparticle concentration in the coolant with feasible pressure drop, to ensure the best possible cooling performance of the device.

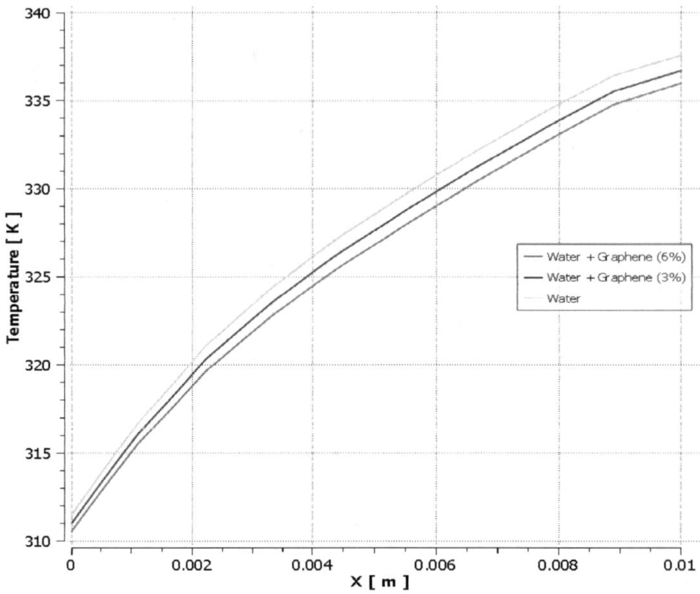

Figure 6. MCHS Bottomwall Temperature Gradient along Flow Length (X) for different fluids at inlet velocity 1.3 m/s.

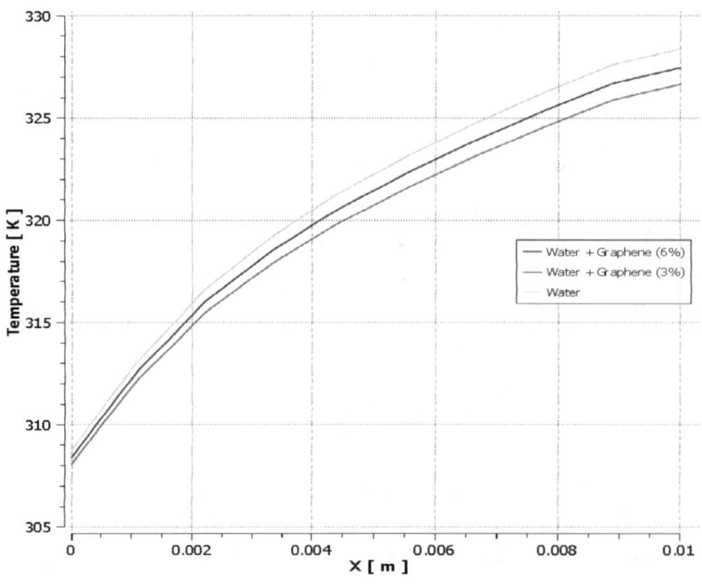

Figure 7. MCHS Bottomwall Temperature Gradient along Flow Length (X) for different fluids at inlet velocity 2.5 m/s.

6. References

[1] Tuckerman D B and Pease R F W 1981 High-performance heat sinking for VLSI *IEEE Electron Device Lett.* 2 126–9
[2] Qu W and Mudawar I 2002 Experimental and numerical study of pressure drop and heat transfer in a single-phase micro-channel heat sink *Int. J. Heat Mass Transf.* 45 2549–65
[3] Li J, Peterson G P and Cheng P 2004 Three-dimensional analysis of heat transfer in a micro-heat sink with single phase flow *Int. J. Heat Mass Transf.* 47 4215–31
[4] Jung J Y and Kwak H Y 2008 Fluid flow and heat transfer in microchannels with rectangularcross section *Heat Mass Transf.* 44 1041–9
[5] Chai L, Xia G D and Wang H S 2016 Numerical study of laminar flow and heat transfer incross section *Heat Mass Transf.* 44 1041–9 *Appl. Therm. Eng.* 92 32–41
[6] Choi S U S and Eastman J A 1995 Enhancing Thermal Conductivity of Fluids with Nanoparticles *ASME International Mechanical Engineering Congress & Exposition*
[7] Lee J and Mudawar I 2007 Assessment of the effectiveness of nanofluids for single-phase and two-phase heat transfer in micro-channels *Int. J. Heat Mass Transf.* 50 452–63
[8] Sadiq Al-Baghdadi M A R, Noor Z M H, Zeiny A, Burns A and Wen D 2020 CFD analysis of ananofluid-based microchannel heat sink *Therm. Sci. Eng. Prog.* 20 100685
[9] Cai Z, Tian M and Zhang G 2020 Experimental Study on the Flow and Heat Transfer of Graph - ene-Based Lubricants in a Horizontal Tube *Processes* 8 1675
[10] Baby T T and Ramaprabhu S 2011 Enhanced convective heat transfer using graphene dispersed nanofluids *Nanoscale Res. Lett.* 6 1–9
[11] Akhavan-Zanjani H, Saffar-Avval M, Mansourkiaei M, Sharif F and Ahadi M 2016 Experimen -tal investigation of laminar forced convective heat transfer of Graphene-water nanofluid inside a circular tube *Int. J. Therm. Sci.* 100 316–23
[12] Tharayil T, Asirvatham L G, Ravindran V and Wongwises S 2016 Thermal performance of miniature loop heat pipe with graphene-water nanofluid *Int. J. Heat Mass Transf.* 93 957–68
[13] Dinçer İ and Zamfirescu C 2016 Appendix B: Thermophysical Properties of Water *Drying Phenomena: Theory and Applications* (John Wiley & Sons, Ltd.) pp 457–9
[14] Sandhya M, Ramasamy D, Sudhakar K, Kadirgama K and Harun W S W 2021 Enhancement of the heat transfer in radiator with louvered fin by using Graphene-based hybrid nanofluids *IOP Conf. Ser. Mater. Sci. Eng.* 1062
[15] Xuan Y and Roetzel W 2000 Conceptions for heat transfer correlation of nanofluids *Int. J. Heat Mass Transf.* 43 3701–7
[16] Farsad E, Abbasi S P, Zabihi M S and Sabbaghzadeh J 2011 Numerical simulation of heat transfer in micro channel heat sinks using nanofluids *Heat Mass Transf.* 47 479–90

Experimental Analysis using an Innovative Catalytic Converter coated with Nano-particles for Pollution Control from Automobiles

Authors. Dr. Mukesh Thakur[1] and Dr. Rohan Senanayake[2]

Authors' Affiliations. [1]Principal, NMDC DAV Polytechnic, Dantewada (C.G.), India.
[2]Associate Professor, Lincoln University, Malaysia.

Abstract. Air pollution control has become the area of interest due to the ever increasing air pollution problem from the automobiles. Several researchers have conducted several experiments to control the air pollution concentration from automobiles by using several techniques but this area has a wide scope for improvement. This research paper is based on an experiment conducted on a Four Stroke Spark Ignition engine test rig using an Innovative catalytic converter which was coated with nano-particles to estimate its effectiveness in air pollution control. The results of the experiment conducted clearly indicate that the Innovative design of catalytic converter is effective in air pollution control from automobiles.

Keywords. Automobiles, catalytic converter, nano-particles, pollution.

1. Introduction

Glencross et al. [1] described that the air pollutants are very dangerous to the immune system of human beings. Air pollutants can disrupt the immune response and trigger health issues. So, the air pollutants should be treated so that they do not create health problems for human beings.

Fattorini et al. [2] analysed the role of air pollutants in the spread of the novel corona virus in Italy. They described that the presence of air pollutants in large concentrations in an area is harmful and can aid in spreading of the Covid disease causing breathing problems. So, control of air pollutants is necessary to ensure the availability of clean and pollution free breathing air for all human beings.

Twigg [3] described that the oxidation of the exhaust emissions emitted from the tail pipe of automobiles by a suitable catalyst can be very helpful in decreasing their concentration. The catalyst which is coated on the surface of a suitable catalytic converter aids the reaction between the exhaust emissions and oxygen thereby converting the Carbon monoxide (CO) and Hydrocarbons (HC) in to Carbon dioxide (CO_2) and water vapour.

Dey and Dhal [4] proposed a method for control of Carbon monoxide (CO) concentration from exhaust of automobiles using catalytic converter coated with Copper oxide catalyst. The results of the method depicted that the concentration of air pollutants from automobiles can be decreased by using a suitable catalyst on the surface of a catalytic converter.

Feldheim et al. [5] elaborated the advances in the synthesis, characterization and applications of nano-particles in different areas. They described that nano-particles have evolved as an innovative technology for an effective solution of various prevalent problems, one of them being the air pollution and its control. Metal nano-particles have superior properties compared to other bulk materials mainly due to their active surface area which is very helpful for reducing the exhaust emissions concentrations emitted from automobiles.

Kishore and Krishna [6] performed experiments for determination of exhaust emissions concentrations from Copper coated Spark Ignition engine using a blended fuel. From the exhaustive experimentation, they demonstrated that use of copper may be very helpful in controlling the concentrations of harmful exhaust emissions. Since, copper is cheap and easily available, so, it is an excellent option to be used in the control of exhaust emissions concentrations emitted from the Spark Ignition engine automobiles.

Heiter et al. [7] conducted an experimental investigation for determination of exhaust emissions from automobiles using methyl esters. From the experimental results, it was evident that if some treatment methodology is used for reducing the exhaust emissions concentrations, it would lead to a cleaner and greener atmosphere.

Murali et al [8] conducted a comparative analysis on the use of a Copper coated catalytic converter for controlling the exhaust emissions concentrations from a Two Stroke and Four Stroke Spark Ignition engine. From the results of the experimental analysis, it was clear that Copper coated catalytic converter is very helpful in control of exhaust emissions from Spark Ignition engine automobiles.

Dey and Mehta [9] presented a detailed investigation on control of exhaust emissions from automobiles using metal catalyst. They explained that the metal catalysts are very effective in controlling the concentration of harmful exhaust emissions from the automobiles. So, metal catalyst can be employed in the catalytic converters of automobiles and that will result in a cleaner environment.

Dey and Dhal [10] presented an experimental study on controlling the harmful exhaust emissions from automobiles using the copper oxide catalysts on a catalytic converter. They explained that this method would be very effective to keep a check on the concentration of Carbon Monoxide emitted from the automobiles. The reason for the superior performance of Copper oxide catalysts are their superior properties which are very helpful to convert the harmful Carbon Monoxide in to Carbon dioxide.

Metal nano-catalysts are being used in automobile emission control systems and in other pollution control and treatment applications to facilitate petroleum extraction and production and to produce chemicals and chemical products. Catalysts usually have two principal roles in nano-technology areas: (i) In macro quantities, they can be involved in some processes for the preparation of a variety of other nano-structures like quantum dots, nano-tubes, etc. (ii) Some nano-structures themselves can serve as catalysts for certain chemical reactions.

2. Experimentation

The experimentation for evaluation of the utility of an innovative catalytic converter in the control of air pollution from automobiles consists of several steps. They are as follows:

2.1. Design and Fabrication of an Innovative Catalytic Converter

In the first step, a design of the catalytic converter named as “Innovative design” was prepared. The design was prepared according to the design recommendations. Thereafter, an innovative catalytic converter was fabricated on the basis of the design.

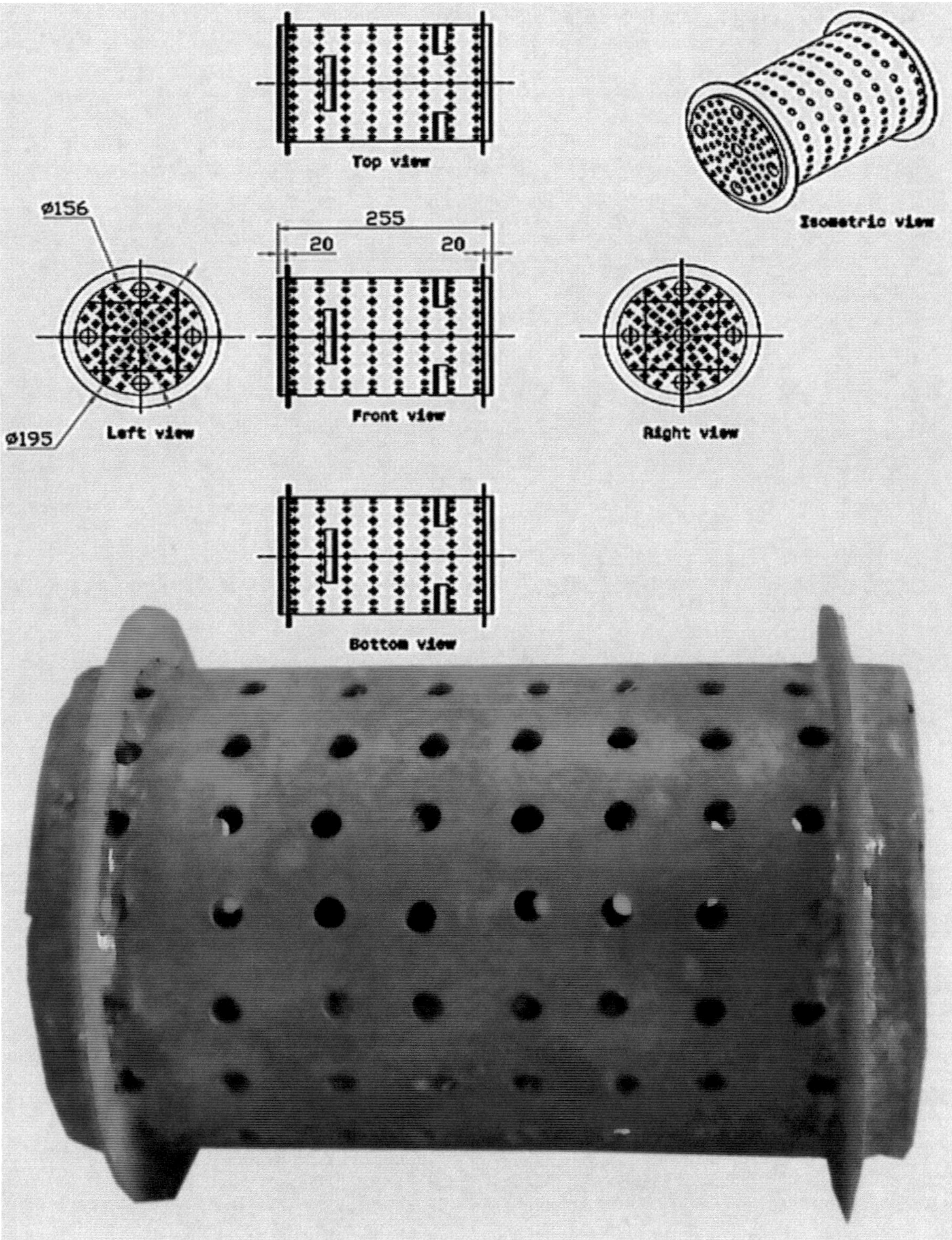

Figure 1. Innovative Catalytic Converter

2.2. Selection of a Four Stroke Spark Ignition Test Rig

In the second step, a suitable Four Stroke Spark Ignition engine test rig was selected for experimentation.

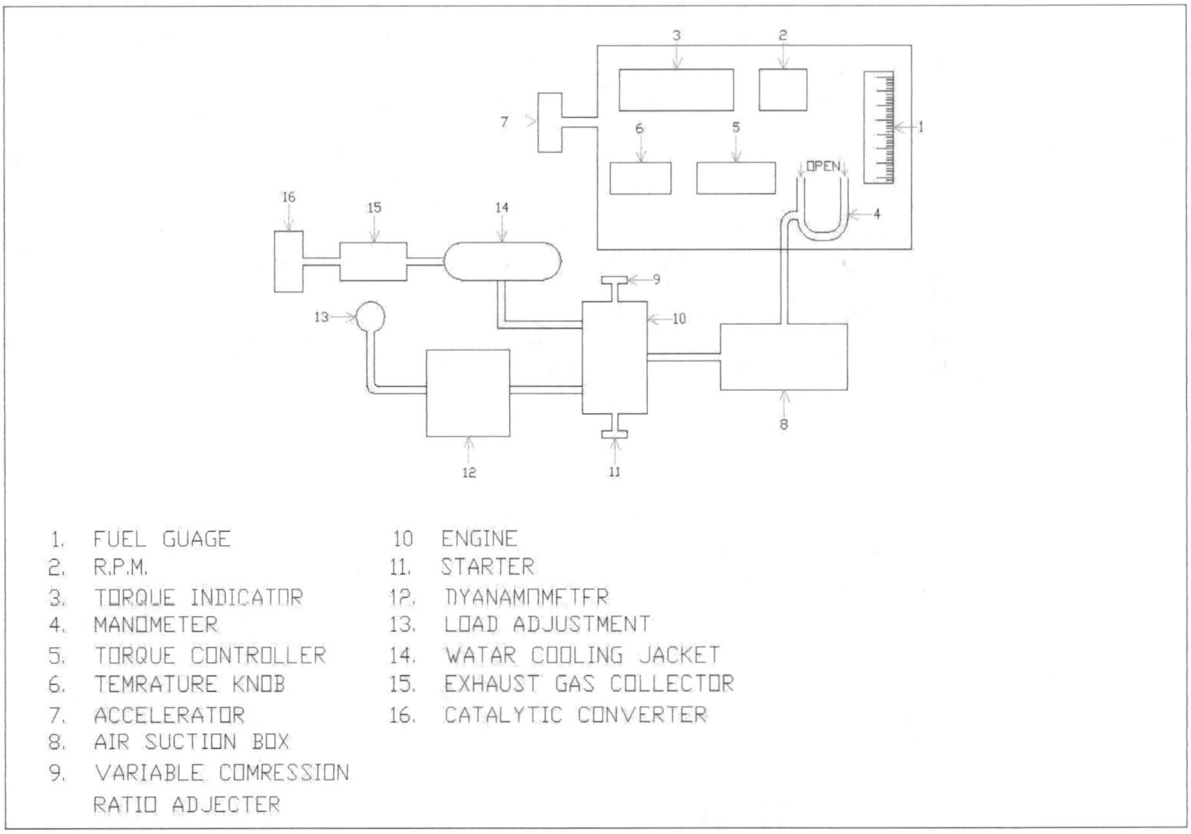

Figure 2. Four Stroke Spark Ignition Engine test rig

After selection of the test rig, the fabricated catalytic converter was coated with Copper nano-particles using drop casting method. The heat treatment was carried out using a muffle furnace.

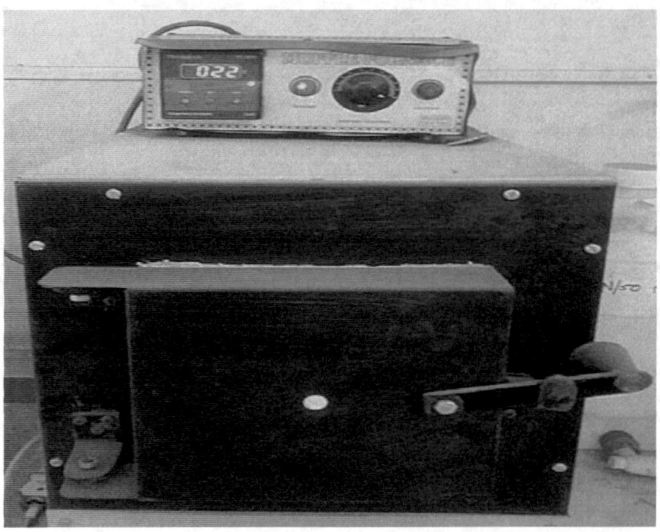

Figure 3. Heat treatment in muffle furnace

The specifications of the muffle furnace are as follows:

Inside Chamber Size	150 x 150 x 149mm (6" x 6" x 6"), 3.6L
Standard Working Temperature	1600° C (continuous)
Maximum Working Temperature	1700° C (< 1 hours)
Heating Rate	0 ~20 °C/min (≤10°C/min recommended)
Temperature Accuracy	+/- 1° C

Then, the experiment for measurement of exhaust emissions before and after using catalytic converter was conducted. A multi gas analyzer was used for measurement of exhaust emissions concentration.

Figure 4. Multi gas analyzer

The specifications of NPM MGA 2 gas analyzer are as follows:

Model Name/Number	NPM-MGA-2
Brand	NETEL
Usage/Application	VECHICLE POLLUTION TESTING
Product Type	MULTIGAS ANALYZER
Power Supply	12 DC and 230V AC
Analysis Time	Quick
Color	Gray
Automation Grade	Automatic
Country of Origin	Made in India

IOP Conf. Series: Materials Science and Engineering 1206 (2021) 012009 doi:10.1088/1757-899X/1206/1/012009

The experimental procedure is as follows:
Connect the instrumentation power input plug to a 230 V, 50 Hz single phase AC supply. Now all the digital meters namely, rpm indicator, temperature indicator display the respective readings. Fill up the petrol to the fuel tank mounted side of the panel. Check the lubricating oil level in the oil sump.
Start the engine with the help of given attachments. Allow the engine to stabilize the speed, i.e., 1500 rpm to 2200 rpm by adjusting the accelerator. Apply ¼ load, i.e. slowly vary the load attachment.
Note down all the required parameters mentioned below:
(a) Speed of the engine in rpm.
(b) Load from spring balance.
(c) Time taken for 10 cc of fuel consumption.
(d) Manometer readings.
Then, Load the engine step by step with the use of load attachment such as,
(a) $\frac{1}{4}$ load
(b) $\frac{1}{2}$ load
(c) $\frac{3}{4}$ load
(d) Full load

3. Results and Discussion

After conduction of the experiment, the exhaust emissions concentrations were compared before and after using the innovative design catalytic converter. The comparison is shown in the following figures:

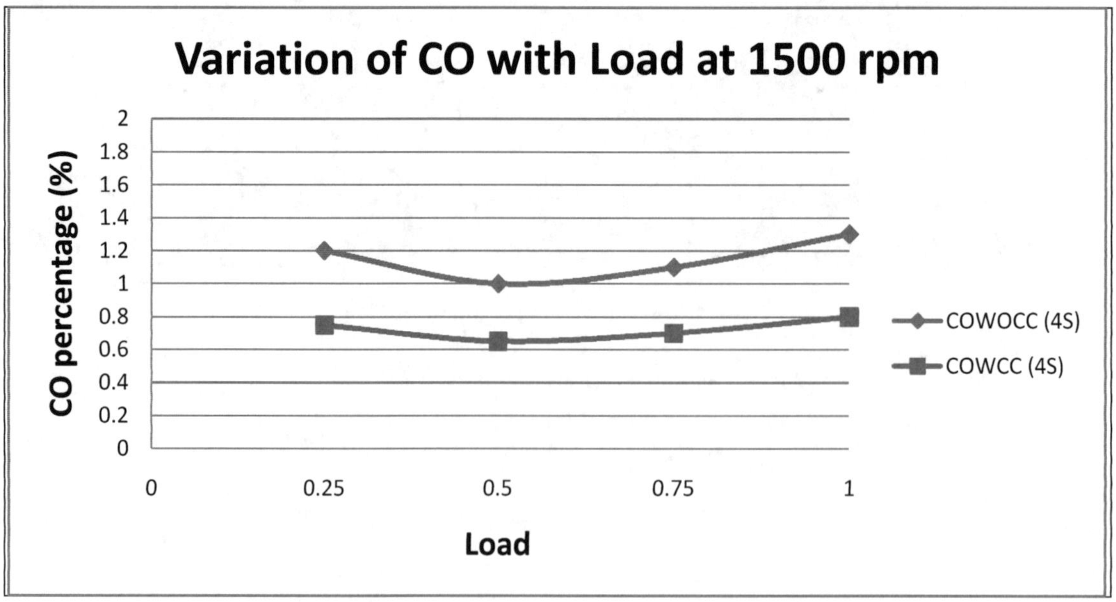

Figure 5. CO percentage Without Catalytic Converter (WOCC) and With Catalytic Converter (WCC) at 1500 RPM

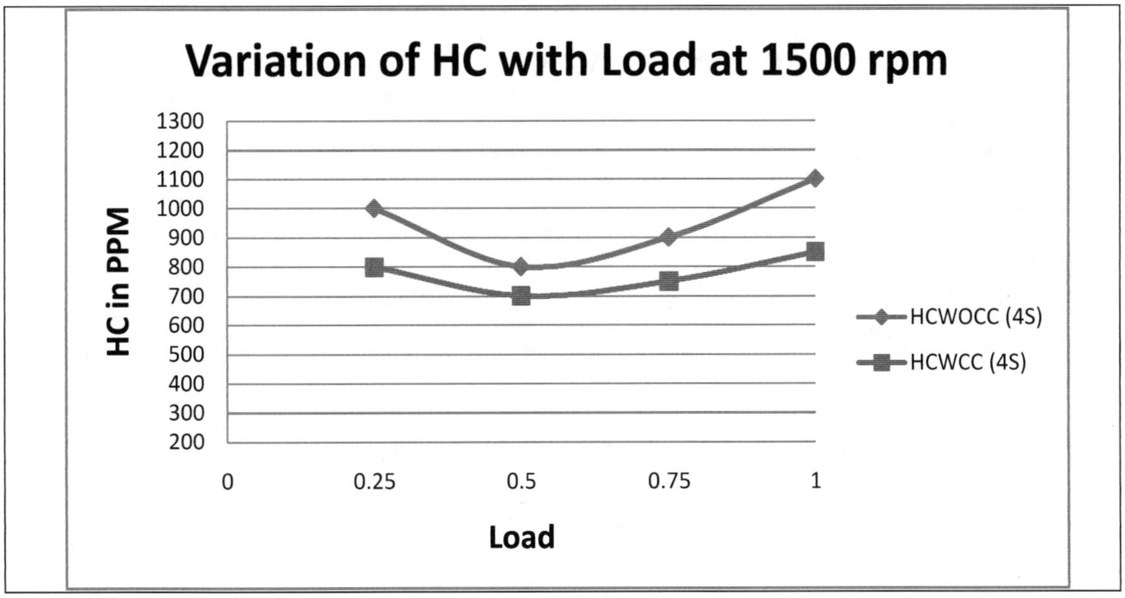

Figure 6. HC PPM Without Catalytic Converter (WOCC) and With Catalytic Converter (WCC) at 1500 RPM

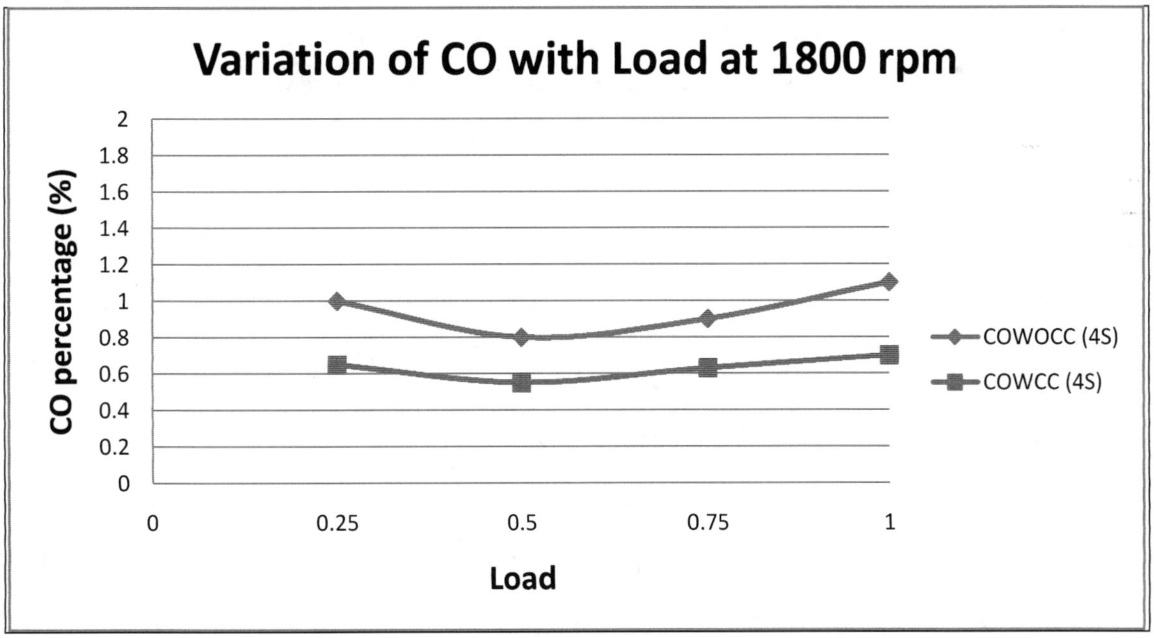

Figure 7. CO percentage Without Catalytic Converter (WOCC) and With Catalytic Converter (WCC) at 1800 RPM

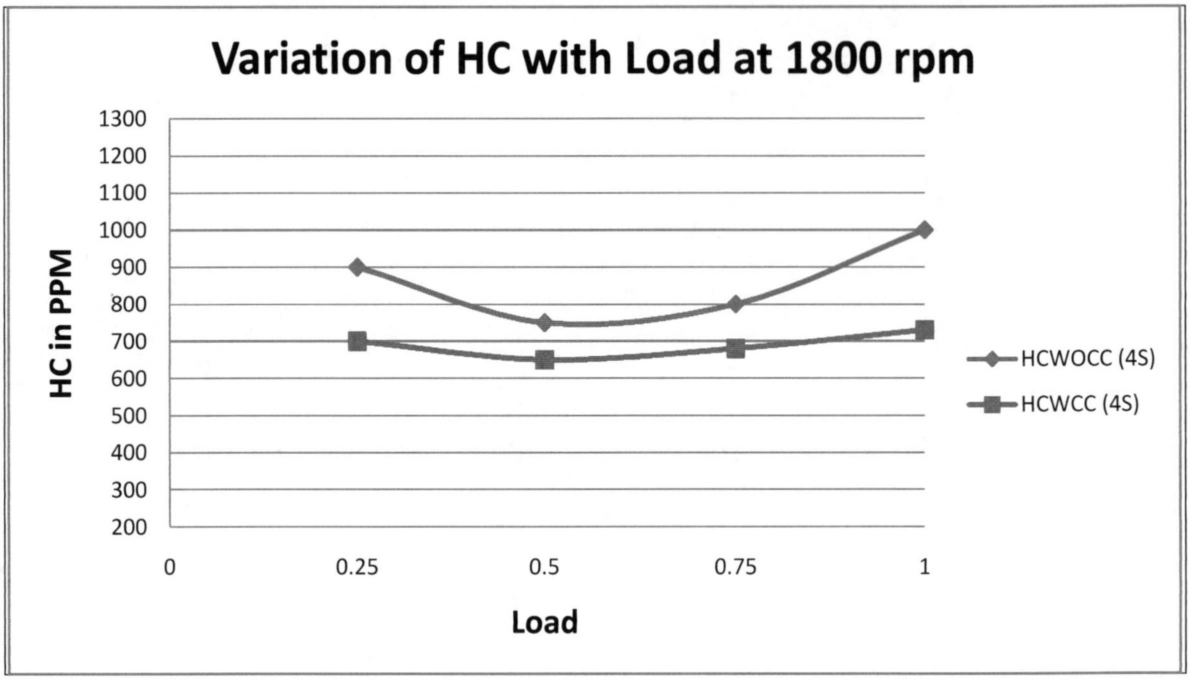

Figure 8. HC PPM Without Catalytic Converter (WOCC) and With Catalytic Converter (WCC) at 1800 RPM

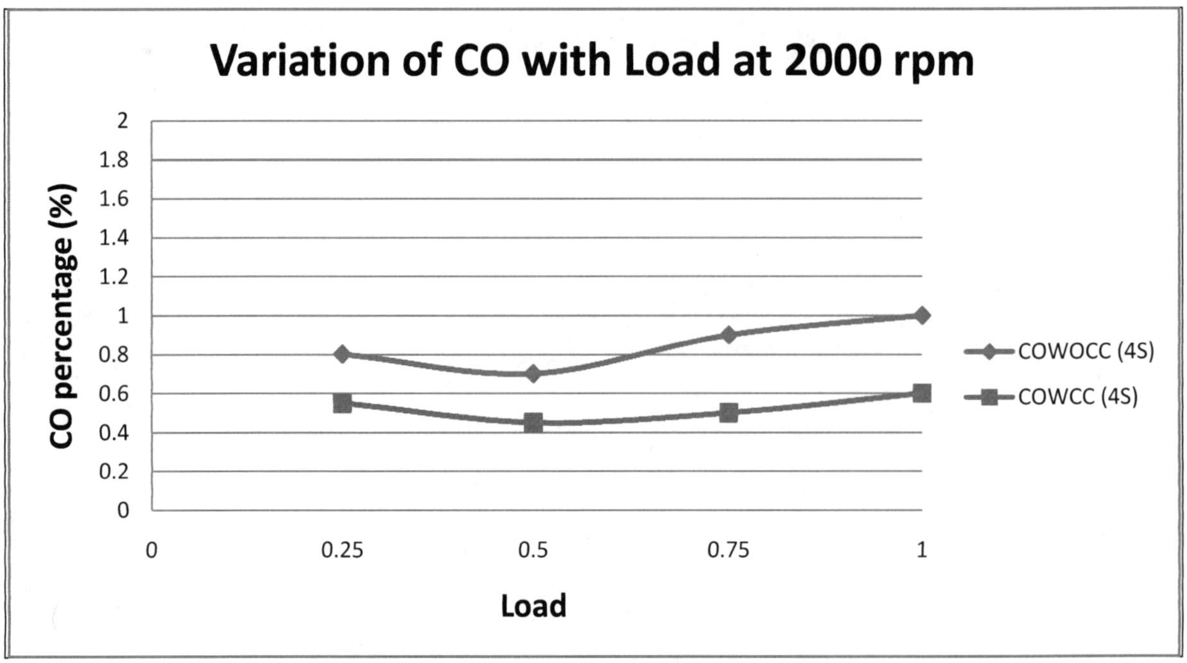

Figure 9. CO percentage Without Catalytic Converter (WOCC) and With Catalytic Converter (WCC) at 2000 RPM

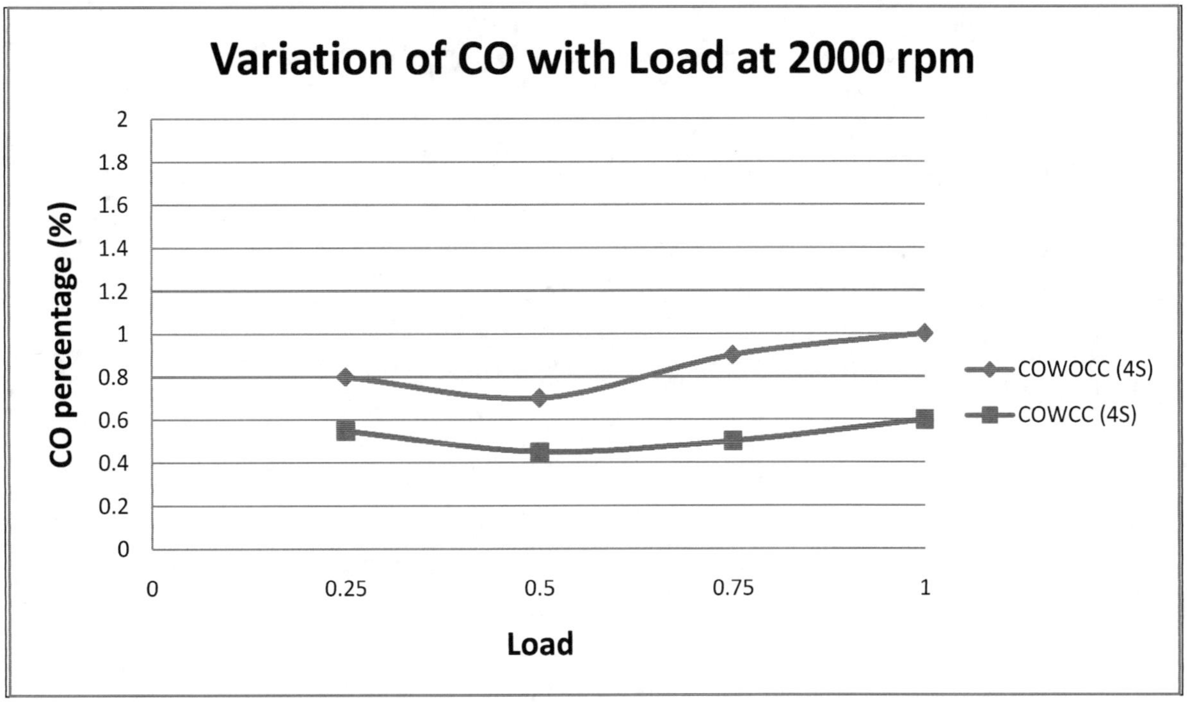

Figure 10. HC PPM Without Catalytic Converter (WOCC) and With Catalytic Converter (WCC) at 2000 RPM

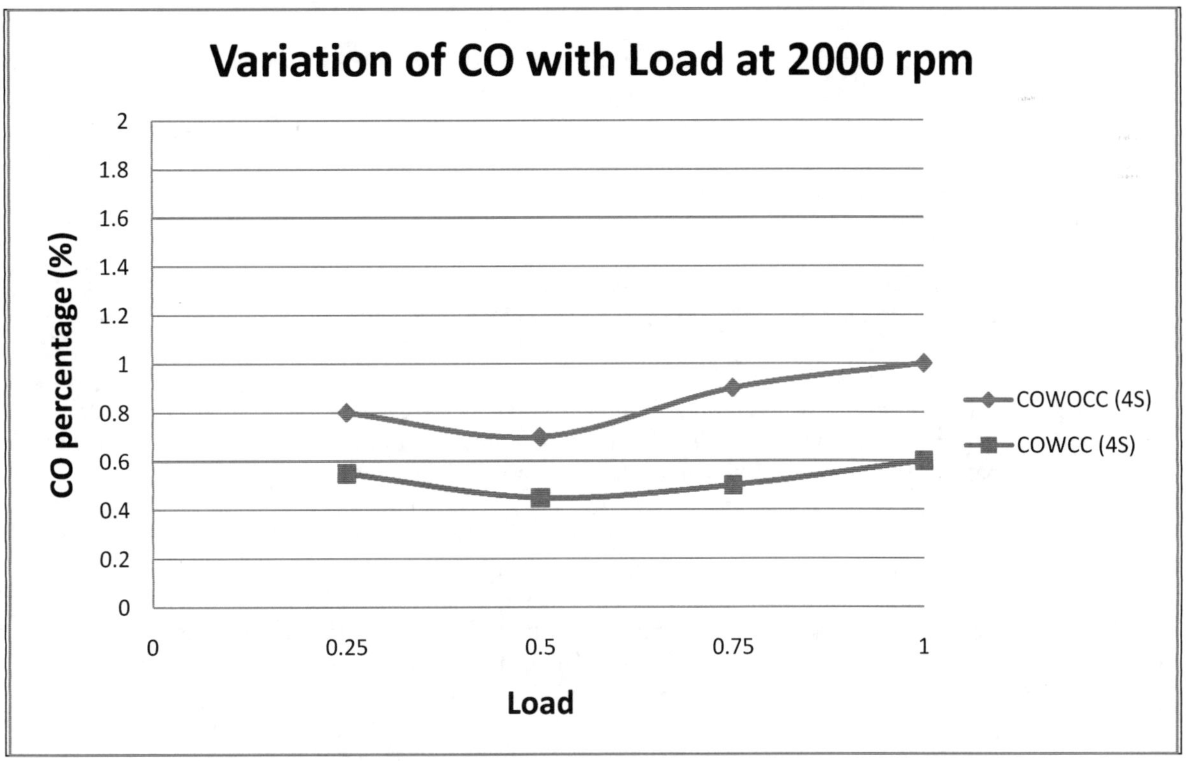

Figure 11. CO percentage Without Catalytic Converter (WOCC) and With Catalytic Converter (WCC) at 2200 RPM

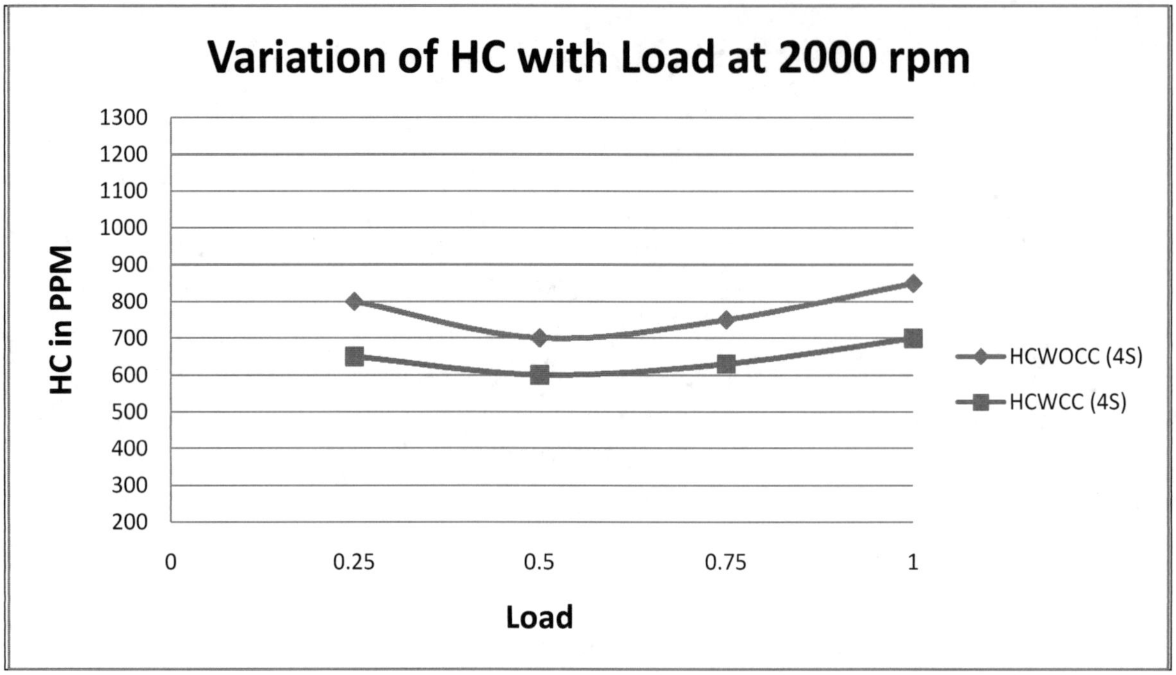

Figure 12. HC PPM Without Catalytic Converter (WOCC) and With Catalytic Converter (WCC) at 2200 RPM

4. Conclusion

This research paper highlights the use of an innovative catalytic converter coated with nano-particles for pollution control from Spark Ignition engine automobiles. From the results, it is evident that the proposed method is effective for pollution control from Spark Ignition engine automobiles. This research paper also highlights the use of Copper as an effective option for use in catalytic converter. Since, the proposed approach in this research paper is based on the Post pollution control, so, it is easy for implementation in automobiles and cheaper as compared to design modification based methods.

The idea behind the work is to create a structure that exposes the maximum surface area of catalyst to exhaust stream, also minimizing the amount of catalyst required. The exhaust gases pass through a bed of catalyst and the catalytic action takes place at surface of copper which are porous and the higher catalytic activity towards the oxidation of CO and HC could be due to the higher catalytic surface area of small nano-particles. The proposed method is very effective in the prevention of environmental pollution contributed from two-wheeler automobiles. It involves the use of copper nano-particle that is cheaper than the platinum, palladium and rhodium nano-particles used in automobiles.

In the current scenario, the pollution control is a crucial area for research and development. Lots of work has been done by the researchers in this connection, but the area is still lacking for an efficient technique which can provide highly robust emission control for all the operating conditions of vehicles. This research work opens a gateway to study the changes in the concentration of exhaust emissions due to the nano-material copper coating. The modeling will help in understanding the mathematical nature of the process and simulation will help in predicting the results with ease.

5. References

[1] D A Glencross, T R Ho, N Camina, C M Hawrylowicz and P E Pefeffer 2020 *J. Free Radical Biology and Medicine* **151** 56-68
[2] D Fattorini, F Regoli 2020 *Environmental Pollution* **264** 114732
[3] Twigg 2006 *J. Catalysis Today* **117** 407-418
[4] S Dey and G C Dhal 2020 *Materials Today Chemistry* **17** 100282
[5] Feldheim D L and Foss C A 2004 *Applications of Physics and Advanced Materials Science Processes* **78** 73-79
[6] Kishore K and Krishna M 2008 *Journal of Scientific and Industrial Research* **67** 543-548
[7] Heiter L A and Ravi R 2012 *International Journal of Environmental Sciences* **3** 23-29
[8] Krishna M M, Kishor P, Murthy A, Gupta S and Narsimha K 2012 *International Journal of Scientific and Technology Research* **1** 85-90
[9] S Dey and N S Mehta 2020 *Resources, Environment and Sustainability* **2**
[10] S Dey and G.C. Dhal 2020 *Materials Today* **17**

IOP Conf. Series: Materials Science and Engineering 1206 (2021) 012010 doi:10.1088/1757-899X/1206/1/012010

Applying analytical hierarchy process for addressing the agile manufacturing drivers

Rohit Sharma[1] and **Ubaid Ahmad Khan**[2]

[1] M.Tech Student, Department of Mechanical Engineering, Shambhunath Institute of Engineering & Technology, Prayagraj, India.
[2] Associate Professor, Department of Mechanical Engineering, Shambhunath Institute of Engineering & Technology, Prayagraj, India.-uk.mnit@gmail.com

Abstract. In order to incorporate agile manufacturing (AM) in materials and systems, the manufacturing sectors have drivers to face obstacles. Agility is generally accepted for satisfying diverse consumer demands as a new strategic principle in the automotive industry. There has now been a prerequisite for evaluating AM in industry. An organization's effectiveness relies on their ability to find and pay special attention to the crucial success drivers to achieve a high level of efficiency. This paper suggests a number of Agile Manufacturing Drivers (AMDs) to evaluate AM that is deemed suitable to the production industry. In order to prioritise performance drivers, the analytical hierarchy process (AHP) approach is used to summarise the perspective of an expert. The proposed AMDs are believed to encourage and assist the manufacturing sector in producing agile products to achieve higher efficiency so as to improve competition.
Keywords: Agile manufacturing (AM), Agile manufacturing drivers (AMDs), Analytical hierarchy process (AHP), Consistency Ratio (C.R.), Consistency Index (C.I.), Random Index (R.I.).

1. Introduction

The definition of agile manufacturing (AM) was first published by the Iacocca Institute at Lehigh University in the 21st century Enterprise Strategy report [1].

AM has already been taken into account by analysts, administrators or consultants. Today businesses continue to restructure and refurbish to meet challenging consumers' challenges, who shift their needs quickly [2]. Customer trends and technical needs push suppliers to implement innovative business models to be competitive on the market. Since "shift" seems to be one of the key features, companies intend to follow (AM), since it rapidly embraces evolving environments. Agility allows companies to compete in an unpredictable and constantly changing business climate [3]. Given the outcome of economy of scale and mass production in uncomplicated plants, it is important to create more versatile and sensitive organisations [2][4]. Companies now consider speed, pricing and versatility, rather than costs that are stressed to meet the specific demands of consumers and industries, to address market uncertainty [5].

The Agility System offers an industry forum that enables them to take advantage of current and evolving business opportunities [6] [7]. The four principal elements of AM must be well understood before its introduction, including AMDs, strategic capacities, agility providers and agility ability [8]. In order to adapt to the reforms, some AMDs are expected to compel an organisation to reconsider its current approach, to implement an agility strategy and to acknowledge the need for agility. The agility suppliers do not always rely and need to be centred on multiple classifications by the policymakers. Strong venture capital, high technology and high integrated products sectors are the automotive industries. In this unpredictable and evolving market, the AM framework serves as a strategic initiative to respond to dynamic demand of customer with efficient costs, for the benefit of a machine agility, used the informative diagram of control and a table model and a decision tree model. So we may conclude that agile production allows the company to succeed in this age of constant and competitive consumer shifts. [9].

The research did not however lead to a study of the interrelationship between AMDs, [10] noted that few components contribute successfully to improving agility from a range of facilitators. A number of facilitators must also be investigated and their relationship investigated. Furthermore, it is important to prioritise AMDs. Example the performance of car firms relies largely on the manufacturer and dealer relationship, which is why the complexity of the partnership is examined by a methodology with a broad number of device variables, which serve as driving forces [11]. It was widely believed that there are challenges in managing complicated situations or structures for people or organisations of policymakers. The ambiguity of the problems or structures is caused by a wide range of elements and relationships between them. ISM provides a greater view of the structure of a function and provides valuable guidance to generate a structure graphical representation [12]. The processing of the AMDs organised by ISM takes the form of a successful AHP operation, which is their individual contribution to the realisation of agile development. The study helps managers to develop policies which maintain their individual value in the effective implementation of AM in view of complex interdependence among drivers. This will help decide whether to take preventive action to tackle these drivers.

2. Literature Review

Following the development of the concepts of agile production, the researchers started conducting studies to determine AMDs to help businesses apply AM principles. These research has been published in literature on a variety of occasions. In these research papers the agility defined by the authors is listed in Table 1.

Table 1 Identified AMDs

S.No	AMDs	[13]	[14]	[15]	[16]
1	Responsive Supply Chain			✓	
2	Top management support	✓		✓	
3	Training and Education			✓	
4	Customer response adoption				✓
5	Concurrent engineering		✓	✓	
6	Re-configurable layout		✓		
7	Appropriate Hardware			✓	
8	Rapid prototyping		✓		
9	Organisational Culture	✓			
10	Customer Involvement	✓			
11	Knowledgeable Workers			✓	
12	Employee involvement				✓
13	Proper database management			✓	
14	Automation				✓
15	Virtual Enterprise		✓	✓	
16	IT Integrations			✓	✓

An summary of the Table 1 material indicates that rapid prototyping and virtual enterprise are also referred to as agile production facilitators. Rapid prototyping helps fast prototyping development so that firms can reconfigure the goods to meet their customer's needs prior to actually producing the product on the shop floor. In addition, the virtual enterprise allows the company to unite other companies temporarily to take benefit from changing markets. The virtual enterprise willingness to make use of other companies' capability to incorporate them into its product factory to satisfy the complex demands of customers' manufacturing process. At this point, researchers assessed the potential of the IT integration model to improve the virtual enterprise and configurable abilities of companies to gain agility assets in the last few years.

The IT integration model has the ability to help organisations, through the use of IT tools, acquire the information they need for agile enterprises within the minimum period. In addition to rapid prototyping and virtual company management, AMD is often required for the storing and use of previous works in the proper database management. Critical thinking would suggest that the output monitoring for product creation is complicated without the help of proper database management. As previously mentioned, they all support other AMDs and have their relationship with one another in the ISM model and are assessed by AHP for their efficacy in agile production.

3. AHP Methodology

AHP is an "analytical hierarchy process" originating in 1977 by Satty. For the outcome of ISM, AHP is used. The ISM model is designed to establish the hierarchy of AMDs defined (Figure 1). Since that model, the ISM results in a hypothetical hierarchy that requires quantitative analysis for the efficacy of its percentage. ISM is used to build relationships between AMDs, as it is an uninterrupted, logical methodology [17]. Thus, the ISM model's soundness can be checked using AHP. In the 70s AHP had its origins and was tremendously employed by Thomas L. Saaty, who was present for the decisions for intricate situations when human consciousness, views and effects were taken together [18]. Their results were long-term. In order to examine and differentiate the AHP application in an independent way, it begins with a problem which is divided into a hierarchy of criteria. By comparing the available options, alternatives or variables for each selected benchmark or criteria in the same way, decision makers can properly review them. In order to enter subjacent information, the comparison can use existing data from the criteria and human judgments [19].

At the strategic level of planning AHP has been widely used to take complex decisions [20]. AHP converts the most frequently real comparisons, into an analysed and comparable numeric number. The weight of each AMDs ensures that each element within the defined hierarchy is properly evaluated [21]. The primary uniqueness of the AHP methodology when compared with other methodologies is this opportunity to change empirical evidence into mathematical modelling.

After all comparisons are made, relative weights are developed between each criterion and the probability calculations for each alternative are calculated. This probability assesses potential for alternative to implement expected targets. More likely alternative is, more likely final goal will be reached. Comparison is made in parallel to determine each alternative's comparative importance. The judge must show his opinion in this method that he re-holds the importance of a single comparison on pairs (Fig 2). A variable parallel comparison will allow the variable to calculate priority and meaning weights. When we compare two components, we obey the straight forward rule suggested by Saaty. [21].

As priorities are only recognised if a consistency check is obtained from consistent matrices. The Consistency Index (CI) has been found by Saaty [21], obtained through the own value method:

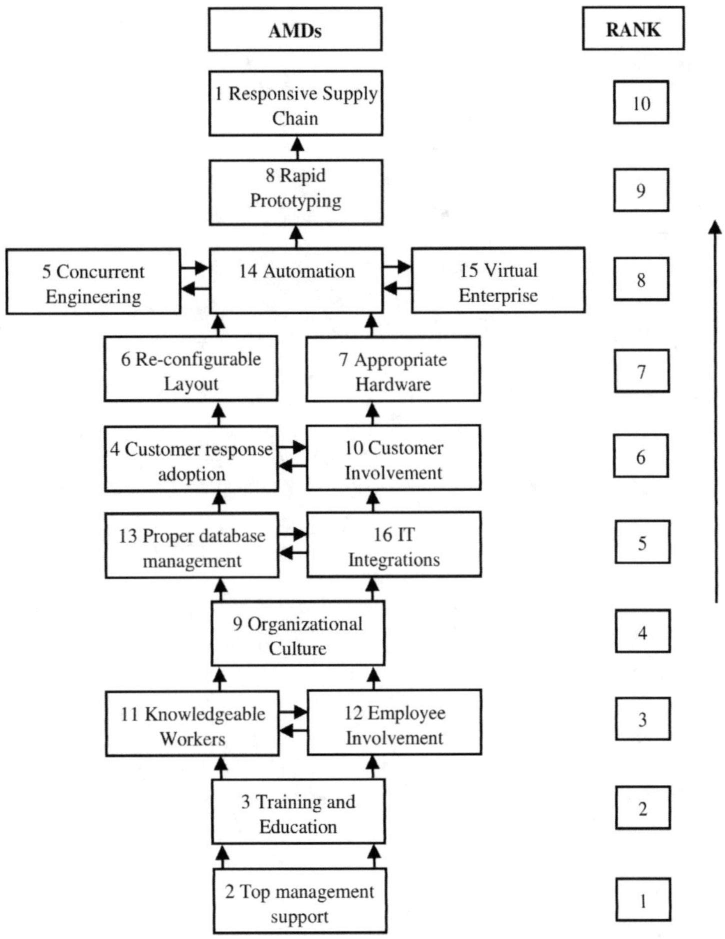

Figure 1 ISM model

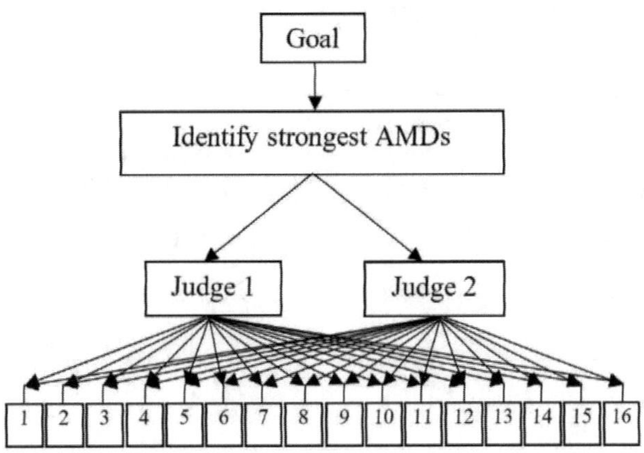

Figure 2 AHP Model

Judge-1

We categorize ten AMDs ranks by ten groups from D1 to D10 in this pair-scale comparison.

Table 2 Groups of AMDs

Groups	D1	D2	D3	D4	D5	D6	D7	D8	D9	D10
AMDs	2	3	11,12	9	13,16	4,10	6,7	5,14,15	8	1
Intensity of Importance	1	1.89	2.78	3.67	4.55	5.44	6.33	7.22	8.11	9
Reciprocal	1	0.53	0.36	0.27	0.22	0.18	0.16	0.14	0.12	0.11

Table 3 Pairwise comparision matrix

AMDs	1	2	3	4	5	6	7	8	9	10	11	12	13	14	15	16
1	1	0.11	0.12	0.22	0.36	0.27	0.27	0.53	0.16	0.22	0.14	0.14	0.18	0.36	0.36	0.18
2	9	1	1.89	5.45	7.22	6.33	6.33	8.11	3.67	5.45	2.78	2.78	4.56	7.22	7.22	4.56
3	8.11	0.53	1	4.56	6.33	5.45	5.45	7.22	2.78	4.56	1.89	1.89	3.67	6.33	6.33	3.67
4	4.56	0.18	0.22	1	2.78	1.89	1.89	3.67	0.36	1	0.27	0.27	0.53	2.78	2.78	0.53
5	2.78	0.14	0.16	0.36	1	0.53	0.53	1.89	0.22	0.36	0.18	0.18	0.27	1	1	0.27
6	3.67	0.16	0.18	0.53	1.89	1	1	2.78	0.27	0.53	0.22	0.22	0.36	1.89	1.89	0.36
7	3.67	0.16	0.18	0.53	1.89	1	1	2.78	0.27	0.53	0.22	0.22	0.36	1.89	1.89	0.36
8	1.89	0.12	0.14	0.27	0.53	0.36	0.36	1	0.18	0.27	0.16	0.16	0.22	0.53	0.53	0.22
9	6.33	0.27	0.36	2.78	4.56	3.67	3.67	5.45	1	2.78	0.53	0.53	1.89	4.56	4.56	1.89
10	4.56	0.18	0.22	1	2.78	1.89	1.89	3.67	0.36	1	0.27	0.27	0.53	2.78	2.78	0.53
11	7.22	0.36	0.53	3.67	5.45	4.56	4.56	6.33	1.89	3.67	1	1	2.78	5.45	5.45	2.78
12	7.22	0.36	0.53	3.67	5.45	4.56	4.56	6.33	1.89	3.67	1	1	2.78	5.45	5.45	2.78
13	5.45	0.22	0.27	1.89	3.67	2.78	2.78	4.56	0.53	1.89	0.36	0.36	1	3.67	3.67	1
14	2.78	0.14	0.16	0.36	1	0.53	0.53	1.89	0.22	0.36	0.18	0.18	0.27	1	1	0.27
15	2.78	0.14	0.16	0.36	1	0.53	0.53	1.89	0.22	0.36	0.18	0.18	0.27	1	1	0.27
16	5.45	0.22	0.27	1.89	3.67	2.78	2.78	4.56	0.53	1.89	0.36	0.36	1	3.67	3.67	1

Table 4 Calculation of priority weight (W) and eigenvector λmax

AMDs	SQUR(P)	ω	Aω	λ=Aω/ω
1	0.24032608	0.0103947	0.17885652	17.2064448
2	4.57056714	0.19768914	3.38304393	17.1129476
3	3.55331495	0.15369028	2.57684005	16.766446
4	0.9423191	0.04075780	0.66954633	16.4274396
5	0.44416857	0.01921146	0.31532210	16.4132226
6	0.64731231	0.02799797	0.45881281	16.387358
7	0.64731231	0.02799797	0.45881281	16.387358
8	0.3165713	0.0136925	0.22906919	16.7294766
9	1.91679426	0.08290643	1.37031066	16.5283999
10	0.9423191	0.04075780	0.66954633	16.4274396
11	2.64451225	0.11438216	1.8944650	16.5625918
12	2.64451225	0.11438216	1.8944650	16.5625918
13	1.36080174	0.05885828	0.9690927	16.4648490
14	0.44416857	0.01921146	0.31532210	16.4132226
15	0.44416857	0.01921146	0.31532210	16.4132226
16	1.36080174	0.05885828	0.9690927	16.4648490
				λ_{max} =17.20644481

Judge-2

In this example, in the saaty scale rating by the judge, 10 AMDs rank are categorized into five groups, from 1 to 9. (Tables 5, 6 and 7).

Table 5 Groups of AMDs

Groups	D1	D2	D3	D4	D5
AMDs	2,3	9,11,12	4,10,13,16	5,14,15,6,7	1,8
Intensity of Importance	1	3	5	7	9
Reciprocal	1	0.33	0.2	0.14	0.11

Table 6 Pairwise Comparison matrix

AMDs	1	2	3	4	5	6	7	8	9	10	11	12	13	14	15	16
1	1	0.11	0.11	0.2	0.33	0.33	0.33	1	0.14	0.2	0.14	0.14	0.2	0.33	0.33	0.2
2	9	1	1	5	7	7	7	9	3	5	3	3	5	7	7	5
3	9	1	1	5	7	7	7	9	3	5	3	3	5	7	7	5
4	5	0.2	0.2	1	3	3	3	5	0.33	1	0.33	0.33	1	3	3	1
5	3	0.14	0.14	0.33	1	1	1	3	0.2	0.33	0.2	0.2	0.33	1	1	0.33
6	3	0.14	0.14	0.33	1	1	1	3	0.2	0.33	0.2	0.2	0.33	1	1	0.33
7	3	0.14	0.14	0.33	1	1	1	3	0.2	0.33	0.2	0.2	0.33	1	1	0.33
8	1	0.11	0.11	0.2	0.33	0.33	0.33	1	0.14	0.2	0.14	0.14	0.2	0.33	0.33	0.2
9	7	0.33	0.33	3	5	5	5	7	1	3	1	1	3	5	5	3
10	5	0.2	0.2	1	3	3	3	5	).33	1	0.33	0.33	1	3	3	1
11	7	0.33	0.33	3	5	5	5	7	1	3	1	1	3	5	5	3
12	7	0.33	0.33	3	5	5	5	7	1	3	1	1	3	5	5	3
13	5	0.2	0.2	1	3	3	3	5	0.33	1	0.33	0.33	1	3	3	1
14	3	0.14	0.14	0.33	1	1	1	3	0.2	0.33	0.2	0.2	0.33	1	1	0.33
15	3	0.14	0.14	0.33	1	1	1	3	0.2	0.33	0.2	0.2	0.33	1	1	0.33
16	5	0.2	0.2	1	3	3	3	5	0.33	1	0.33	0.33	1	3	3	1

Table 7 Calculations for priority weight (W) and eigenvector λmax

AMDs	SQUR(P)	ω	Aω	λ=Aω/ω
1	0.25027698	0.01053413	0.18034246	17.1198120
2	4.44197579	0.18696240	3.19032015	17.0639662
3	4.44197579	0.18696240	3.19032015	17.0639662
4	1.1472026	0.04828566	0.79420693	16.4480881
5	0.50543546	0.02127373	0.34847135	16.3803535
6	0.50543546	0.02127373	0.34847135	16.3803535
7	0.50543546	0.02127373	0.34847135	16.3803535
8	0.25027698	0.01053413	0.18034246	17.1198120
9	2.41938808	0.1018318	1.68888656	16.5850523
10	1.1472026	0.04828566	0.79420693	16.4480881
11	2.41938808	0.1018318	1.68888656	16.5850523
12	2.41938808	0.1018318	1.68888656	16.5850523
13	1.1472026	0.04828566	0.79420693	16.4480881
14	0.50543546	0.02127373	0.34847135	16.3803535
15	0.50543546	0.02127373	0.34847135	16.3803535
16	1.1472026	0.04828566	0.79420693	16.4480881
		$\sum \omega = 1$		**λ_{max} =17.11981204**

3.1. Checking Consistency Ratio

Matrix is rejected if the consistency ratio is C.R.>0.1. Matrix is compatible and therefore suitable as each matrix, CR < 0.1, therefore (Table 8). Based on the acceptability of the matrix C.R. < 0.1. Random Index (R.I.) for n=16 is 1.5978.

Table 8 Consistency ratio

Judges	λ_{max}	$Consistency\ Index\ (C.I.) = \frac{\lambda_{max-n}}{n-1}$	Consistency Ratio (C.R.) $= \frac{C.I.}{R.I.}$
Judge-1	17.20644481	0.08042965	0.05033775
Judge-2	17.11981204	0.07465414	0.04672308

3.2. Effectiveness of AMDs

We perform multiplication of AMDs matrix by individual judges to determine the effectiveness of each AMD (Fig. 4, Table 9).

$$\begin{Bmatrix} \text{Judge 1} & \text{Judge 2} \\ 0.010395 & 0.010534 \\ 0.197689 & 0.186962 \\ 0.15369 & 0.186962 \\ 0.040758 & 0.048286 \\ 0.019211 & 0.021274 \\ 0.027998 & 0.021274 \\ 0.027998 & 0.021274 \\ 0.013693 & 0.010534 \\ 0.082906 & 0.101832 \\ 0.040758 & 0.048286 \\ 0.114382 & 0.101832 \\ 0.114382 & 0.101832 \\ 0.058858 & 0.048286 \\ 0.019211 & 0.021274 \\ 0.019211 & 0.021274 \\ 0.058858 & 0.048286 \end{Bmatrix} \text{X} \begin{Bmatrix} 1/2 \\ 1/2 \end{Bmatrix} = \begin{Bmatrix} 0.01046 \\ 0.19233 \\ 0.17033 \\ 0.04452 \\ 0.02024 \\ 0.02464 \\ 0.02464 \\ 0.01211 \\ 0.09237 \\ 0.04452 \\ 0.10811 \\ 0.10811 \\ 0.05357 \\ 0.02024 \\ 0.02024 \\ 0.05357 \end{Bmatrix}$$

Figure 3 Matrix multiplication

Table 9 AHP Result

S. No	AMDs	Rank by ISM	Effectiveness by AHP (%)
1	Responsive Supply Chain	10	1.046444
2	Top management support	1	19.23258
3	Training and Education	2	17.03263
4	Customer response adoption	6	4.452174
5	Concurrent engineering	8	2.02426
6	Re-configurable layout	7	2.463586
7	Appropriate Hardware	7	2.463586
8	Rapid prototyping	9	1.211334
9	Organisational Culture	4	9.236914
10	Customer Involvement	6	4.452174
11	Knowledgeable Workers	3	10.8107
12	Employee involvement	3	10.8107
13	Proper database management	5	5.357198
14	Automation	8	2.02426
15	Virtual Enterprise	8	2.02426
16	IT Integrations	5	5.357198

4. Result and Discussion

Currently, any manufacturing company must embrace AM to succeed and fulfil consumer demand. The principal purpose of the analysis is to measure, contrary to the title which determines the efficacy of AMDs through the use of AHP, the results of the ISM. There are four AMDs in our efficiency quantification analysis that seriously impact the effective execution of AM with a large percentage of efficiency are top management support, training and education, knowledgeable workers and employee involvement.

5. Conclusion

The results achieved will contribute to implement AMDs for the effective application of AM after achieving the percentage effectiveness of AMDs. For the theoretical and administrative dimensions of AM the effects of AHP may be implied, as they help managers to remove AMDs for proper functioning of AM by taking certain measures depending on the driver percentage effectiveness. This research further allows one to consider the principles and drivers of AM by applying diverse revisions methodologies such as ISM and AHP. A wide range of AM is available in future research directions as manufacture moves towards creativity and personalization, and so AMDs should be researched to execute AM successfully.

6. References

[1] Goldman, L. & Nagel, Roger & Preiss, Kenneth. (1995). *Agile Competitors and Virtual Organizations-Strategies for Enriching the Customer.*

[2] BUNCE, P. and GOULD, P., 1996, *From lean to agile manufacturing.* IEE Colloquium (Digest), **No. 278,** pp. 3/1-3/5.

[3] Richards, C.W. (1996) *'Agile manufacturing: beyond lean?'*, Production and Inventory Management Journal, **Vol. 37,** No. 2, pp.60–64.

[4] Moore, S.M.R. (1995) *'Agility is easy, but effective agile manufacturing is not'*, IEEE Colloquium on Agile Manufacturing, pp.3/1–3/4, DOI: 10.1049/ic:19951099.

[5] Yusuf, Y., Sarhadi, M. and Gunasekaran, A. (1999) *'Agile manufacturing: the drivers, concepts and attributes'*, International Journal of Production Economics, **Vol. 62,** Nos. 1–2, pp.33–43.

[6] Poolton, J., Ismail, H. S., Reid, I.R. and Arokiam, I.C. (2008) *'Implementation of an agile marketing approach for the manufacturing-based SME'*, International Journal of Entrepreneurship and Small Business, **Vol. 6,** No. 2, pp.264–279.

[7] Browne, J., Sackett, J. and Wortmann, J. (1995) *'Future manufacturing systems towards the extended enterprise'*, Computers in Industry, **Vol. 25,** No. 3, pp.235–254.

[8] Sharifi, H. and Zhang, Z., 2001. *Agile manufacturing in practice - Application of a methodology.* International Journal of Operations & Production Management, **21(5/6),** pp.772-794.

[9] Haider, A. and Khan, U., 2020. *Finding the Percentage Effectiveness of Agile Manufacturing Barriers: An AHP Approach.* Lecture Notes in Mechanical Engineering, pp.133-145.

[10] Garbie, I.H., Parsaei, H.R. and Leep H.R. (2008) *'A novel approach for measuring agility in manufacturing firms'*, International Journal of Computer Applications in Technology, **Vol. 32,** No. 2, pp.95–103.

[11] Chen, S-P. and Wu, W-Y. (2010) *'A systematic procedure to evaluate an automobile manufacturer-distributor partnership'*, European Journal of Operational Research, **Vol. 205,** No. 3, pp.687–698

[12] Vivek, S.D., Banwet, D.K. and Shankar, R. (2008) *'Analysis of interactions among core, transaction and relationship-specific investments: the case of offshoring'*, International Journal of Operations Management, **Vol. 26,** No. 2, pp.180–197.

[13] Aldahmash, A., Gravell, A. and Howard, Y., 2017. *Using Factor Analysis to Study the Critical Success Factors of Agile Software Development.* Journal of Software, **12(12),** pp.957- 963.

[14] Kumar, D., Agarwal, S.P., Mishra, A.K. and Singh, S.K., *An ISM Approach to Analysis of Enablers of Agile Manufacturing.*

[15] Hasan, M., Shankar, R., Sarkis, J., Suhail, A. and Asif, S., 2009. *A study of enablers of agile manufacturing.* International Journal of Industrial and Systems Engineering, **4(4),** p.407. 10

[16] Ajay Guru Dev, C. and Senthil Kumar, V., 2016. *Analysis on critical success factors for agile manufacturing evaluation in original equipment manufacturing industry-an AHP approach.* Chinese Journal of Mechanical Engineering, **29(5),** pp.880-888.

[17] Kaswan, M. S., & Rathi, R. (2019). *Analysis and modeling the enablers of Green Lean Six Sigma implementation using Interpretive Structural Modeling.* Journal of Cleaner Production, **231,** PP.1182–1191.

[18] Bhushan, N., & Rai, K. (2004). *Strategic decision making: Applying the analytic hierarchy process.* New York: Springer.

[19] Saaty, T.L., 2008. *Decision making with the analytic hierarchy process.* International Journal of Services Sciences, **1 (1),** 83–98.

[20] Sharma, B. P; Singh, M. D., & Kumar, A. (2012). *Knowledge sharing barriers: An integrated approach of ISM and AHP.* In 2012 International Conference on Information and Knowledge Management (ICIKM 2012) IPCSIT **(Vol. 45).** Singapore: IACSIT Press.

[21] Saaty, T. L. (1980). *The analytical hierarchy process.* New York, NY, U.S.A: McGraw-Hill International.

Ridge Gap Waveguide based band pass filter for Ku-band Application

Neetirajsinh J Chhasatia[1], Jitendra P Chaudhari[2], Amit V Patel[3]
[1,2,3]V.T. Patel Department of Electronics & Communications, Chandubhai S Patel Institute of Technology,
Charotar University of Science and Technology, Changa, Anand, Gujarat, India 388421
[1]neetiraj.n.c@ieee.org, [2]jitendrachaudhari.ec@charusat.ac.in,
[3]amitvpatel.ec@charusat.ac.in

Abstract. This paper describes a simple, low loss, compact tuneable band pass filter based on ridge gap waveguide (RGW) technology for the Ku-band applications. This is achieved by keeping the height of the air gap in the gap guide structure equal to the thickness of the substrate or base of the structure. The resonant frequency and electromagnetic (EM) field distribution of the structure is investigated. This filter is designed by inserting the proposed ridge in the cut-off region of the gap waveguide. The frequency of tuning has been carried out using the slot created on ridge, which generates capacitive effect. Experimental results of the manufactured structure show an insertion loss of approximately 0.15 dB and a return loss of 16.38 dB over 4.5% relative bandwidth in Ku-band. The structure, put forwarded here, has been designed and optimized in the CST microwave studio environment and simulated results are validated by experimental results. The size of the structure is 64.65 mm x 64.65 mm x 7 mm.

Keywords. Dispersion Diagram, Insertion Loss, Kurtz band application, Return Loss, Ridge Gap Waveguide Technology.

1. Introduction

Now days, it is expected to have a high-density technique with substantial process of fabrication at lower cost, which offers widespread solutions for future millimetre wave applications. To meet this expectation, an extensive research in this area has been carrying out in antenna/propagation/microwave communities. Techniques started providing the improved results, kicked off with SIW technology [1], followed by one of the famous low loss TFMS - thin film microstrip line [2], Low Temperature Co-Fired Ceramic [3] and many more. These methods have their own pros and cons. However, as market has not been sticking at a single point, a gap waveguide technology is booming now-a-days designing the various high frequencies components or devices. This is the concept of a structure, contains metallic base, pins and a lid. This periodic structure of pins creates artificial magnetic conductor – predominately provides high impedance, which helps preventing the electromagnetic wave propagations for particular frequency region. This historic theory is presented in [4] with reference to the simulations. On later stage, based on the experimental results, idea was confirmed in [5]. Aforementioned literature validates the electromagnetic band gap characteristic of the structure, which can be advantageous for the application at high frequencies.

This technology provides us the promising results in an absence of dielectric material in any form, diminishing losses pertaining to the radiation & dielectric material, which are associated predominately with microstrip based/SIW structures [15].

The point that makes the structure an easy and straightforward is – no electric contact between two plates. It is the main pros in fabricating the components at millimetre waves compared to conventional tech designs. In [6], three different types of gap waveguides are demonstrated viz. – Ridge Gap Waveguide, Groove Gap Waveguide and microstrip Gap Waveguide.

In the year of 2009[7], Kildal has given the concept of ridge gap waveguide to the microwave researcher's community. Three-dimensional view of the structure is shown in the Fig. 1. To have an internal view of the structure, upper lid is removed.
This concept is an early extension of the concept of hard and soft surfaces. PMC surface can be felt by arranging the nails of quarter wavelength with two-dimensional placement. Upper lid made up of the same material as the ground plane, covers the structure. Ridge is place in between the columns of nails, behaves as PEC which allows signal to be guided along with it [8-10]. Surrounding nails behave like PMC surface, which prevent leakages from the sides of the structure. This is the proof of existence of the band gap, which prohibiting wave propagation inside the periodic cells. Researchers are exploring higher frequency bands to design the microwave components, and gap waveguide is the best suited guiding structures to meet the design expectations.

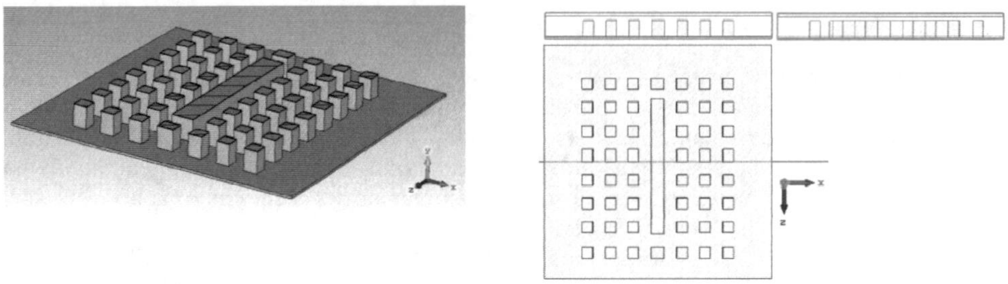

Figure 1. Three-dimensional view of the structure providing Insights of the periodic structure

These guiding structures are well known for carrying microwave signals specifically in the form of quasi-TEM mode [11, 12]. Even if time lapsed, post announcement this, it is still being snatching the focus of the communities due to its own advantages. Blockage of side leakage is the substantial pros of the artificial PMC surroundings over the micro strip lines.
Key part of designing is to construct the smallest part of structure – a unit cell, which acts as the PMC surface. Structure of this unit cell characterizes the operating bandwidth, by forming the artificial magnetic conductor surfaces. Discussions have been depicted on the proposed unit cell in the subsequent sections of this paper. Dispersion diagram – one of the key aspects, connected with this unit cell, has also been illustrated. People are putting rigorous efforts in this area, exploring various shapes of unit cells to achieve good bandwidth ratio, exploring the ways for a transition between the standard connecters and ridge gap waveguide [13, 14].
Here, a proof of concept of tuneable band pass filter based on ridge gap waveguide technology is put forward. Started with unit cell first, band stop region is identified and post that whole structure is simulated in CST microwave studio. The same structure is also fabricated and results are compared in Simulations & Results section. Agreement between simulated and measured results confirms the validity of all simulated results.
At this point, the scope of this paper is to demonstrate the capability and manufacturing flexibility of the proposed ridge gap waveguide filter only.

2. Unit cell & stop band

Unit Cell: The desired filter is implemented using Gap Waveguide technique; hence, gap waveguide structure must be designed first. Aforesaid structure is designed in such a way that the frequency band gap incorporates the operating frequency band of the filter. This two-dimensional structure provides us the stop band for the electromagnetic waves.
By examining a part of the structure – called a unit cell with periodic boundary conditions, stop band can be calculated between different modes of frequencies. Unit cell structure is shown in the Fig. 2.

Various parameters viz. – width of pins (a), height of pins (h), distance of upper plates from pins i.e., airgap (a_g), period of alignment of pins (p) – are the crucial to consider while designing the structure, as they control the frequencies of operation, size of band, impedance and so on. Band gap can be defined by the upper and lower bounds of frequencies and respectively, can be controlled by varying the height of pins and air gap distance – distance of lid/upper plate from pins. For the proposed design, appropriate dimensions of the structure are calculated to create required band gap, based on parametric study.

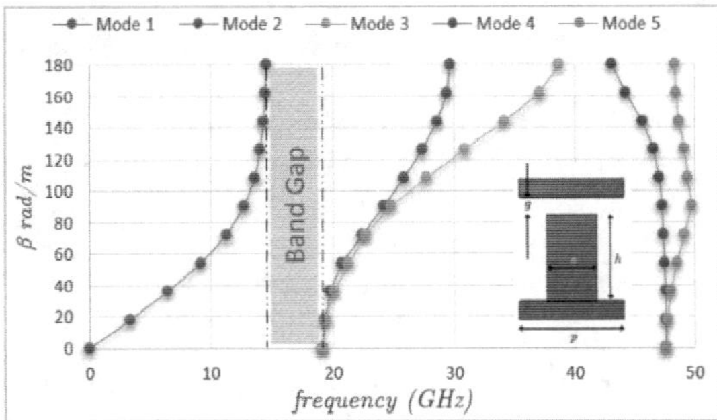

Figure 2. The dispersion diagram for several different modes *(p=6.5 mm, h=4.5 mm, a=3 mm, g=1 mm)*.

With reference to the Fig. 2, the band gap frequency range is from 14.5-19 GHz and its relative bandwidth of (f_{max}/f_{min}) – 2.24. By introducing the proper defects within the structure, filter is designed by developing apt sized ridge.

Such a structure electrically represented as the resonant parallel LC circuit, which has impedance equal to the impedance of the parallel LC circuit. Generally, impedance of parallel LC circuit hit the peak if the frequency of structure reached at resonant frequency.

3. Simulation Results

Based on discussions in previous section, RGW structure is designed. Table 1 contains the parametric view of the same. The proposed structure has been constructed in Computer Simulation Technology (CST) microwave studio software as shown in Figure 3. The same software is used for simulation purpose. The electric field distribution inside the proposed stricter is shown in Figure 4. The scattering parameters are shown in Figure 7. They show that over the expected bandwidth, return loss observed better than 15 dB, and the insertion loss is better than 1 dB for almost the same bandwidth.

Pin Height *(h)*	**Period *(p)***	**width *(a)***	**Air gap *(g)***	**Thickness of base**	**Thickness of upper metal lid**	**Desired Stop Band**
λ/4	λ/3	λ/6	λ/20	λ/2	λ/20	14.5-19 GHz

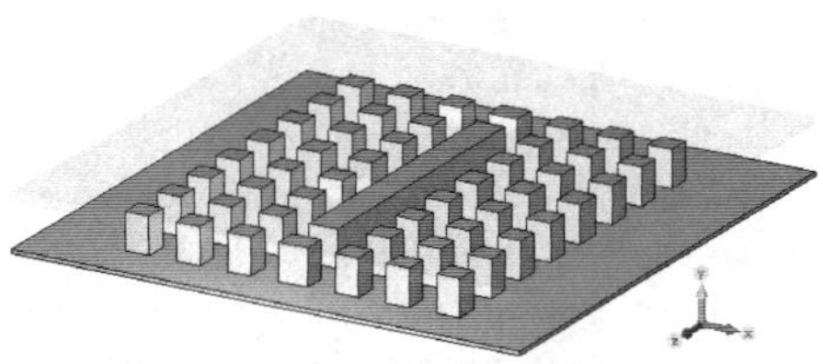

Figure 3. Perspective view of the complete geometry including bed of nails underneath.

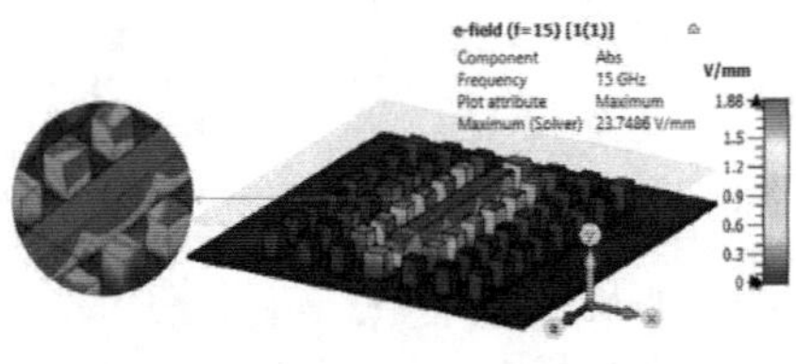

Figure 4. The E-field distribution in ridge gap waveguide filter.

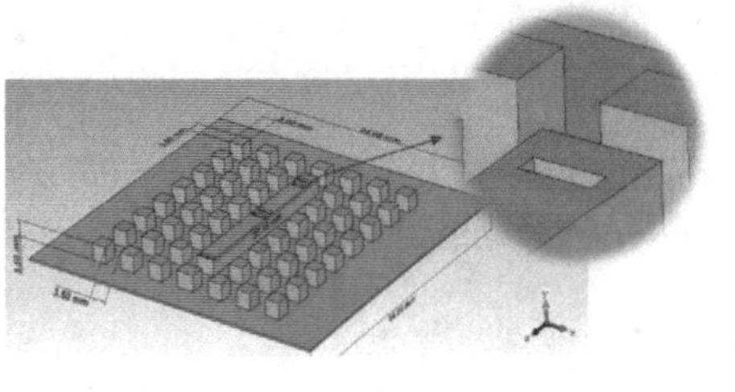

Figure 5. Structure with slot on the ridge.

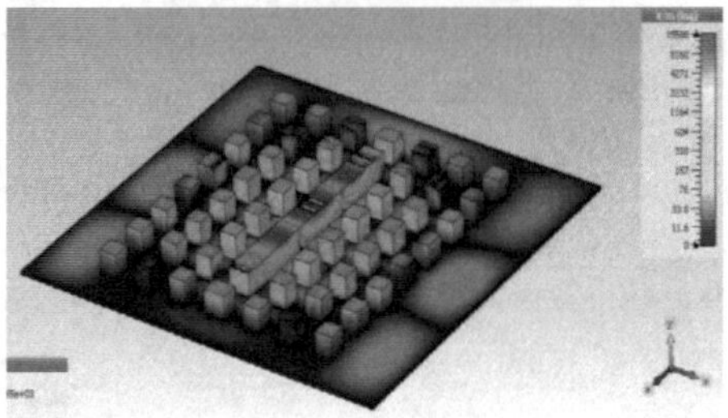

Figure 6. The E-field distribution in ridge gap waveguide filter.

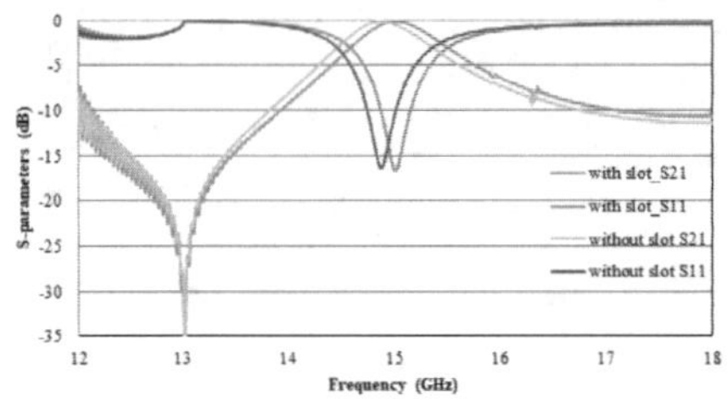

Figure 7. Simulated frequency response of ridge gap waveguide filter.

Figure 8. Structure under measurement - complete geometry including bed of nails underneath.

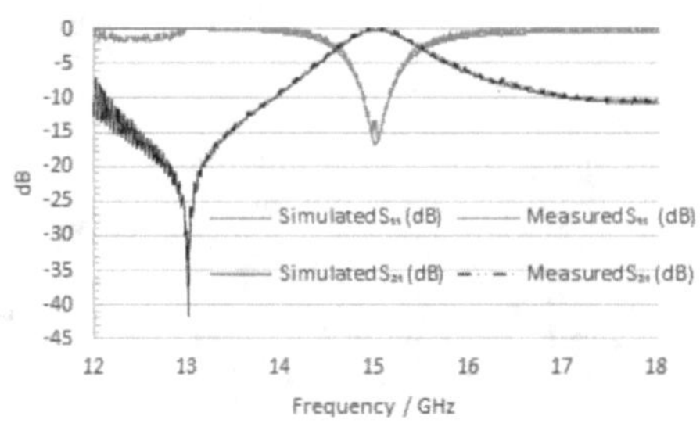

Figure 9. Simulated and measured response of the filter with a slot on ridge.

Second part of the design is to explore tuning capability of the structure. By introducing the slot on the ridge, gives us the capability to fine-tune the filter in the operating frequency band(s), keeping the scattering parameters almost similar. Figure 5 shows the structure with a slot on a ridge of the gap waveguide structure. Figure 6 shows the field distribution inside the structure which is confine to the ridge at the desired frequencies and structure will not let the flux to leakage from the sides. Figure 7 shows the comparison between the simulated results of the structure without a slot on a ridge and with a slot on a ridge of the proposed structure. Figure 8 is the measurement setup of the fabricated filter. The optimized response of the filter is plotted in Figure 9 along with the measured results. The measured results in Figure 9 show an acceptable agreement with the analyses in general as a first prototype trial. Some additional loss are observed in the measurements due to the possible presence of additional airgap between the layers.

4. Conclusion

This paper has been proposed the structure, purely based on the concept of perfect magnetic conductor, having a symmetric geometry, and is realized using gap waveguide technology. This geometry causes considerably less fringing fields around the ridge located in the centre. This structure contains fewer numbers of periodic cells on the both sides of the ridge, which reduces the structure size. Simulation results confirmed the theoretically expected band obtained via the Eigen mode solver in CST. The simulated results for the filter are capable to meet the stringent commercial specifications. However, fabrication tolerance associated with commonly used manufacturing techniques is to be investigated more in depth. At last, structure is fabricated, tested, and observed good agreement between measured and simulation results.

5. References

[1] Ke Wu, "Towards system-on-substrate approach for future millimeter-wave and photonic wireless applications," 2006 *Asia-Pacific Microwave Conference*, 2006, pp. 1895-1900, doi: 10.1109/APMC.2006.4429778

[2] G. Six, G. Prigent, Eric Rius, G. Dambrine and H. Happy, "Fabrication and characterization of low-loss TFMS on silicon substrate up to 220 GHz," in *IEEE Transactions on Microwave Theory and Techniques*, vol. **53**, no. 1, pp. 301-305, Jan. 2005, doi: 10.1109/TMTT.2004.839915

[3] J. Hasch, E. Topak, R. Schnabel, T. Zwick, R. Weigel and C. Waldschmidt, "Millimeter-Wave Technology for Automotive Radar Sensors in the 77 GHz Frequency Band," in *IEEE Transactions on Microwave Theory and Techniques*, vol. **60**, no. 3, pp. 845-860, March 2012, doi: 10.1109/TMTT.2011.2178427

[4] P. Kildal, "Three metamaterial-based gap waveguides between parallel metal plates for mm/submm waves," 2009 *3rd European Conference on Antennas and Propagation*, 2009, pp. 28-32.

[5] S. Kildal, A. Uz Zaman, E. Rajo-Iglesias, E. Alfonso, A. ValeroNogueira, "Design and experimental verification of ridge gap waveguides in bed of nails for parallel plate mode suppression", *IET Microwaves, Antennas & Propagation*, Vol. **5**, No 3, pp. 262-270, March 2011

[6] Kildal P-S. Three metamaterial-based gap waveguides between parallel metal plates for mm/submm waves. In: *3rd European Conference on Antennas and Propagation EUCAP*; 2009

[7] P.-S. Kildal, E. Alfonso, A. Valero-Nogueira, and E. Rajo-Iglesias, "Local metamaterial-based waveguides in gaps between parallel metal plates," *Antennas and Wireless Propagation Letters,IEEE*, vol. **8**, pp. 84-87, 2009

[8] M. Sharifi Sorkherizi and A. A. Kishk, "Fully Printed Gap Waveguide With Facilitated Design Properties," in *IEEE Microwave and Wireless Components Letters*, vol. **26**, no. 9, pp. 657-659, Sept. 2016

[9] L. Kumar and M. S. Parihar, "A Compact Reconfigurable Low-Pass Filter With Wide-Stopband Rejection Bandwidth," in *IEEE Microwave and Wireless Components Letters*, vol. **28**, no. 5, pp. 401-403, May 2018

[10] Y. Shi, J. Zhang, M. Zhou, W. Feng, B. Cao and W. Che, "Miniaturized W-Band Gap Waveguide Bandpass Filter Using the MEMS Technique for Both Waveguide and Surface Mounted Packaging," in *IEEE Transactions on Circuits and Systems II: Express Briefs*, vol. **66**, no. 6, pp. 938-942, June 2019

[11] D. Sun and J. Xu, "A Novel Iris Waveguide Bandpass Filter Using Air Gapped Waveguide Technology," in *IEEE Microwave and Wireless Components Letters*, vol. **26**, no. 7, pp. 475-477, July 2016

[12] M. Ali and A. Sebak, "Compact Printed Ridge Gap Waveguide Crossover for Future 5G Wireless Communication System," in *IEEE Microwave and Wireless Components Letters*, vol. **28**, no. 7, pp. 549-551, July 2018
[13] H. Raza, J. Yang, P. Kildal and E. Alfonso Alós, "Microstrip-Ridge Gap Waveguide–Study of Losses, Bends, and Transition to WR-15," in *IEEE Transactions on Microwave Theory and Techniques*, vol. **62**, no. 9, pp. 1943-1952, Sept. 2014
[14] A. U. Zaman, P. Kildal and A. A. Kishk, "Narrow-Band Microwave Filter Using High-Q Groove Gap Waveguide Resonators With Manufacturing Flexibility and No Sidewalls," in *IEEE Transactions on Components, Packaging and Manufacturing Technology*, vol. **2**, no. 11, pp. 1882-1889, Nov. 2012
[15] P. Kildal, E. Lier and J. A. Aas, "Artificially soft and hard surfaces in electromagnetics and their application," 1988 *IEEE AP-S. International Symposium, Antennas and Propagation*, pp. 832-835 vol.**2**, 1988

Design and Analysis of Release Mechanism using Shape Memory Alloy for Spacecraft

Shrutika Dahake[1,*], Nilesh Awate[2], Rupesh Shelke[2]

[1]PG Scholar, Department of Mechanical engineering, G H Raisoni Collage of Engineering Nagpur, India
[2]Professor, Department of Mechanical Engineering, G H Raisoni Collage of Engineering Nagpur, India

shrutikadahake6@gmail.com[1,*], nilesh.awate@raisoni.net[2], rupesh.shelke@raisoni.net[2]

Abstract: Shape memory alloy (SMA) is used as a smart material. It is also called as Nitinol and intelligent material because it regains their original shape after deformation. It works on the principle of shape memory effect which regain their original shape after passing through different temperature phases. SMA get some more attention in incoming years due to their superior and unique properties. This paper presents design and maximum stress analysis of release mechanism using SMA. It also covers the capability and efficiency of notch bolt of Frangibolt mechanism. Its design, calculation and analysis has done in CAD software and simulation software.

Keywords. Shape Memory Alloys, Nitinol, Frangibolt, Release Mechanism.

1. Introduction

During the launch of a space vehicle, Antenna and solar panels are initially folded to reduce its size. Once the spacecraft reaches its orbit, it needs to be deployed to perform its function. Design constraints viz. simple, compact in size, lightweight, low shock, reusable and easy to reset have been considered while designing the Release Mechanism.

Release mechanism is actuated by using Shape Memory Alloy (SMA) which works on the Shape Memory Effect (SME) i.e., When SMA wire is heated above its transformation temperature, it will transform from Martensite phase to Austenite phase. i.e. it will regain its parent shape. This release mechanism can be used for the deployment of solar arrays and antennas of small size satellites. Shape memory alloy (SMA) having some native properties, specially regain its original form by increasing or decreasing temperature. In 1938 Mooradian and Greninger observe the shape memory effect for (Cu-Sn).In 1961 Buehler and Wiley invented a Nickel-Titanium alloy called NiTi-NOL[1] . In Nitinol Ni contain 55 wt% and the rest is Ti[2].

2. Aerospace applications

In 1960s, Raychem developed Cryofit connectors for F-14 airplane hydraulic circuit. It is the first industrial SMA applications in aeronautics.[1].Because of unique properties of SMA, it has huge

application in aerospace such as actuators [3], sealer, release or deployment mechanism, vibration dumper, manipulators and the pathfinder applications and structural connectors [1].
In space craft, SMAs used as low shock release mechanism. Best example is Cryofit. CryoFit in which coupling is made from SMA material. Applications of Cryofit are such as tubes used in aircraft, hydraulic lines, and fuel lines. When these tubes are broken, they often have to be repaired by people working in confined conditions. In this situation, it is very difficult to use welding and brazing because these processes require bulky equipment. Hence, CryoFit coupling is used. Because they release slowly with minimum vibration. Cryofit and Frangibolt [4] are some example of small device which shape memory alloys material is used.

3. Types of Hold Down release Mechanism (HDRM)
There are two basic types of hold down release mechanism, Pyrotechnic HDRM (Explosive HDRM) and Non-Pyrotechnic HDRM (Non-Explosive HDRM).

3.1. Pyrotechnic HDRM.
In this pyrotechnic HDRM, shaft is made inside of the bolt that specifically made to break at certain location along its length as shown in the fig.1 [4]. This location typically has a smaller diameter than the rest of the bolt inside the shaft. A low sensitivity and high density explore taking place on the top of that small amount of low-density. If this is placed finally, the filament bridge wire initiator is placed above the low density where it's at time to separate the parts that are held together by the boat and illogical current is passed through a pyramid inside the initiator. It heats up quickly and the ignites small amount of pyrotechnic powder. The Powder burns rapidly creating lots of gas at high temperature and pressure. This causes the low-density explosive to detonate. These detonations create a shock wave and then the final explosion inside of the bolt causes the part of the bolt that has a smaller diameter to fill first because it's the weakest part of the bolt. The bolt breaks apart cleanly causing the two parts to separate.

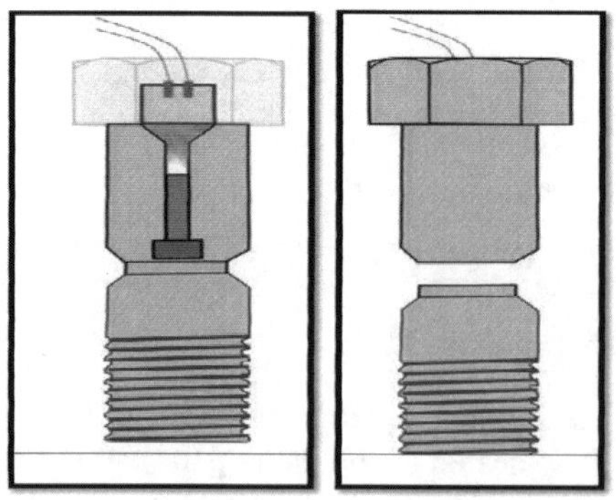

A.Before Actuation B. After Actuation

Fig.1 Pyrotechnic HDRM

This sequence of events is overly complicated and may wonder why the high-density exclusive is not connected directly. The main reasons are safety, reliability and accuracy. The high-density explosive is made less sensitive to prevent it from exploding due to force encountered during the mission because of this small explosive is required to trigger it. This comes from the low density is used to trigger the expressive and exploding bridge wire is used because it can reliably detonate the low-density explosive accurately in a certain amount of the time after activation. This timing is important during deployment because the one second delay means that a spacecraft will be many km away from where an event was to be triggered. Another configuration for the pyrotechnic hold down release mechanism is one where the expanding gas generated by the explosive is used to push to hemisphere of threaded coupling apart. These will lose the bolt that screwed into the coupling. A compressed spring can pull the parts away. pyrotechnic HDRM suffers from one measure issue. The detonation of exclusive produce shockwaves when they are triggered. This shockwave is transmitted through out the spacecraft and can potentially damage electronic circuit boards optical device alignment and other sensitive component. For reducing this shockwave was one of the reasons for the deployment of a second type of HDRM.

3.2. Non pyrotechnic HDRM

Non pyrotechnic HDRM is the split spool initiator. In this design a groove has curved into the middle of cylinder as shown in fig. 2A. the cylinder is split in half lengthwise but is then held together by wrapping a tensile wire around it one end of this wire is fixed to the frame of the HDRM and the other end is hooked to linked wire inside the groove of the cylinder is the pleasure pin which is being pressed into the groove by spring loaded attachment inside, this attachment is threaded coupling to which the voges attached as shown in fig 2. This coupling is also split lengthwise but held together by protrusions inside of the attachment to trigger this HDRM and release the bolt a high current is passed through the linked wire this causes is to vaporise and release one end of the tensile wire. Now with the only one side held down tensile wire unwind and is unable to hold it to hemispheres of the cylinder together under pressure from the spring. the attachment along the plunger pin is now pushed forward, now that the cylinder is no longer blocking the movement of the plunger pin as a consequence of the attachment moving forward, the protrusions that kept the two halves of the threaded coupling to heater are now moved forward. This Dennett result in the two halves of the coupling moving apart and releasing the bolt just like with the pyrotechnic design. The apparent complexity of this design is related to safety and reliability.

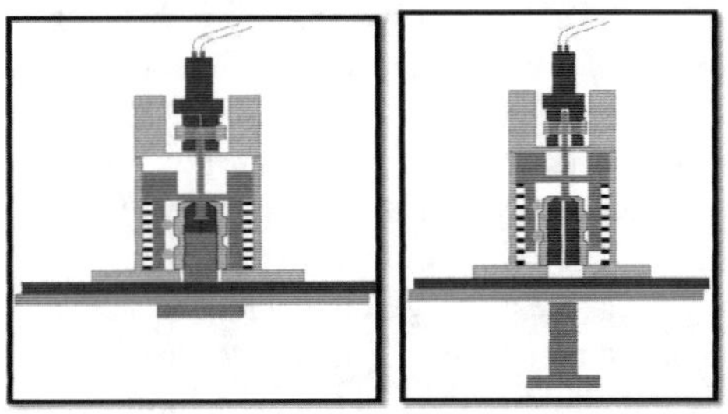

A. Before Actuation B. After Actuation

Fig.2 Non- Pyrotechnic HDRM

since this design does not contain any chemical reaction. The delay from trigger to release is more consistent. The rest of the mechanism is there to redirect the energy of the spring to make the connection between the bolt and coupling fail cleanly as spacecraft become bigger and more complicated innovative ways of folding a space craft of fate inside a rocket are being created.

So, while holding release mechanisms do not take about frequency when it comes to spacecraft systems. Hence non-pyrotechnic HDRM is always better option than pyrotechnic HDRM. SMA based and paraffin-based mechanism are the two types of non-pyrotechnic HDRM.

IOP Conf. Series: Materials Science and Engineering 1206 (2021) 012012 doi:10.1088/1757-899X/1206/1/012012

4. Functional requirement of release mechanism in today's era

- It should be of non-pyrotechnic type
- Capable of moderate preload
- Design should be simple, compact lightweight in construction
- In flight model Capable of functional testing
- Less shock than explosive devices
- Easy to reset

5. Comparison Between Pyrotechnic and Non-pyrotechnic HDRM

Name of parameter		Pyrotechnic HDRM	Non-pyrotechnic HDRM	
		Pyrocutter	SMA Based	Parafin based
1.	Source shock generation	Very High (20000 to 6000g)	Low (4000 g to 0g)	Low (100 g to 0g)
2.	Preload Capacity	Very High	High	High
3.	Power Requirement	Low	Low	High
4.	Out gassing	Moderate to Ni	Nil	Nil
5.	Time of actuation (min-sec)	Low	Low	High

In the present study the SMA have been extensively used and lot of developmental activities is in progress. In this paper we are going to discuss on SMA based release mechanism.

6. SMA Based mechanisms

Shape Memory Alloy (SMA) works on the principle of shape memory effect (SME). In SME previously deformed alloy can be made to recover its original shape simply by heating. When the alloy is stressed and it returns to the original shape without heating then it is called Super Elasticity (SE). Material is in the austenite phase at high temperature and in the martensite phase at low temperature. When a material is cooled it changes phase from austenite to martensite, but this martensite is in the form of twinned martensite. When a material is loaded the material gets deformed and its form changes from twinned martensite to de-twinned martensite. The maximum deformation is limited to 8%. When heat is applied SMA recovers its original form. Martensite is converted back to austenite by heating. Fig.3 shows a stress-strain curve of it. [4] Some of the examples of the SMA alloys are Ni-Ti, Ti-Nb, Ti-Nb-Sn, Ti-Nb-Al, Ti-Nb- Ta, Ti-Nb-Pt, Cu–Al–Be, Cu–Zn–Al, etc. [4] Frngibolt and Pin-puller are the example of SMA based hold down release mechanism.

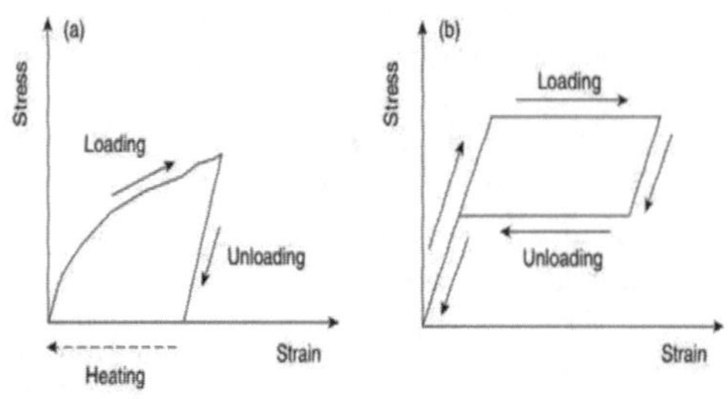

A.Shape memory effect B. Super elasticity

Fig.3.: Stress-strain curves [1]

6.1. One way shape memory effect (OWSME)

During heating the shape changes, this transformation is called the one-way shape memory effect. OWSME deliver more economical solution. Diverse training method was proposed [2] such as spontaneous and load assisted induction.

6.2. Two way shape memory effect (TWSME)

SMA's ability to recover a native shape when heated, and then return to an alternative shape when cooled down its transformation temperature is called as two-way shape memory. Mild heating in two-way systems destroys SME. Two-way SMAs are a more dynamic mechanism than SMAs used in one-way systems.

6.3. Super thermo elastic effect

The basic behavior is the same as in the case of TWSME: some of the mechanical and physical properties belongs between thermal conductivity, young's modulus, thermal expansion and electrical resistivity [5]

7. Frangibolt

The Frangibolt is a simple release device consists of four major component groups. (A) Notched bolt element and matched nut (B) SMA actuator (C)Heater and insulation (D) Joint materials and other hardware. The basic idea of the frangibolt is to utilize shape memory effect to generate forces that will break the notch bolt. When a current is applied through the heater, the SMA cylinder elongates and fractures the bolt element, in this way the separation of two or elements takes place.

7.1. Bolt element

- Design of bolt element of frangibolt [6]

Diameter= 6.35mm
Length= 50mm
1.11mm deep notch with 0.25 radius

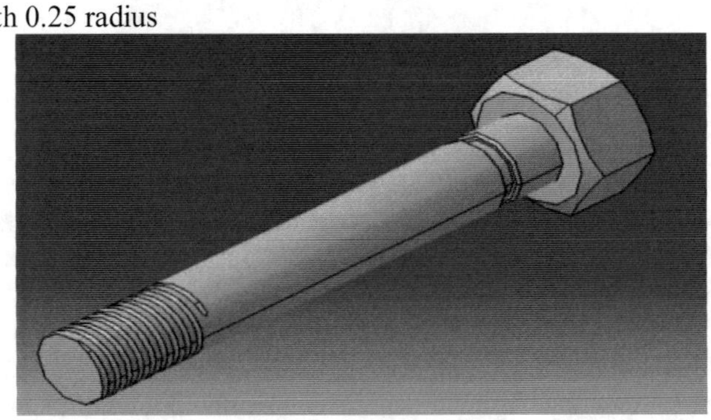

Fig.4 CAD Model of bolt

- Material Selection

Titanium 318 (6Al- 4V) material is selected because it possesses good fatigue properties with corrosion resistance, great strain recovery beyond its ultimate tensile limit.

Table 1: Properties of the Titanium used in the Ansys model [7-8]

Name of property	Value
Density	4429 kg/m^3
Coefficient of thermal expansion	8.789e-5 C^{-1}
Young's Modulus	1.11e11 Pa
Poisson's ratio	0.3387
Bulk Modulus	1.074e11Pa
Shear Modulus	3.713e10Pa
Ultimate tensile strength	985-1090 MPa[7]
Elastic Modulus	106 GPa
Rapture elongation	4.2%

7.2. SMA Actuator

- Design of SMA actuator of frangibolt

Length = 25mm
Inside diameter = 6.6mm
Outside diameter = 12.7mm

Fig.5 CAD model of actuator

- Material Selection

Shape memory alloy material is selected because it generates larger strain recoveries and higher forces as compare to other material and it maturely manufactured to meet aerospace specification. SMA can produce as much as 400 MPa.

Table 2: Properties of the SMA used in the Ansys model [6, 9-10]

Name of property	Value
Hardening Parameter = 85MPa	85MPa [9]
Reference Temperature = 296K	296K [9]
Elastic Limit	192MPa [10]
Temperature Scaling Parameter	5.6 [10]
Maximum Transformation Strain	0.04
Martensite Modulus	6000MPa
Thermal Conductivity	18 W/m ^{0}c
Young Modulus	90000 MPa [6]
Phase transformation coefficient	1.25 Pa

7.3. Design of joint material and other hardware

Nut, washer is designed as requirement of bolt and all this material made up by SS-304

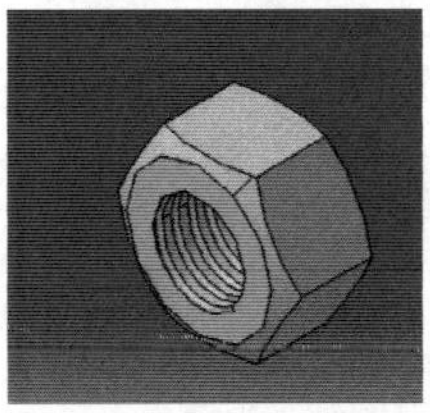

Fig.6 CAD model of nut

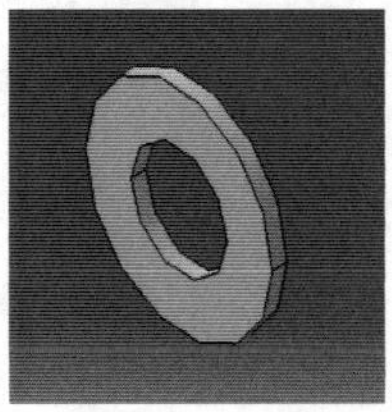

Fig.7 CAD model of washer

SS-304 material is already preloaded in Ansys 2020R2

Table 3: Properties of the SS-304 used in the Ansys model

Name of property	**Value**
Density	7850 kgm-3
Young Modulus	2E+11 Pa
Poissons ratio	0.3
Tensile yield strength	2.5E+08 Pa

7.4. Assembly

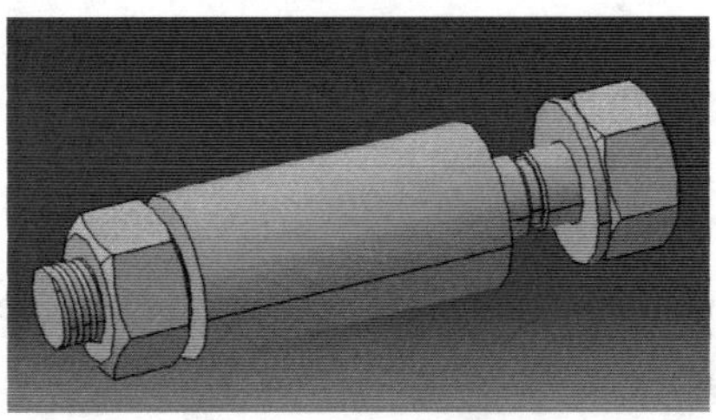

Fig.8 CAD model of assembly of frangibolt

7.5. Analysis

Titanium Bolt preload gives benefits the operation of actuator. By rising preload value, the SMA actuator stroke wanted to break the bolt. So firstly, we have to calculate notch capacity of bolt, because notch play very important role in frangibolt device. According to this we can calculate how much maximum force can sustain a bolt.

7.6. Finite Element Model

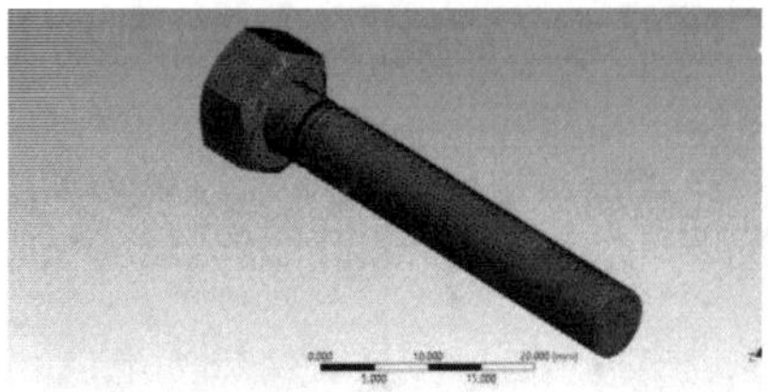

Fig. 9 Meshing model using Ansys Workbench

Tetrahedron meshed was used for discretization. Average quality of element was found to be a 0.86945. The number of elements used for analysis is approximately 150,000.

7.7.. Boundary Condition

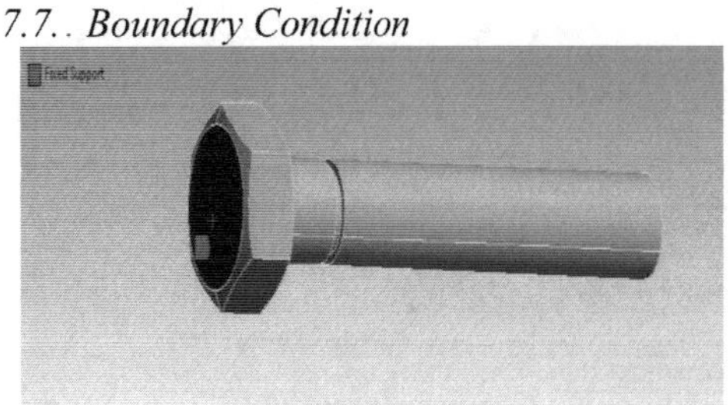

Fig. 10 Fixed support

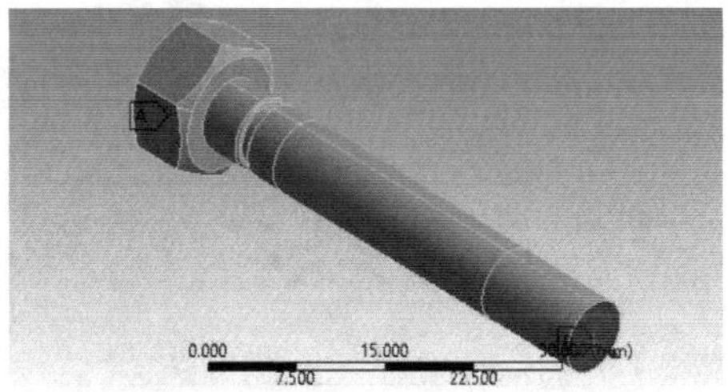

Fig. 11 Pulling Force

The head of nut is fixed. Fixed constraints are as shown in Fig.10. and pull force applied on another end shown in fig.11

7.8. Forces

Equivalent stresses are determined using structural analysis

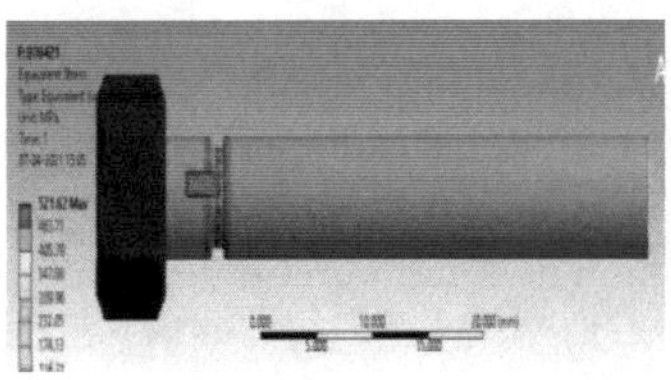

a. P= 8 KN

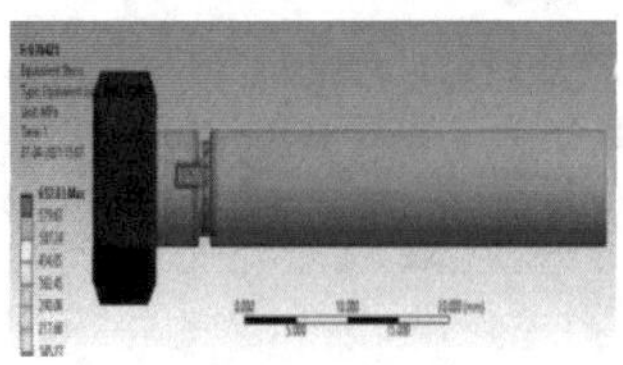

b. P= 10 K

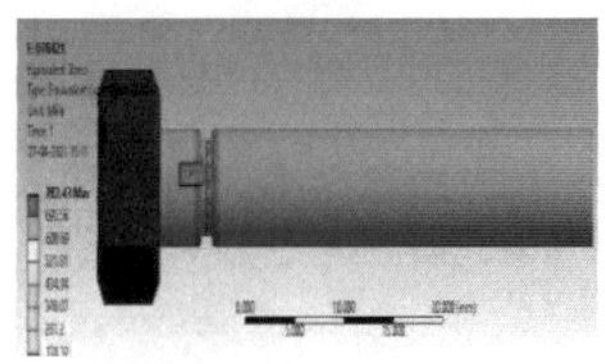

c. P= 12 KN

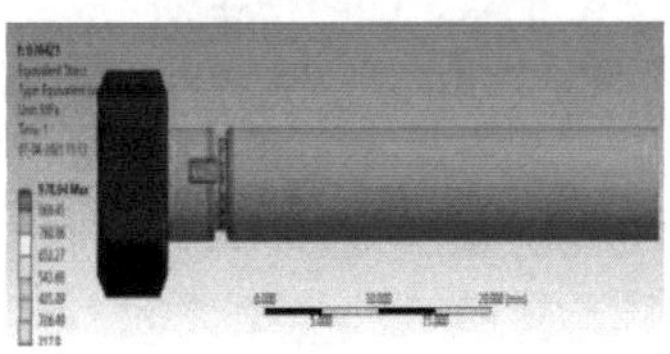

d. P= 15 KN

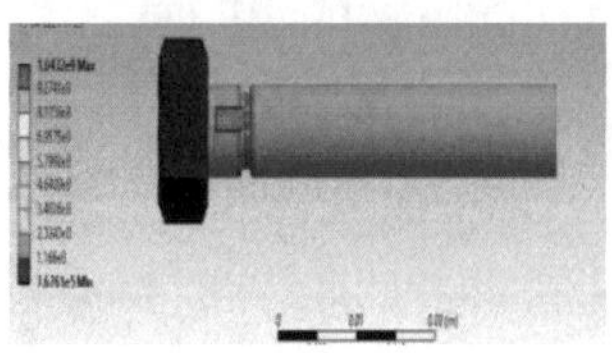

e. P= 16 KN

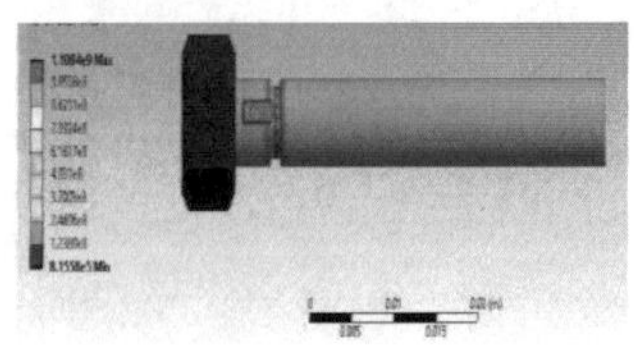

f. P= 17 KN

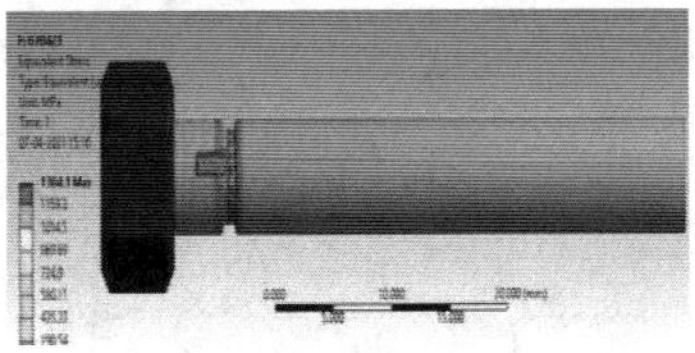

g. P= 18 KN

Figure 12. Equivalent stresses acting on frangibolt

After analysis we can consider that minimum 17 KN force gives is required to break the notch

7.9. Force on Assembly
After applying 17 KN force on assembly it generates 1108.4 Mpa equivalent stress. The SMA actuator produce more force than this, then only release mechanism will work. So, we can conclude that 17 KN minimum load required for release mechanism

7.10. Effect of applying various load on bolt
Fig.13 shows that effect of applying various load on bolt

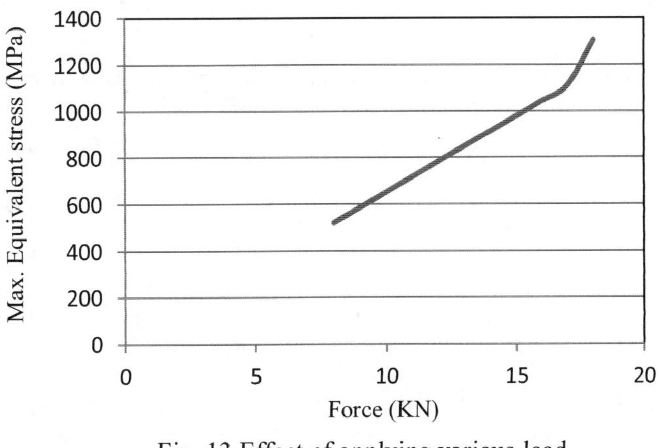

Fig. 13 Effect of applying various load

8. Conclusions

During the analysis, the value of minimum force required to break the notch of bolt element is recorded. A non-explosive release mechanism is designed and tested. The release of a notch bolt is done by SMA wire. SMA wire works the principle of the Shape Memory Effect (SME). When SMA material is loaded it gets deformed. When it is heated it regains its original parent shape. This will generate force, which is used for the actuation of the release mechanism. The testing of the release mechanism is done. Test shows that it will sustain a preload of P =17 KN. In the future, higher preload capacity, quick release, low shock, lightweight, higher reliability release devices can be designed. This analysis will give major role to contribution of improving frangibolt design.

References

[1] A. G. Shivasiddaramaiah, U. S. Mallik, R. Mahato, C. Shashishekar, L. Shivaramu, and S. Prashantha, "Evaluation of biocompatibility of Cu-Al-Be-Mn quaternary shape memory alloys using antibacterial test by agarwell diffusion method," *Mater. Today Proc.*, vol. 17, pp. 61–69, 2019, doi: 10.1016/j.matpr.2019.06.401.

[2] S. Miyazaki, "My Experience with Ti–Ni-Based and Ti-Based Shape Memory Alloys," *Shape Mem. Superelasticity*, vol. 3, no. 4, pp. 279–314, 2017, doi: 10.1007/s40830-017-0122-3.

[3] H. S. Randhawa, G. Pearce, R. Hepton, J. Wong, I. F. Zidane, and X. Ma, "An investigation into the design of a device to treat haemorrhagic stroke," 2019, doi: 10.1177/0954411919890698.

[4] A. D. Johnson, "Non-Explosive Separation Device," vol. 100, no. 362, p. 625, 1992, [Online]. Available: https://patentimages.storage.googleapis.com/08/4b/25/b2e425eeeeca67/US5119555.pdf.

[5] J. Mohd, M. Leary, A. Subic, and M. A. Gibson, "A review of shape memory alloy research , applications and opportunities," *Mater. Des.*, vol. 56, pp. 1078–1113, 2014, doi: 10.1016/j.matdes.2013.11.084.

[6] D. Matthew, "Shape Memory Alloy Fracture as a Deployment Actuator In the Graduate College," 2021.

[7] S. A. Amin and A. Y. Hassan, "Numerical And Experimental Study Of Shape Effect Behavior Of Nitinol Wire," vol. 23, no. 02, pp. 1–11, 2019.

[8] S V Dusane, M K Dipke and M A Kumbhalkar, "Analysis of Steering Knuckle of All Terrain Vehicles (ATV) Using Finite Element Analysis", IOP Conference Series: Material Science and Engineering, Volume 149, 012133.

[9] P. Manuel, C. Lopes, and M. Savi, "Modeling And Simulation Of A Shape Memory Release Device For Modeling And Simulation Of A Shape Memory Release Device For Aerospace Applications Pedro Manuel Calas Lopes PACHECO 1," no. January, 2000.

[10] E. Farber, J. Zhu, A. Popovich, and V. Popovich, "Materials Today : Proceedings A review of NiTi shape memory alloy as a smart material produced by additive manufacturing," *Mater. Today Proc.*, no. xxxx, pp. 1–7, 2020, doi: 10.1016/j.matpr.2020.01.563.

Aerodynamic analysis of aircraft model using indigenously developed wind tunnel facility

D Makhija[1], S V Jain[1], A M Achari[1] and K Ghosh[1]

[1] Mechanical Engineering Department, Institute of Technology, Nirma University, Ahmedabad, India

17bme027@nirmauni.ac.in, sanjay.jain@nirmauni.ac.in, madhusudan.achari@nirmauni.ac.in, kunal@iitk.ac.in

Abstract. This paper presents a design of force balance setup that can measure lift force acting on the aircraft model. The setup was developed indigenously and installed in an open circuit low-speed wind tunnel. It mainly consists of two components viz. a traverse mechanism that can hold the model in the test section at different angles of attack and air speeds and a supporting frame to hold the traverse mechanism over it. The spring balances are used to obtain lift force readings at different angles and air speeds. The experimental and numerical investigations were done in the wide range of Reynolds number (range: 0.55 to 1.12 lakh) and angle of attack (range: -6° to 20°). The results are presented in terms of pressure contours, velocity contours, pressure coefficient and lift coefficient. From the experiments it was found that value of lift coefficient increases with angle of attack and stalling occurs at 18° for all the air speeds. However, in the numerical results the stalling was observed little earlier than 18° angle of attack. The experimental results were compared with CFD results and an average relative error of 18% was observed which may be due to assumption of 2-D airfoil in CFD analysis.

Keywords: aerodynamics, aircraft, airfoil, CFD, lift, wind tunnel

1. Introduction

Wind tunnels are large cross-section tubes of various shapes with air moving inside them. Air is blown around an object in the wind tunnel, which can be considered as the moving object in the stagnant air as there is a relative motion between the object and air. These tunnels are used to mimic the conditions of an object in flight. Aero-dynamists, engineers, and scientists use wind tunnels to study the aerodynamic characteristics of different objects, from jet wings to car windshields. By getting a better understanding of the way air moves around or through objects under test, designers and manufacturers can devise and create faster, more reliable, safer, and more efficient products. Low-speed wind tunnels move air inside at a very low Mach number (M), with speeds of air in the test section up to 480 km/h (~ 134 m/s i.e., M = 0.4) [1].

In wind tunnels, test scale models of aircraft and spacecraft can be tested while some wind tunnels are large enough to accommodate full-size versions of vehicles. Considering the case for testing an aircraft model, the main motive is to test the wings of the aircraft because wings are responsible to generate the lift force required by the aircraft to fly. The magnitude of lift will depend on the kind of airfoil used in the wing and most commonly National Advisory Committee for Aeronautics (NACA) airfoils are used and its nomenclature is based on the numbering that decides the quality and performance of that airfoil.

Li et al. [2] used a low speed closed circuit wind tunnel for experimentation in which a support system was used to hold the model. For force measurement, a six-component sting-type strain gauge balance in conjunction with a signal amplifier was used. Mizoguchi et al. [3] used an open-circuit wind tunnel to test wings of different aspect ratios. In order to measure aerodynamic coefficients three-component balance was used. A strut was used to hold the model in the inverted position at different angles of attack (AoA). Laitone [4] tested the performance of different wings at low Reynolds number by measuring lift and drag forces and for this purpose, a sensitive two-component beam balance was designed. The wing was hung vertically from the beam balance mounted on the top of the tunnel.

Mueller and Batill [5] conducted experiments to study the separation characteristics for two-dimensional (2-D) airfoil at different Reynolds number (Re) between 40,000 to 400,000. A highly sensitive external strain gauge platform balance was used, and it was mounted with an aircraft model in the tunnel. Saleh et al. [6] proposed a design of multi-component strain-gauge external balance with trial-and-error method using finite-element simulations. This gauge was manufactured, and calibration was done for lift, drag, and pitching moment and interference error was found to be less than 2%.

Sahin and Acir [7] performed a numerical analysis of NACA 0015 airfoil in ANSYS Fluent to determine lift and drag coefficients. A semi-circular-rectangular shaped flow domain was selected and Spalart Allmaras viscous model was used. Muftah [8] used similar shaped domain for analysis of NACA 2414 airfoil in COMSOL Multiphasic Design modeler. Spalart Allmaras viscous model was used to obtain contours for velocity and pressure. Kulshreshtha et al. [9] carried out the numerical study for three airfoils i.e., NACA 2412, NACA 2414, and NACA 2415 using the k-epsilon viscous method and compared their lift and drag coefficients. Sharma [10] investigated four airfoils numerically in Fluent at low Reynolds number to calculate force coefficients and compared them to find the airfoil with the best performance. Spalart Allmaras viscous model was chosen for the analysis.

2. Experimental work

In the present study, the experiments were performed in an open-circuit low-speed wind tunnel with an aircraft model in the Reynolds numbers range of 55413 to 112505 and angles of attack range of -6° to 20°.

2.1. Wind tunnel

The experiments were performed in the subsonic open circuit wind tunnel (suction type) which was indigenously developed at Mechanical Engineering Department, Nirma University and it is shown in figure 1. The wind tunnel consists of an entry section, settling chamber, convergent section, test section, diffuser section, and fan at end of the wind tunnel. To minimize the turbulence level, a honeycomb structure is provided at starting of the settling chamber. The size of the test section is 381 mm × 254 mm (15" × 10"). The fan is incorporated with a variable frequency drive (VFD) to generate different air speeds corresponding to the frequency given. The calibration of the wind tunnel was carried out by using pitot-tube and micro manometer to measure air speeds at different locations inside the test section at various fan speeds. In the present study, experiments were carried out at four different air speeds i.e., 11.88 m/s, 16.4 m/s, 20 m/s, and 24.12 m/s and corresponding Reynolds number.

IOP Conf. Series: Materials Science and Engineering 1206 (2021) 012013 doi:10.1088/1757-899X/1206/1/012013

Figure 1. Open-circuit wind tunnel used for experimentation.

All the measuring instruments were calibrated before putting in use. The experimental uncertainty in the lift coefficient was estimated in accordance with the procedure given by Kline and McClintock [11] and the relative uncertainty in the lift coefficient was found to be in the range of $\pm 5\%$.

2.2. Aircraft model

In order to perform the experiments, three dimensional (3-D) aircraft model was developed as shown in figure 2.

Figure 2. Aircraft model used for experimentation.

It was fabricated using mild steel and teak wood and to hang the model strings were attached to the wings and fuselage. The details of the wing profile are provided in table 1.

Table 1. Wing specifications

Wing	Details
Profile	NACA 2414
Material	Teak wood
Wingspan	340 mm
Chord length	72 mm
Planform area	23800 mm^2

2.3. Traverse Mechanism

The traverse mechanism consists of a stainless-steel tube of 1 x 1 inch^2 cross-section and bolts. It was required to move the wing of the model, keeping the tail at constant height, so that the angle of attack of the wing can be varied. Two holes were drilled on the front and top faces which were at the same distance as of string on the aircraft wing. Holes on the top face accommodate bigger bolts having a slot along their length. Smaller bolts were added perpendicularly in these slots from the front face, and this will restrict the rotational motion of the bigger bolt and allows only vertical translational motion. With this arrangement angle of attack of the aircraft model in the wind tunnel test section could be changed. Apart from this, two bolts were welded at the end of the bigger bolt to hold spring balance using a nut. Figures 3 and 4 show a 3-D CAD model and photograph of the traverse mechanism respectively.

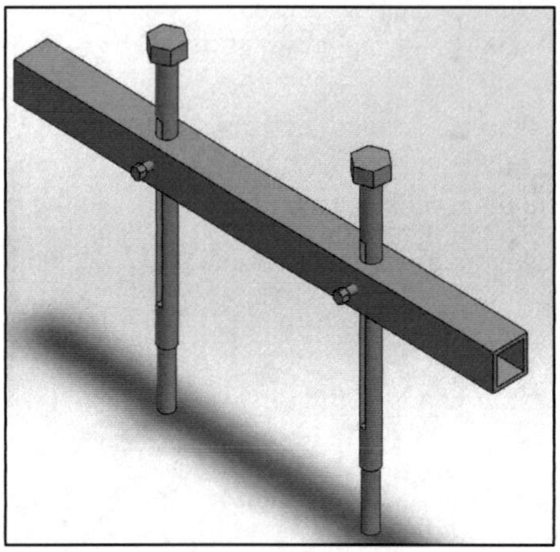

Figure 3. 3-D CAD model of traverse mechanism.

Figure 4. Traverse mechanism.

2.4. Supporting frame

Supporting frame was required to hold the traverse mechanism and it can accommodate the mechanism when placed over the wind tunnel roof. The dimensions of the stand were decided based on the dimensions of the test section of the wind tunnel and aircraft model wing. Figure 5 shows the supporting frame used for the experimentation.

Figure 5. Supporting frame.

2.5. Experimental setup

A 3-D CAD model of the complete assembly comprised of the aircraft model in the test section, traverse mechanism, and frame is shown in figure 6.

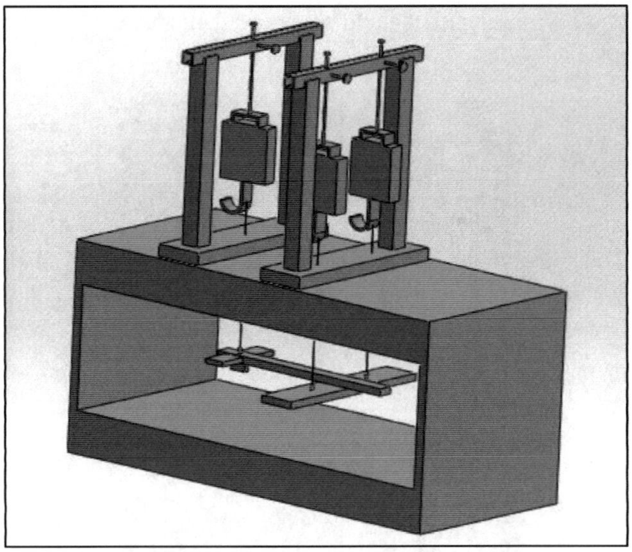

Figure 6. 3-D CAD model of the experimental setup.

It can be observed that the aircraft model is hanged upside-down because in this arrangement lift force will be generated downwards and it will create tension force in the strings. This tension force was measured using the spring balance. Figure 7 shows the test section of the wind tunnel.

Figure 7. Test section.

When air blows in the tunnel it tends to push the model backward and this will include a drag component in the force readings. In order to overcome this and to prevent the backward motion of the aircraft model, a C-section was attached behind the aircraft model with the tunnel roof as shown in figure 8.

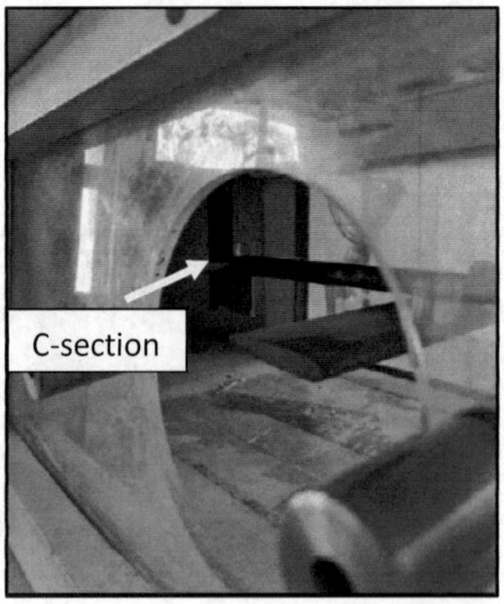

Figure 8. C-section in test section.

2.6. Experimentation
Experiments were performed to obtain the lift force (L) generated by the aircraft model when it was in the wind tunnel at a different angle of attack and air speeds. Air speed in the tunnel depends on the speed of the fan installed at end of the tunnel. Fan speed depends on the frequency of the Variable Frequency Drive (VFD) and this frequency was already calibrated with the air speed in the wind tunnel.

Experiments were performed for angle of attack varying from -6° to 20° with a step of 2° each. At each angle, the wind tunnel was operated at four different speeds viz. 11.88 m/s, 16.4 m/s, 20 m/s, 24.12 m/s and based on these speeds, Reynolds number (Re) was found as 55413, 76496, 93288 and 112505 respectively. Three force scales are used to measure the force acting on the aircraft model when the tunnel is running, and readings are added to get the total force for the wing. This sum also includes the weight of the model which is subtracted from this sum to obtain the net lift force. The lift coefficient and pressure coefficient were calculated using equations (1) and (2).

$$C_L = \frac{L}{\frac{1}{2}\rho A v^2} \quad (1)$$

$$C_p = \frac{p - p_\infty}{\frac{1}{2}\rho A v^2} \quad (2)$$

where, ρ is density of air (kg/m^3), A is area (m^2), v is air speed (m/s), p is static pressure (N/m^2), p_∞ is free stream pressure (N/m^2).

2.7. Experimental results

The lift force acting on the model was obtained at different angles of attack and air speeds and calculations were carried out to obtain the value of the lift coefficient (C_L). Values of C_L are plotted against AoA at different Re as shown in figure 9.

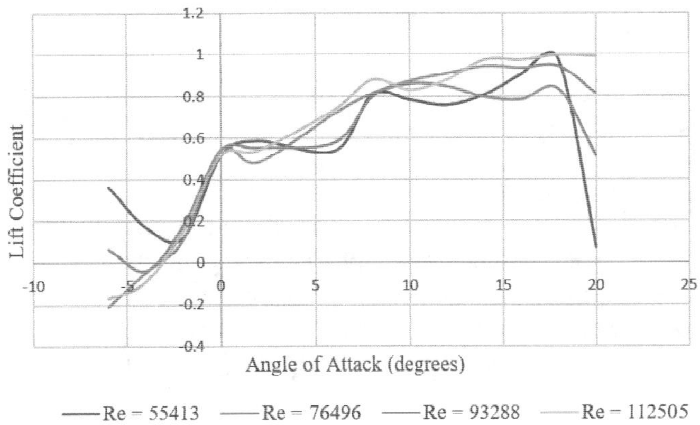

Figure 9. Variation of lift coefficient with angle of attack at different Re.

From figure 9, it can be observed that C_L increases with Re and it also increases with AoA up to a point called stall angle and reduces after this. It was found that the stall occurs at around 17° AoA for all the Reynolds numbers. At 0° AoA value of C_L is greater than zero and it becomes zero at negative AoA, this is one of the characteristics of the cambered airfoil profile. Among the previous studies, similar results were also reported by Sahin and Acir [7] for NACA 0015 profile, by Muftah [8] for NACA 2414 airfoil profile and by Kulshreshtha et al. [9] for NACA 2412 profile.

3. Numerical investigations

The two-dimensional numerical investigations of NACA 2414 airfoil were carried out by computational fluid dynamics (CFD) analysis software ANSYS Fluent to obtain lift coefficient at different AoA and Re. These results will be comparable to the experimental results because the wing used for the experimental work is of uniform cross-section throughout its span. CFD analysis is performed for 2-D NACA 2414 from AoA -4° to 16° with a step of 4° each to obtain the value of lift Coefficient (C_L) at four different Re corresponding to the experimentation.

3.1. Computational domain and grid generation

For the analysis, a semi-circular rectangular 2-D domain was selected which had a radius of 15C (C is chord length) and a rectangular portion length of 20C. The computational domain was discretized using the hexahedral grid. In order to study the effect of grid size, a grid independence test was performed by varying the edge sizing of the mesh from 0.0035 m to 0.05 m and it was found that total grid elements of around 250,000 were appropriate and results do not vary on the further grid refinement. Based on the grid independence test, edge sizing of 0.005 m was chosen for the further simulations and corresponding number of grid elements were 258664. The computational domain and the mesh are shown in figure 10 (a) and (b).

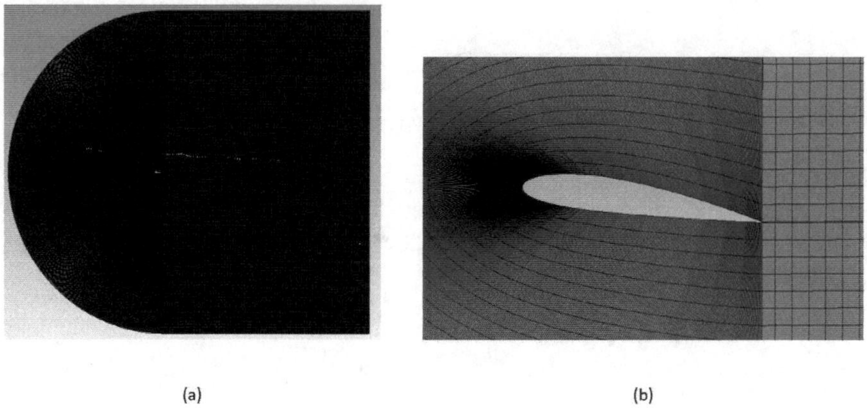

Figure 10. (a) 2-D computational domain and (b) mesh used for CFD analysis.

3.2. Solution methodology
In order to capture the turbulence, realizable k-epsilon viscous model was selected for the numerical analysis because it can give solution convergence even at higher AoA and it is quite accurate method for flows involving boundary layers and separation. The semi-circular portion was chosen as velocity inlet boundary condition whereas airfoil was given no-slip shear condition. The coupled scheme was selected for pressure-velocity coupling and default settings for spatial discretization were kept. An absolute criterion for residual monitors was considered as 10^{-6} and initial values for calculation were defined based on the velocity input.

3.3. Results and discussion
A comparison is made between the values of lift coefficient obtained from CFD analysis and experimental values for different values of Reynolds number to check for the error and deviation in the results. Figures 11 and 12 show plot of comparison corresponds to Re = 55413 and Re = 93288 respectively. It is observed that in CFD results stall occurs earlier i.e., at 9° as compared to experimental results i.e., at 16°. Also, the experimental values of lift coefficient are found to be greater than CFD values at both the Re with an average relative error of 18%. This error can be due to the fact that the experiments were performed for a 3-D aircraft model whereas CFD analysis was carried out for a 2-D airfoil profile. Moreover, turbulence in the wind tunnel, the least count error of the force balance device, swaying motion of the model, etc. may also lead to some error.

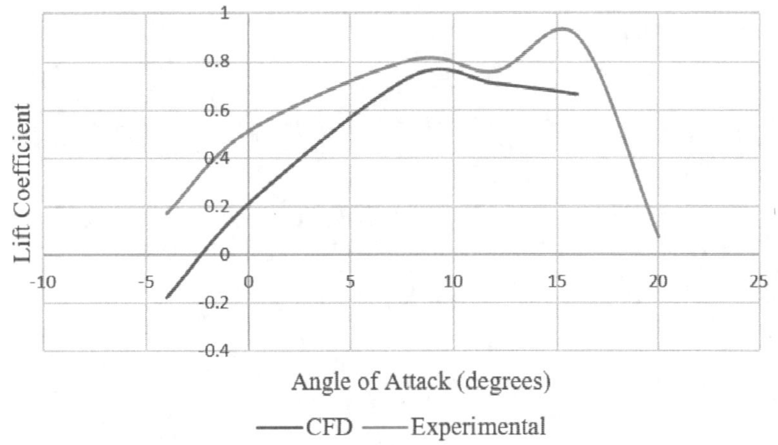

Figure 11. Comparison of CFD and experimental results for C_L vs. AoA at Re = 55413.

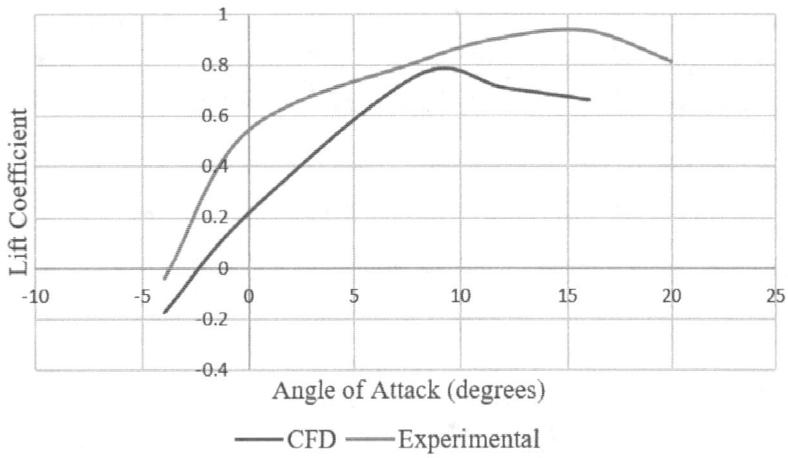

Figure 12. Comparison of CFD and experimental results for C_L vs. AoA at Re = 93288.

From the CFD analysis, velocity and pressure contours were obtained for the airfoil at different angles of attack for a given Reynolds number. Figure 13 depicts velocity contours at Re = 93288 at different AoA. It can be observed that velocity is relatively higher at the top edge of the airfoil as compared to the bottom edge except at α = -4°. This created low pressure at the top edge and high pressure at the bottom edge. According to Bernoulli's theorem, as a result of this pressure difference, an upward lift force is generated. At α = -4°, the situation gets reversed, velocity is higher at the bottom edge and lower at the upper edge. This creates a low-pressure zone at the bottom and produces a downward lift force.

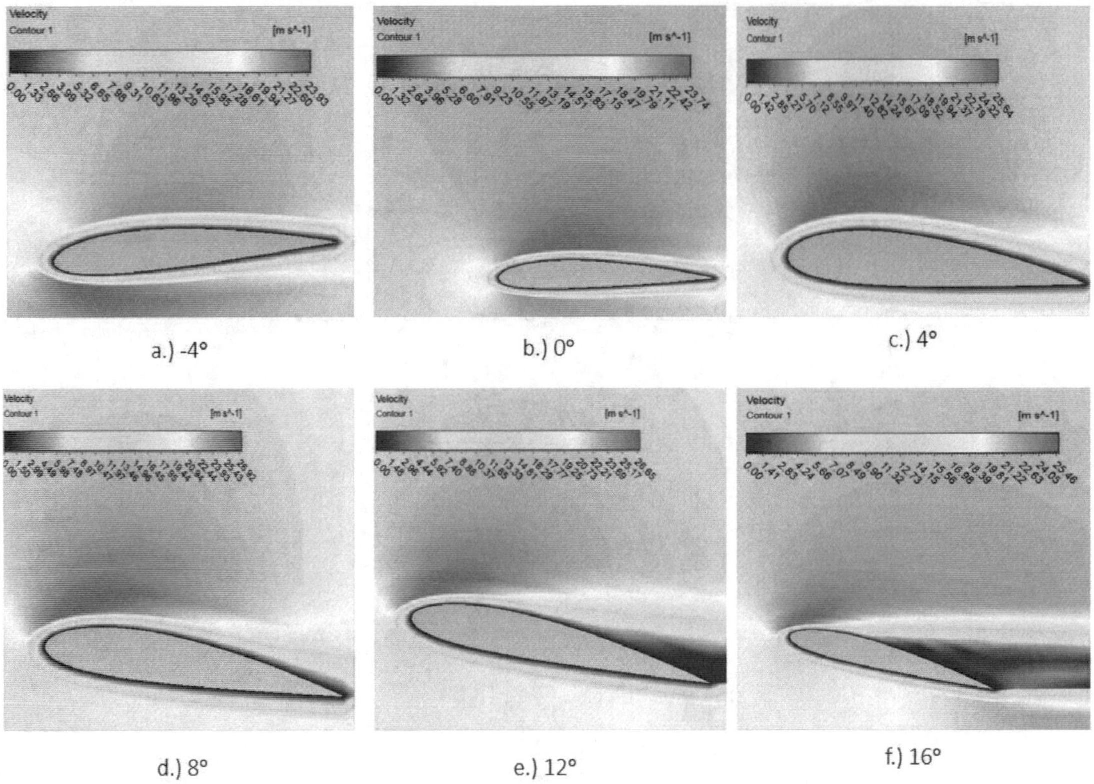

Figure 13. Velocity contours at different angles of attack at Re = 93288.

It can be seen that the flow separation starts after $\alpha = 8°$ and as the angle increases the separation point starts shifting towards the upper edge When α reaches 16°, the flow is separated from around 50% of the chord length of the airfoil and due to this lift force reduces significantly.

The pressure contours were obtained at different angles of attack at Re = 93288 as shown in figure 14. As discussed, in all the contours (except $\alpha = -4°$), velocity is found higher at the top edge, and this forms a low-pressure zone near the top region. Due to this an upwards lift force is generated. After $\alpha = 8°$, as AoA is increased the pressure difference across the airfoil is reduced and this resulted in decreases in the magnitude of the lift force. In case of $\alpha = -4°$, there is a low-pressure zone at the bottom edge and because of this direction of the lift force gets reversed (i.e., downwards).

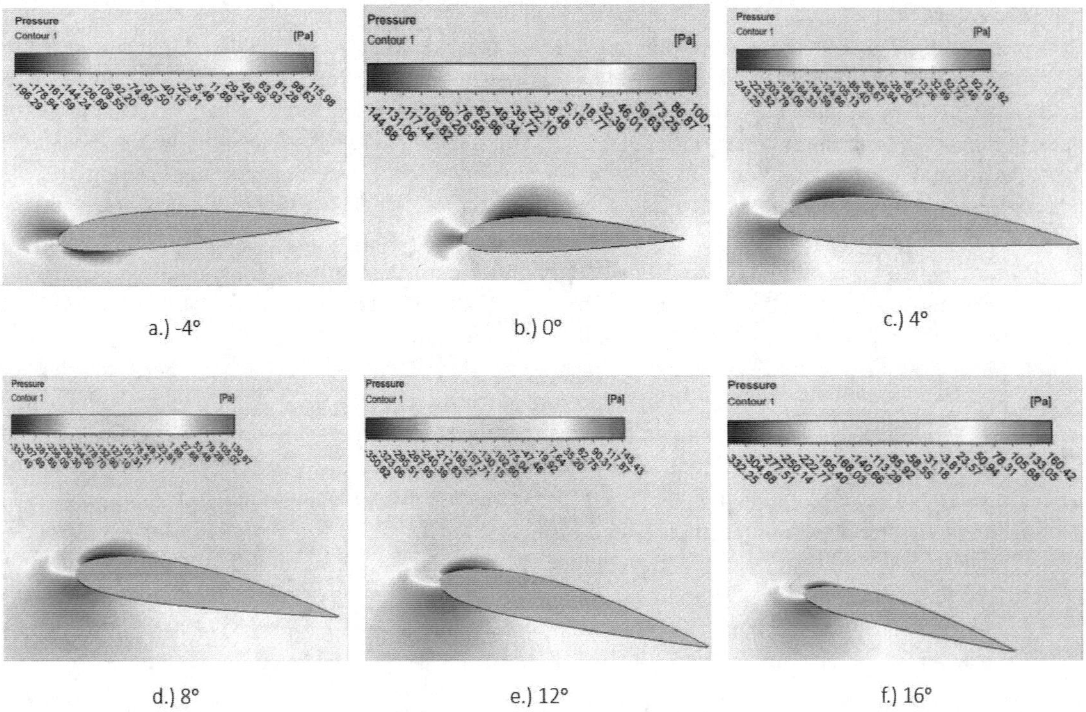

Figure 14. Pressure contours at different angles of attack at Re = 93288.

The variation of pressure coefficient (C_p) obtained from the numerical analysis along the upper and lower surfaces of the airfoil at Re = 93288 for different AoA is shown in figure 15. The direction of the y-axis is reversed so that the upper edge remains up in the plot as it will be subjected to the negative value of C_p. Higher the difference in the value of C_p between upper and lower edges, higher will be the value of lift force acting on the airfoil and this can be observed from these plots.

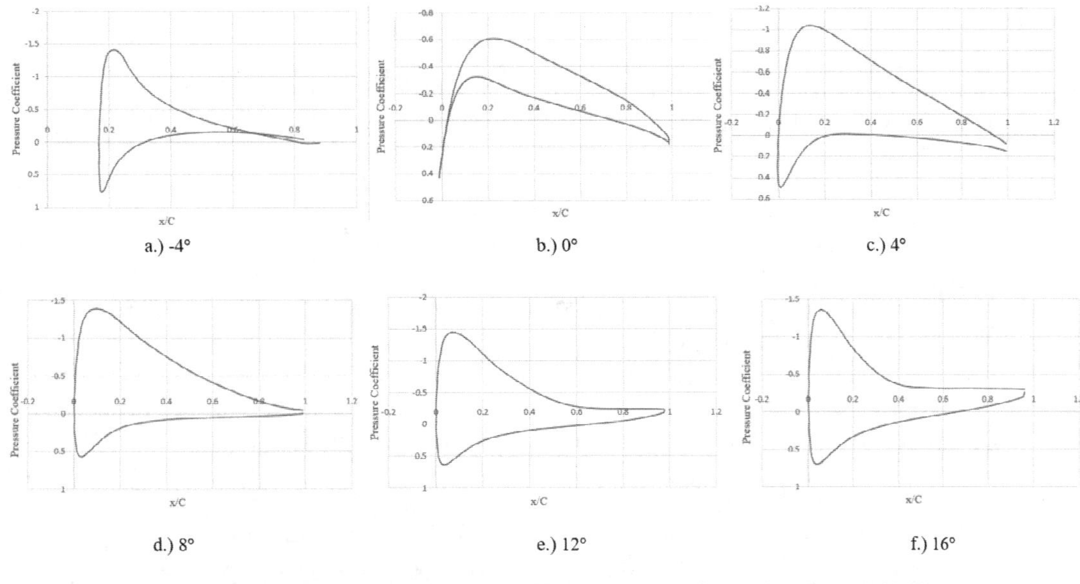

Figure 15. Pressure contours at different angles of attack at Re = 93288.

Among the previous works, similar results were also reported by Kamas and Omar [12] for NACA 2414 airfoil at different speeds and by Madhavan [13] for NACA 2414 airfoil.

4. Conclusions

In the present study, a design and development of force balance setup were presented that can measure lift force acting on the aircraft model. The setup was developed indigenously and installed in an open circuit low-speed wind tunnel facility. It consists of a traverse mechanism that can vary the angle of attack of the model and supports the model in the wind tunnel with help of strings with the force balance. The experimental and numerical investigations were done in the Reynolds number range of 0.55 to 1.12 lacks and angle of attack range of -6° to 20°. The major conclusions drawn from the study are as under:

- From the grid independence test, it was found that an edge size of 0.005 m and the corresponding number of grid elements of 258664 were appropriate for the computational analysis.
- From the velocity contours correspond to Re = 93288, it was observed that flow gets detached from the airfoil surface after 8°, i.e., stall occurs after this point.
- From pressure contours correspond to Re = 93288, it was observed that there was higher pressure at the upper surface and low pressure at the lower surface for AoA = -4°, which created a negative lift. However, for positive AoA, there was lower pressure on the upper surface and higher pressure on the lower surface which induces positive lift. Moreover, the magnitude of pressure difference and hence the lift force increased with AoA up to stall condition.
- Comparison between experimental and CFD results for C_L at Re = 55,413 and Re = 93288, showed that stall occurs a little early in case of CFD analysis and the relative average error of 18% was observed. This error can be due to the fact that the experiments were performed for a 3-D aircraft model whereas CFD analysis was carried out for a 2-D airfoil profile.
- The average relative uncertainty for the lift coefficient was found to be in the range ±5%.

5. References

[1] nasa.gov/audience/forstudents/k-4/stories/nasa- knows/what-are-wind-tunnels-k4
[2] Li Y, Ding X and Wang Y 2009 Force measurement system of a low-speed wind tunnel Proc. 9th Int. Conf. Electronic Measurement & Instruments p 300
[3] Mizoguchi M, Kajikawa Y and Itoh H 2016 Aerodynamic characteristics of low-aspect-ratio wings with various aspect ratios in low Reynolds number flows *Trans. Japan Soc. Aero. Space Sci.* 59 56-63.
[4] Laitone E V 1997 Wind tunnel tests of wings at Reynolds numbers below 70000 *Exp. fluids* 23 5 405-409.
[5] Mueller T J and Batill S M 1982 Experimental studies of separation on a two-dimensional airfoil at low Reynolds numbers *AIAA J 20* 4 457-463.
[6] Saleh A R, Setoodeh A R and Gholamzadeh M 2016 A novel multi-component strain-gauge external balance for wind tunnel tests: Simulation and experiment *Sensors Actuators A Phys.* 247 172-186
[7] Şahin I and Acir A 2015 Numerical and experimental investigations of lift and drag performances of NACA 0015 wind turbine airfoil *Int. J. Mater. Mech. Manuf. 3* 1 22-25
[8] Muftah A 2019 CFD Modeling of airfoil of wind turbine under different effect of operating conditions *SUSJ Journal 9* 1 27-43
[9] Kulshreshtha A, Gupta S K and Singhal P 2020 FEM/CFD analysis of wings at different angle of attack *Mater. Today: Proc.* 26 1638-43.
[10] Sharma S 2016 An aerodynamic comparative analysis of airfoils for low-speed aircrafts *Int. J. Eng. Res.* 5
[11] Kline S J and McClintock F A 1953 Analysis of uncertainty in single-sample experiments *Mech. Eng. J. 75* 3
[12] Kamas T and Omar M 2010 A two-dimensional assessment of a cambered and a symmetric spoiler at various angles of attack and speeds: a CFD study *J. Future Eng. Tech.* 3 5
[13] Madhavan N 2019 Design and analysis of NACA 2414 airfoil *Int. J. Sci. Res. Rev.* 8 168

Design and analysis of smoke flow visualization apparatus for wind tunnel

D Raval[1], S V Jain[1], A M Acharil[1] and K Ghosh[1]

[1] Mechanical Engineering Department, Institute of Technology, Nirma University, Ahmedabad, India

15mmet23@nirmauni.ac.in, sanjay.jain@nirmauni.ac.in, madhusudan.achari@nirmauni.ac.in, kunal@iitk.ac.in

Abstract. In the present study, the design and analysis of smoke generator are done for the low-speed wind tunnel. The wind tunnel fan is fitted with the Variable Frequency Drive to produce the wind speed in the range of 3 to 32 m/s with fan speed of 150 to 1500 rpm. The design of smoke generator was done according to Preston Sweeting mist generator principle corresponding to the free stream velocity of 3 m/s. A controlled smoke generator consisting of kerosene reservoir, controlled heater, blower, liquid column height adjustment mechanism, valves etc. was designed and fabricated. The smoke generator produced the smoke at the rate of 154 cm^3/s which was close to the design flow rate of 149 cm^3/s. To supply the required quantity of smoke in the wind tunnel, the smoke rake of NACA 0010 profile was developed and installed in the rapid contraction section of the wind tunnel to achieve the streamlined flow. The parametric studies were done on the smoke generator at different power inputs and its effects were studied on smoke temperature, smoke discharge and boiling time of the kerosene. The flow visualization was carried out on NACA 0015 airfoil model and the images were captured to examine the flow physics around them under different operating conditions.

Keywords: Design, NACA, smoke generator, flow visualization, wind tunnel.

1. Introduction

The smoke flow visualization system is one of the ways to visualize the air flow in the wind tunnel. The need for the smoke visualization technique is necessary to understand that how the flow behaves around or over the body. It gives the idea of the major design procedure of any machine or any corrections needed in the design. Many techniques are available for the flow visualization for the wind tunnel. Be it smoke visualization, tufts, china clay visualization etc. [1]. The smoke visualization is considered appropriate for better understanding of the flow over the object. The smoke flow visualization can be achieved by various means. The technique depends on the type of the visualization and also on the cost procurement of the project. Basically, oil smoke visualization is preferred in the low-speed subsonic wind tunnels, as at the low speeds of around 2 to 3 m/s [1], the smoke does not diffuse in the tunnel and travels with the air in it.

The most important aspect of flow visualization is the condensation, due to which the smoke becomes visible to naked eyes. Usually, the size of the smoke particles is around 0.3 to 0.5 micron. The condensation process plays major role in the flow visualization system. Also, other factors such as density, luminosity etc. play key role for the proper smoke discharge. Also, the smoke particles should be appropriate to control the settling velocity of smoke. The sizes of the smoke particle generated from different sources differ and also depends on the heating instrument used in it. Usually, low particle size results in low settling velocities and large particles results in higher settling velocities. The particle size should be kept small so as to prevent the solid deposition in the flow field [2].

2. Development of smoke rake

The smoke rake is a hollow airfoil shaped component, with the hollow tubes attached at its trailing edge from which smoke is ejected. For the design of smoke rake, NACA airfoil profile is selected which is a symmetrical airfoil. The chord length of the airfoil is selected as 127 mm, as with the increasing chord length the thickness of the airfoil would also increase resulting in the flow separation in the wind tunnel. For the design of the smoke rake, the design parameters considered are (i) the smoke rake must maintain its proper airfoil shape and must free from any dents on the surface (ii) the tubes attached at its end must be properly aligned with the shape of the smoke rake.

Based on the chord length of the NACA 0010 airfoil, the design of the smoke rake was done based on the cross section of the contraction cone of the wind tunnel. The smoke rake has to be installed at the rapid converging section of the wind tunnel where the air flow would accelerate over the symmetric smoke rake such avoiding the adverse pressure gradients and as a result the smoke would travel with the air smoothly without any disturbances. The smoke rake has been installed at the distance of 15" from the test section to the contraction cone. The height of the smoke rake is kept as 14.60". The smoke rake is manufactured from stainless steel because of its high strength and could also withstand high temperatures. The smoke rake is shown in the figure 1.

The fuel used to produce smoke is taken as kerosene because of its easy availability and less toxicity as compared to other fuels. The total volume of the smoke is calculated based on the free stream velocity of 3 m/s. Based on the requirement of smoke for 30 minutes, 2 liters of kerosene capacity is required. Experimental work is also undertaken for the estimation of the smoke discharge time from the fixed amount of the fuel. The smoke rake is installed at the specified junction in the contraction cone, so that the smoke would easily travel with the air flow in the tunnel.

3. Design of smoke generator

3.1 Design principle

In the present study, the smoke generator is designed based on the principle of Preston Sweeting mist generator [1]. Figure 2 shows the schematic diagram of the smoke generator based on this principle. The working of the Preston-Sweeting mist generator is as follows.

- The kerosene is filled in the boiler glass tube which is connected to the kerosene reservoir via U-tube connection.
- A spiral heater is wound around the glass tube, which supplies heat to the glass tube and eventually boils the kerosene.
- The vapor from the kerosene travels through the jet in the mixing chamber where it mixes with the cold air to be supplied by the fan.
- As the hot vapor of kerosene mixes with the air, it condenses and forms the dense white smoke.
- This smoke is guided by the outlet pipe to the separator where it is stored.
- This stored smoke is then released in the wind tunnel through the valve mechanism.
- During the boiling of kerosene no heat loss is taking place.
- The velocity of smoke in the smoke rake is same as the free stream velocity in the wind tunnel.

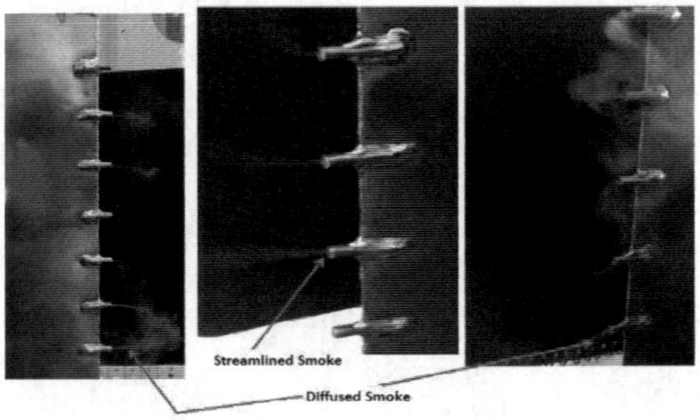

Figure 1. Smoke rake

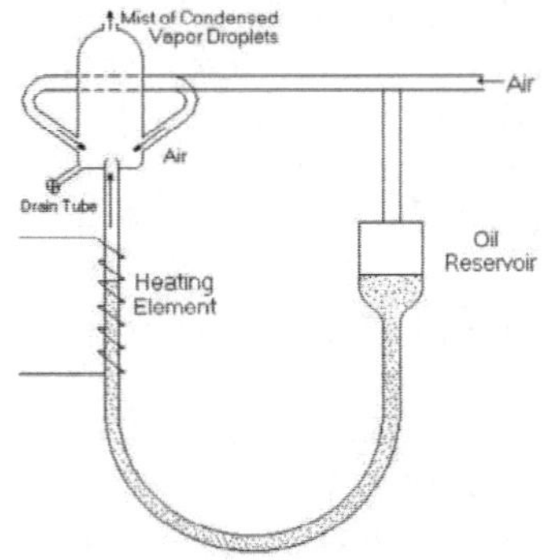

Figure 2. Preston sweeting mist generator

3.2 Design of smoke generator

The design of smoke generator was carried out corresponding to free stream velocity of 3 m/s at the wind tunnel fan speed of 150 rpm. At the speeds higher than 5 m/s to 8 m/s the smoke would diffuse and would result in the failure of the flow visualization [3]. For the installation of the smoke generator in the wind tunnel system, usually the smoke generator is installed at the test section entry so that the smoke gets easily aligned with the flow. But the problem with the installation of the smoke generator at the entry of the test section is that, the flow will not develop and streamlined flow will not be obtained. Hence it is better that it should be installed at the low-pressure section of the wind tunnel. The contraction cone at which the section rapidly converges is the appropriate place for the placing of smoke rake which would supply the smoke from smoke generator. Figure 1 shows the smoke rake. In the smoke rake, 7 cylindrical tubes will be provided with ID and OD of 3 mm and 4 mm respectively with the length of 10 cm. The cross section of a tube was determined as follow:

$$A = (\pi/4) \times d_i^2 \quad (1)$$

Based on the smoke rake to be installed in the wind tunnel the volume of the smoke to be used is calculated as:

$$Q = A \times v \quad (2)$$

where, Q = smoke volume discharge for one pipe, A= Cross sectional area of tube, v = Velocity of 3 m/s.

Hence, such 7 cylinders would contribute to about 148.45 cm^3/s of smoke. Also, the experiment was conducted for about 30 minutes and hence the total volume of the smoke required was 267210 cm^3. Now in order to match the desired smoke volume, the fuel capacity has to be calculated. So, the mass of the smoke can be calculated by multiplying its vapor density with the total volume. Hence the mass of the smoke is calculated as follows:

$$Mass\ of\ smoke = Total\ volume\ of\ smoke\ \times\ vapor\ density \quad (3)$$

Now an assumption is to be done that, no evaporation loss is taking place. Hence the amount of the liquid kerosene evaporated is equal to the amount of smoke generated. As only phase change takes place and mass remains constant, the volume of the kerosene can be found by:

$$Volume\ of\ kerosene = \ Mass\ of\ smoke/Density\ kerosene \quad (4)$$

Hence, for the production of the required amount of smoke, 2 liters capacity of kerosene is needed. The time duration for which the smoke will be liberated from the kerosene for a fixed mass is decided by conducting a small-scale experiment. The smoke rake will be made of symmetric airfoil shaped tube which will be inserted in the direction of flow in the wind tunnel. Small tubes will be inserted at its trailing section of the smoke rake from which the smoke will be discharged. It was installed at the converging section of the wind tunnel because the pressure will constantly reduce that helps the smoke travel through the tube without any disturbance.

4. Development of smoke generator

As discussed earlier, the smoke generator is designed based on the principle of Preston Sweeting Mist generator. Figures 3 and 4 show the smoke generator setup and the wind tunnel developed at Nirma University. The glass tube is subjected to the external heater with the rock wool insulation in order to minimize the heat losses. Flow control valves are also installed in order to control the air flow in the mixing chamber to produce the required amount of smoke. The high temperature resistant silicon tubes are used for supplying the hot vapors and kerosene to the mixing chamber and glass tube. A drain vessel is provided for collecting the condensed kerosene which can be used again. Table 1 shows the major specifications of the smoke generator.

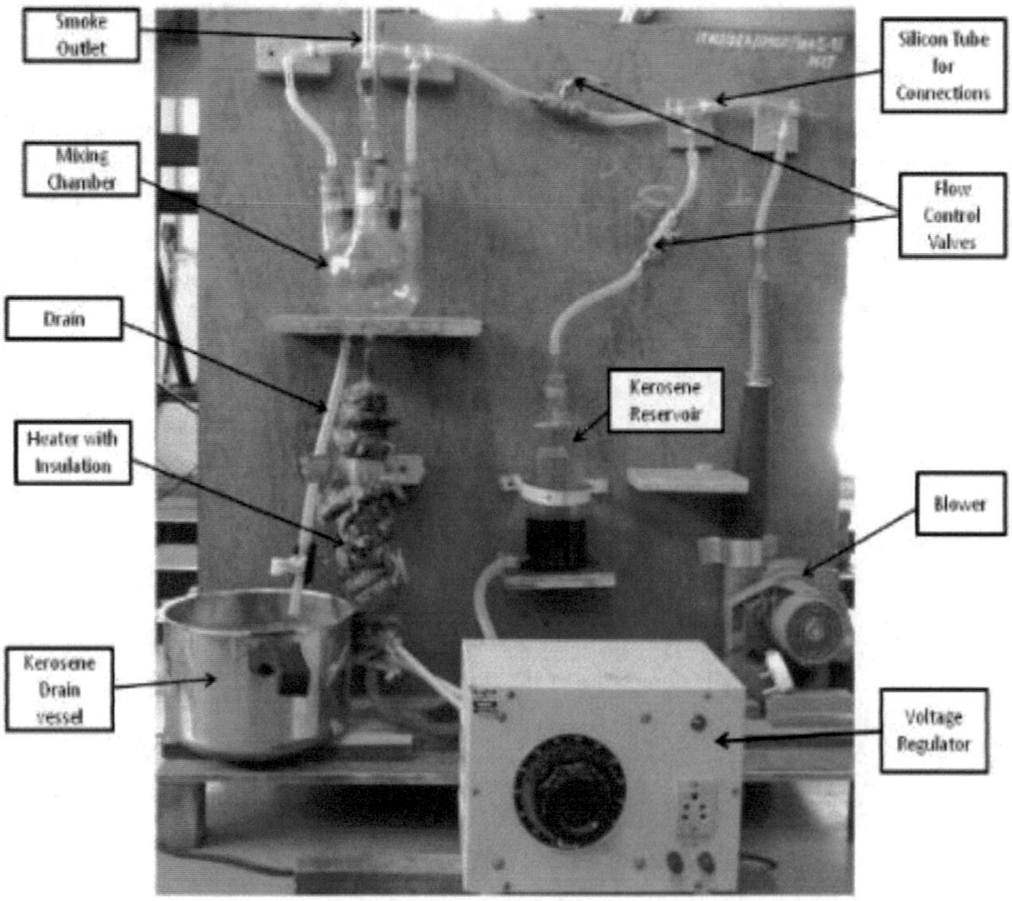

Figure 3. Smoke generator

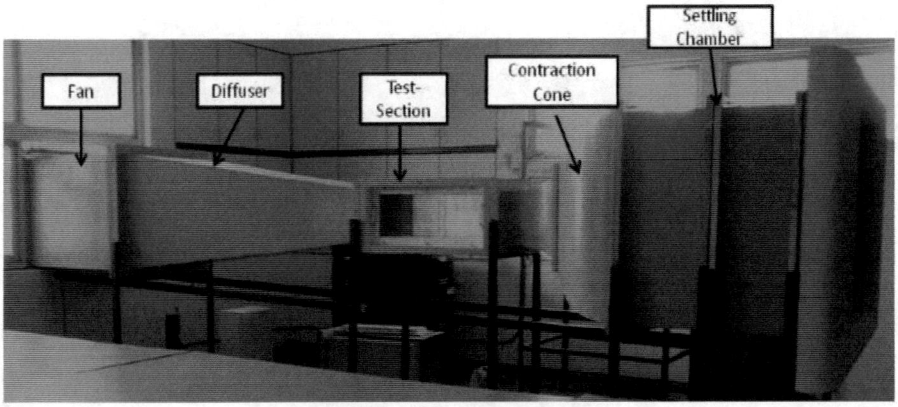

Figure 4. Wind tunnel

Table 1. Major specifications of smoke generator

Components	Major specifications	
Fuel reservoir	Fuel: Kerosene	Capacity: 2 lit
	Material: Borosilicate glass	
Boiler glass tube	ID: 23 mm	OD: 24 mm
Mixing chamber	3 neck type	
Blower	Did: 150 mm m3/min	Capacity: 3.3.
Heater	Tape type	Capacity: 1 kW
Valve	Ball valve	
Pipes	Silicon pipes for connection	

5. Result and discussion

5.1 Flow visualization around an airfoil

The flow visualization is an important aspect of examining the flow physics around the various objects. In the present study, the flow visualization is carried out around NACA 0015 airfoil at different time intervals as shown in Figure 5 [4].

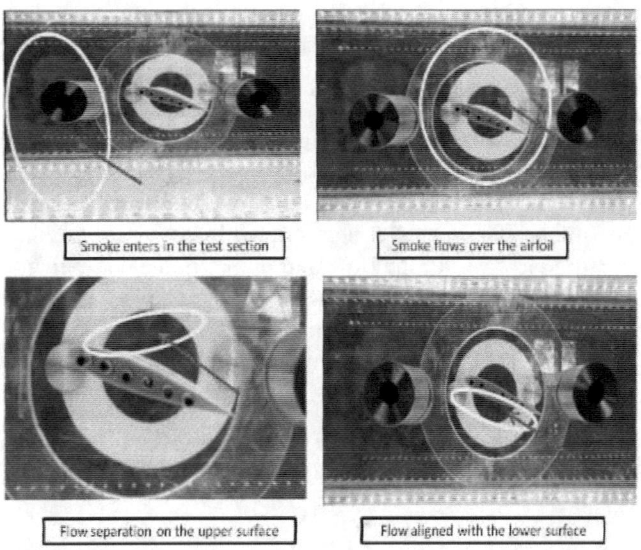

Figure 5. Flow visualization near airfoil at different time

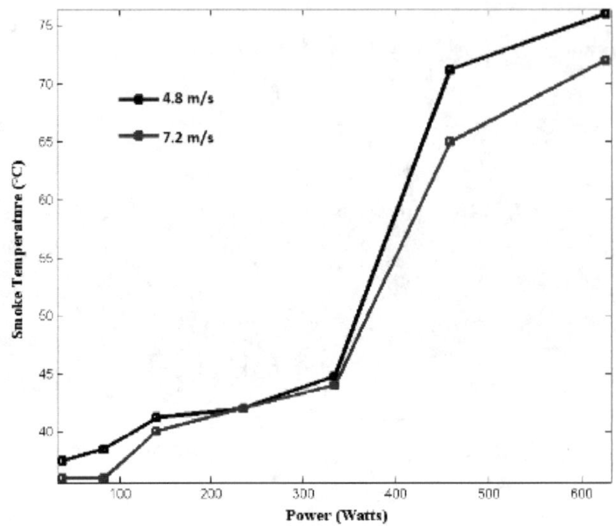

Figure 6. Smoke temperature vs. power inputs

5.2 Parametric study

In the present study, to evaluate the performance of smoke generator parametric studies were done. The effects of boiling time of kerosene, smoke temperature and smoke discharge were studied by varying the power rating at various flow speeds in the mixing chamber. Experimental analysis was carried out in order to measure the smoke temperature and an estimated smoke discharge from the smoke generator. For the experimental purpose (i) a fixed amount of kerosene was taken i.e. 77 ml (ii) valve positions for the air flow were fixed at two points, correspond to the air velocity in the mixing chamber as 4.8 m/s and 7.2 m/s (iii) same amount of kerosene was subjected to different power inputs in order to estimate the smoke temperature and smoke discharge (iv) kerosene boiling time was noted, in order to match the design criteria of boiling the kerosene in 5 minutes.

Figure 6 shows the smoke temperature for same amount of kerosene at different power inputs. It can be seen that at the same power rating, the smoke temperature differs with the air velocity. This is because, at higher flow speeds, the hot liquid rapidly condenses and also gets carried away with the flow resulting in the less smoke temperature. By decreasing the air flow, the smoke temperature increases with increased amount of smoke rate. Further decreasing the air flow results in abnormal smoke generation, which would not allow the smoke to eject out of the glass tube.

Figure 7 shows the estimated smoke discharge from the smoke generator at different power inputs [4]. It can be observed that at lower air speeds the smoke discharge increases as compared to the air flow subjected to higher speeds. Because at the higher air flow rates, the hot vapors condenses rapidly producing less amount of smoke. The smoke generator should not be operated at higher speeds, as the hot liquid particles gets carried away with the flow and damages the equipment. The trend shows that, that at 4.8 m/s velocity and at 550 W, the smoke discharge matches its approximate design criteria of 148 cm³/s. The kerosene boiling time is predicted based on the power inputs. Figure 8 shows the trend of the kerosene boiling time with variation in power input. The design criteria are based on the maximum 5 minutes boiling time of kerosene.

IOP Conf. Series: Materials Science and Engineering 1206 (2021) 012014 doi:10.1088/1757-899X/1206/1/012014

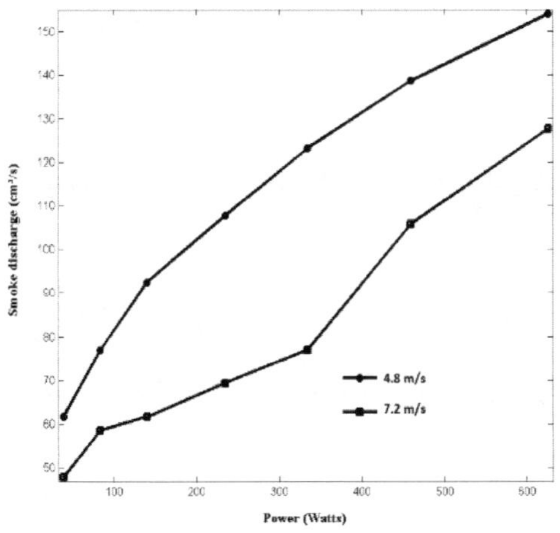

Figure 7. Smoke discharge vs. power inputs

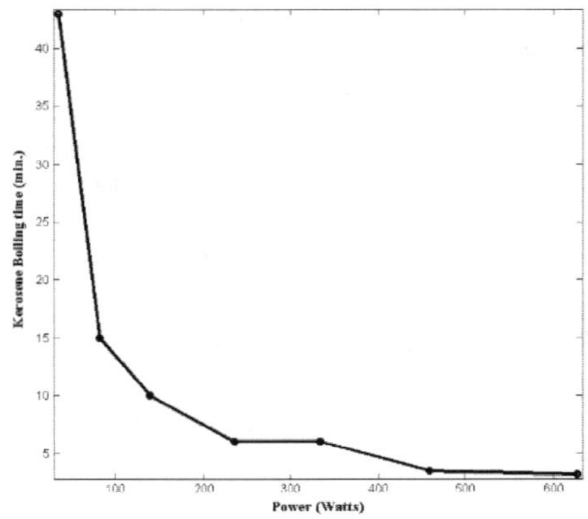

Figure 8. Kerosene boiling time vs. power input

6. Conclusions

In the present study, the design of smoke generator is done according to Preston Sweeting mist generator principle corresponding to the free stream velocity in the wind tunnel as 3 m/s. The parametric studies were done on the smoke generator at different power inputs and its effects were studied on smoke parameters. The smoke generator has produced the smoke at the rate of 154 cm³/s which was very close to the design flow rate of 149 cm³/s. The smoke rake was installed at the rapid contraction section of the wind tunnel to avoid the flow diffusion and it led to streamlined flow. When power input was varied from 32 W to 626 W, the smoke temperature, smoke discharge and boiling time were changed in the range of 37° to 76°C, 61 to 154 cm³/s and 45 to 3 minutes, respectively at an air velocity of 4.8 m/s in the mixing chamber. The flow visualization was carried out on NACA 0015 airfoil model and the images were captured to examine the flow physics around them under different operating conditions.

7. References

[1] Merzkirch, Wolgang. Flow visualization. Elsevier, 2012.
[2] http://www.engineeringtoolbox.com/particle-sizes-d_934.html.
[3] Rathakrishnan, Ethirajan. Instrumentation, measurements, and experiments in fluids, 2007.
[4] Raval D. Design and development of smoke flow visualization apparatus for wind tunnel. M Tech thesis. Nirma University, 2017.

2-Dimensional inline system for peppermint oil extraction

M Anas[1] and A Sharma[2,*]

[1,2] Department of Mechanical Engineering, Integral University, Lucknow 226026, India

anas@iul.ac.in[1], anuragsharma186@gmail.com[2]

Abstract. Peppermint leaves have wide range of applications in medicines, toothpastes, soaps, mouth fresheners, chewing gums etc. However, in India, the traditional way of processing peppermint leaves for extraction of peppermint oil is slow, tedious, labour intensive, hazardous, and causes skin problem as well.
This paper proposes a method of mechanizing the post harvesting process, with least modification in the current apparatus being used by the farmers in the Indian villages. It proposes a non-electric 2-dimensional Mechanized system for movement and compaction of the compact/ hay. The mechanism will help to reduce the number of labourers per unit hay volume and the risk associated with the contact of in-process hay with their skin. Also, a lot of time is wasted in putting the leaves/ hay in the boiler vessel, preparing the hay in a compact form and in removing the processed leaves.
This paper is an attempt to get rid of most of the problems aforementioned and make the process safe, efficient, less labour intensive and more economical for farmers and workers. The mechanism proposed is sustainable and farmer friendly as it does not require any special training for its operation. It will increase the profit of small-scale industries with small capital investment for the setup.

Keywords. Extraction, Inline system, Manual, Mechanization, Menthol, Peppermint oil, Rural technology.

1. Introduction

Peppermint oil is extracted from Mentha piperita and has contents of menthol (45% approx.) and menthone (22%). Peppermint (i.e. Mentha Piperita) belongs to mint family (Lamiaceae), which is widely considered as a medical and aromatic plant and are produced extensively for the medicinal and as food product. It is believed that peppermint originated in the Mediterranean and Northern Africa. It was discovered through the Ebers papyrus, a medical text dating back to 1550 BC, that Egyptians have long known about its stomach calming effects. Its leaves containing essential oils consisting various aromatic chemicals compounds like menthol, menthone, isomenthone and menthofuran which are used in medicines, foods, flavourings, cosmetics and beverages. Tobacco industry is the leading consumer of mint essential oils and consumes about 40% of the total oil produced, followed by pharmaceutical and confectionary industries. Menthol, being the major substance, give the mints major characteristic aromas and flavours. It is used as a flavouring substance in toothpaste, mouthwash, chewing tobacco, mouth fresheners, analgesic balms, cough syrups, chewing gums, candies and tobacco industry. This is due to its fresh minty taste and fragrance and cooling effects due to presence of menthol.[1],[2]

Figure 1. (a) Mentha Piperita **(b)(c)** Peppermint field ready for harvesting

The peppermint oil is reported to have anti-oxidant properties, antibacterial activity and due to this, it is a major constituent of some of the over the counter remedies available in Europe for the irritable bowel syndrome.[3]

There are four varieties of peppermint plants that are cultivated in India - Peppermint, Spearmint, Bergamot mint and Japanese mint. India produces about 75% of the total world mint oil supply by cultivation of Japanese mint on around 60,000 hectares of land. It produces about 12,000 tonnes of menthol mint oil annually. Japanese mint is cultivated in India in around 60,000 ha in tarai districts like Rudrapur, Bilaspur etc. of Uttaranchal and Central U.P. like Barabanki, Moradabad, Bareily, Badaun and Lucknow, besides smaller area in parts of Punjab (Ludhiana and Jallandhar). Along with Japanese mint, India produces Peppermint on an area of about 2,500 ha, Bergamont mint on 1,200 ha and Spearmint on 3,000 ha land which produces 200 tonnes, 150 tonnes and 300 tonnes of oils respectively. Other countries that grow these species and produce oil are China, Brazil, USA, former USSR and Thailand.[4]

Presently, India is one of the largest producer of Mentha oil in the world. As per , The HINDU Business Line report, dated February 21, 2019, India produces 80% of the world's mint supply and exports 75% of its output.[5]

It is widely grown in temperature region of the world. In India it is grown between February to September (cultivated twice and hence extracted twice from single field). The crop gives high yield with regular irrigation and takes about 3.5 months to grow. Peppermint plants are about 30 to 90 cm high and have smooth stems with square cross section. Its leaves can be 4 to 9 cm in length and 1.5 to 4 cm in breath and are of dark green colour with reddish veins, and they have an acute apex and coarsely toothed margins.

Peppermint can be cultivated both in tropical and sub-tropical regions of the world. A mean temperature of 20 to 40 C is required during major part of its growing period and a rainfall between 100 to 110 cm is ideal for its cultivation. Well drained loam or sandy loam soil rich in organic matter having pH between 6 and 8.2 are ideal for its cultivation. Both red and black soil can also be used. In case of acidic soil having pH less than 5.5 is recommended.

1.1. Need of mechanization

The small scale plants are very pathetic and so the old traditional method adopted is. The traditional method of oil production in Indian villages is based on steam distillation of peppermint leaves. The labours put the harvest directly in the boiling vessel and stomp it with their bare feet to compress and compact it for maximum utilization of vessel volume and make each process cycle more economical. But this manual process is hazardous as it makes the labourers come in direct contact with the in-process hay. They have to work inside a 5 ft deep vessel that may have gone through a process cycle before and may still be quite warm. Also, when they stomp the leaves with bare feet, the direct contact with it causes skin irritation. And at last they have to take all the processed hay out of the hot vessel by going inside it and manually throwing it out of the vessel walls with a poorly designed very rough tool, like big fork.

During the time when the distillation process is going on, the vessel is sealed and the workers have to wait for it to end. The farmers have to wait for 4-5 hours i.e. as long as distillation is carried on. Only when the vessel is cleared of the processed leaves, the next cycle can be started again with filling and compacting the leaves.

Most of the villages of Central U.P. like Barabanki, Mahmoodabad, Sitapur, Ramnagar, Bahraich, Gonda were visited. It was found that the whole process is slow and tedious and as mentioned above, hazardous. A lot of time is wasted in preparing the hay in a compact form and putting the leaves/ hay in the boiler vessel. The time is again wasted in removing the processed leaves. This paper is an attempt to get rid of most of the problems aforementioned and make the process safe, efficient, less labour intensive and more economical for farmers and workers.

Figure 2. View of the plant in rural areas and labour putting the leaves in the vessel

Figure 3. Workers compacting the leaves with bare feet.

Figure 4. Vessel is heated to produce steam and Removal of processed leaves/ hay

Figure 5. Fork used for removal of processed leaves and scattering of processed hay over ground for drying

1.2. Literature survey

The old traditional method being adopted by farmers is least available in record. Anas & Abusad (2009) have surveyed and collected a series of problems associated with peppermint industry in rural India and about the many problems faced by the Indian farmers in processing of harvest for extraction of peppermint oil. The problems included unsafe material handling, direct contact of hay with skin, delay in process cycles, compact preparation inside hot vessel, removal of processed hay and many more. They have categorized problems as technical problems, ergonomics problem, economic and management related problems and so on. They also suggested various methods to solve these problems related with handling of the compact and making the whole process mechanized with and without electricity.[6]

Anas et. al. (2020) in their paper "Sustainable Model for Mechanization of Peppermint Oil Extraction Plants for Rural India" discussed about the feasibility of a 3-Dimensional model for movement and compaction of hay. Computerized model was made and motion study was done. They have attempted to reduce the time wasted in filling the drum with hay. They have also made the technique/ process more safe for the farmers It was found that that model could help the farmers to produce more oil in less time with large capital investment. However it was complex method and it required large area and large capital investment.[7]

Braima & Andoh (2016) designed a mechanical system i.e. a prototype for extraction of lemon grass oil using direct steam distillation technique. Tests were conducted to measure the machine performance and quality of oil extracted.[8] Lucchesi et. al. (2004) developed Solvent-free microwave extraction (SFME). Solvent free microwave extraction is an original combination of microwave heating and dry distillation. It offers important advantages over traditional alternatives, namely: shorter extraction times (30 min for SFME method against 4.5 h for hydro-distillation), substantial savings of energy, and a reduced environmental burden (less CO_2 rejected in the atmosphere).[9]

Gavahian et al. (2011) performed Ohmic-assisted hydrodistillation (OAHD) on Zataria multiflora Boiss (Shirazi thyme). OAHD is a combination of ohmic heating and distillation, and could be considered as a novel method for the extraction of essential oils. The method was compared with the quality and quantity of essential oil extracted from traditionally used Hydro-Distillation (HD) technique. In this study OAHD was presented as 'environment friendly' method for essential oil extraction. It was found that OAHD resulted in reduced extraction time as compared to traditional HD and extracted most of the oil from thyme in a time at which the HD method started giving out its first drops. Also the quality of oil extracted from both methods were similar.[10]

However, such methods are very expensive and require technical skills. Hence they are not suitable for Indian context. The mechanized models so far made are not accepted by our farmers because of complex mechanism of handling, high installation cost and lower profits. As a result we see the models have not replaced the old traditional methods of extraction of peppermint oil. The present mechanized model has very less installation cost, can be easily handled and besides that, increase in efficiency will be obtained when implemented. The mechanization of plant will not only make the process faster but also will be very safe to operate.

2. Materials and Methods

As mentioned above, the plants surveyed in rural areas of India run without electricity. The mechanized model hence proposed here is also developed while keeping that in mind and also that very technologically advanced system would be inoperable by the farmers due to lack of technical knowledge.

The model proposed is a 2-dimensional inline mechanized system. The details of the Plant is given in Table 1 and Fig. 6.

Table 1: List of components.

Sl. no.	Component name	Material	Specification/ Capacity
1	I-beam	Structural Steel	SC 140 (140mm x140mm x 33.3Kg/m)
2	I-beam trolley	-	0.5 Ton
3	Chain block	-	0.5 Ton
4	Wire rope	Stainless Steel	3/16 in
5	Hand winch	-	362 Kg
6	Boiling vessel	Alloy Steel	Diameter = 72in Height = 60in
7	Lower compacting container	Mild Steel Frame, Galvanized Steel mesh	Diameter = 71in Height = 36in
8	Upper compacting container	Mild Steel Frame, Galvanized Steel mesh	Diameter = 72in Height = 36in
9	Prepared hay compact	Peppermint hay	Diameter = 70in Height = 30in
10	Base net	Mild Steel Frame, Galvanized Steel mesh	Diameter = 71in Height = 1in
11	Chain Sling	Alloy Steel	3 Legs, 2m
12	Processed hay	Peppermint hay	Diameter = 70in Height = 30in

The trolley moves on an I-beam held at a height of 12 ft. The trolley is connected to the wire ropes of two hand winches present on both columns supporting the I-beam. Hence when a person winds one hand winch and other unwinds the other one, the trolley moves in either directions. The chain block is suspended from the trolley and is used to move the load (compact) in and out of preparation drum and boiling vessel.

The average elbow height of male workers in northern regions of India is 1102mm.[11] Best ergonomic height of crank should be 0.9 times of the elbow height for maximum torque transmission and least discomfort of the worker.[12] Hence the suitable height chosen for hand winch placement on the pillar becomes 992 mm from the ground.

The main stages of working mechanisms are as follows:

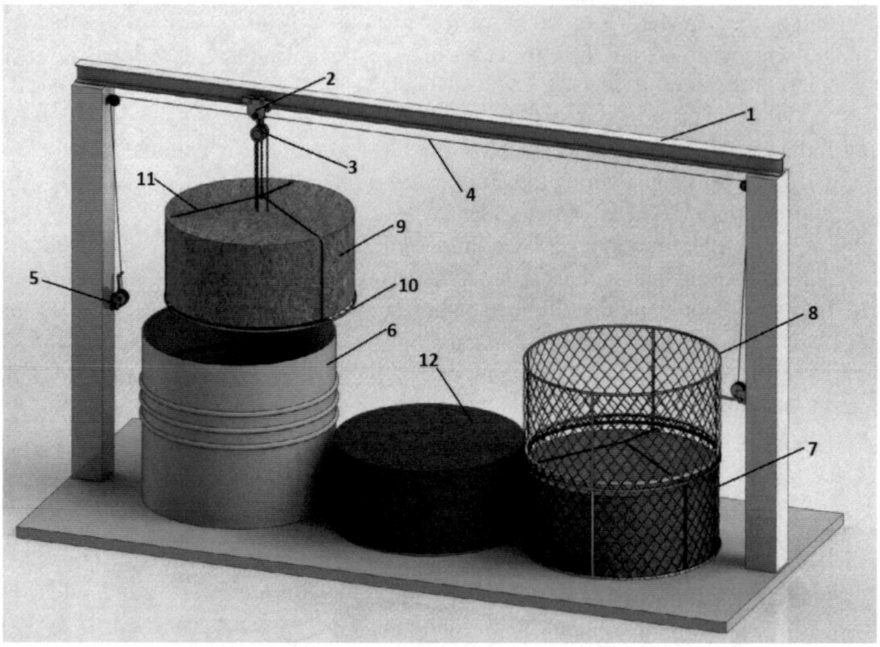

Figure 6. Details of proposed plant components

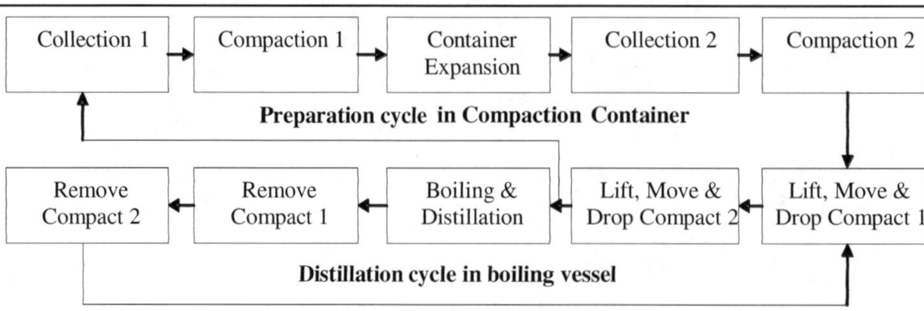

Figure 7. Block diagram of the mechanism's working

1. Leave collection 1: The hay is collected in a net of wire in compaction container that is about the size of boiling vessel.
2. Leaves compaction 1: The workers compact the hay by walking on it. They can't be bare feet. They have to wear long rubber boots to avoid contact of skin with the hay, otherwise it results in skin irritation (Fig 8a).
3. Compaction Drum Expansion: The Compaction drum is telescopic and can be pulled upward when half of it gets filled and compacted. This creates space for preparation second compact (Fig 8b).

4. Leave collection 2: More hay is collected and filled inside the new space generated after the expansion of compaction container.
5. Leaves compaction 2: Similarly, more hay is compacted to form a second compact.
6. Transport of compact 1 & 2: The chain block is used to lift the net that is now filled with compacted hay (Fig 8c). The hand winches on both columns are used to move and align the compact hovering above the boiling vessel (Fig 8d). And again the chain block lowers the compact into the boiler drum. Both the compacts from compaction drum are transported to boiling drum one by one in the same way. Then the vessel lid is fixed over the boiling vessel by chain block and then it is fastened by nut bolts. It is heated at the base, containing water up to 300 mm. The steam so produced extracts the oil from the peppermint leaves. The oil is separated from steam when it is condensed in the condenser and the oil is collected in a container.
7. Removal of the processed compact: The cover is unfastened and lifted by chain block. The chain block further removes the compacts of processed hay one by one, loaded on a rickshaw-trolley, which can be kept beside the boiler drum, and transported over to smoke emission chamber (drier unit) the hay is scattered in the field for faster cooling and drying.

With the addition of a compaction container, the labourers can prepare another round of compacts while two compacts are already in the boiler being distilled. As soon as the first two compacts are removed from the boiling vessel after processing, the pair of prepared compacts can be put inside the vessel for the next round. This would save the otherwise waiting time while the vessel was full and distillation was going on.

3. Results and Discussion

The traditional method adopted in various places is surveyed and it is found that farmers use various sizes of drum (Boiling vessel). However, the preferred size of drum and other details related to plant are as follows:

The working area for entire plant is 7.5 x 3 m^2.

The plant has following major elements

Boiling vessel is of diameter= 6Ft (1.82m) and height= 5Ft (1.52m)

Compaction container of diameter= 6Ft (1.82m) and height= 3Ft (0.92m) (collapsed)

The items required are listed in Table 2 along with their costs and suitable capacities.

Other items that need to be fabricated are:

a) Base net (4 Units)
b) Telescopic Compaction container (with wire mesh walls)

Table 2: List of required items. [13]

Sl.No	Item	Capacity	No.(s)	Avg. Cost
1.	I-beam SC 140 (23Ft)	-	1	Rs.11672@ Rs.50/Kg
2.	I-beam trolley	0.5 ton	1	Rs. 3000
3.	Chain block	0.5 ton	1	Rs. 4000
4.	Hand winch	Min.	2	Rs. 2000x2
5.	Pulley (with mount)	-	2	Rs. 700 x2
	Total cost			Rs. 24072

3.1 Time Calculations

3.1.1 Time taken for lifting and lowering of compacted compact with chain block:

Travel ratio of 0.5 ton chain block = 41:1 (assumed as available in market)

Let's assume that worker moves the hand chain at a constant velocity V_H= 1 m/s

Load chain velocity V_L=1/41 m/s= 0.002439 m/s =0.96024 in/s

Total time to lift or lower load chain by 6 feet = (6x12 in)/(0.96024 in/s)

=74.98s ~ 75s approx.

Total time to lift or lower load chain by 3.5 feet = (3.5x12 in)/(0.96024 in/s)
= 43.74s ~ 44s approx.

3.1.2 *Time taken to move the trolley in one direction:*

Gear ratio of hand winch = 2.9:1 (assumed as available in market)
The trolley travel per revolution of winch handle =75.87 mm/rev (assumed to be const. with each rev.)
Hence, 1 rps on handle = 1/2.9 = 0.3448 rps on drum
The trolley velocity V_T =0.3447x75.87 mm/s =26.16 mm/s
Time to move 3.81m (centre distance of drums) = 145.6s ~ 146 s

3.1.3 *Total time taken in removal of two processed compacts:*

Total time for removal = Time taken to (move from beam centre to vessel centre + lift first compact + move back to beam centre + lower the first compact + move back to vessel + lift second compact + move back to centre + lower second compact)
= 146/2 + 44 + 146/2 + 75 + 146/2 + 75 + 146/2 + 75
= 561s = 9m21s

3.1.4 *Total time taken in transport of two compacted compacts:*

Total time to fill vessel = Time taken to (move from beam centre to container centre + lift first compact + move back to vessel centre + lower the first compact + move back to container centre + lift second compact + move back to vessel centre + lower second compact)
= 146/2 + 44 + 146 + 75 + 146 + 75 + 146 + 44
= 749s = 12m 29s

3.2 *Load determination for hand winch capacity:*

Trolley wheel diameter = 4in = 71 mm
No. of wheels = 4
Mass of Compact = 200kg
Total mass to be hung = Mass of Compact + (Base Net + Chains + Chain Block) = 200 + 50(assumed)
= 250kg
Load per wheel W = (250x9.81)/4 = 613.125 N
Coefficient of rolling friction for steel on steel C_r = 0.001

$$F = C_r W/R \text{ [14]} \tag{1}$$

where, F = Frictional force
C_r = Coefficient of rolling friction
W = Load per wheel (N)
R = Radius of wheel (m)

F = 17.27 N
Total rolling friction on 4 wheels = 17.27x4 = 69.08N
Starting force (2 to 2.5 times of rolling friction) = 69.08x2.5 =172.71N = 17.55 Kgf

3.3 *Load determination for I beam size:*

Beam length L = 23ft = 7.0104m
Total Load P = (Trolley mass + Hung mass)x9.81
= (18 + 250)x9.81 = 2629.08 N
Now,

$$\delta_a = L/500 \text{ [16]} \tag{2}$$

where δ_a = Allowable deflection (m)
This gives δ_a = 0.0140208 m = 14.02mm
Assuming simply supported beam with concentrated load, maximum deflection will occur when trolley is mid span
Hence by beam equation,

$$\delta_{max} = PL^3/48EI \quad (3)$$

where, P = Load on beam centre (N)
L = Beam length (m)
E = Young's Modulus ($N/m^{2)}$
I = Moment of inertia (m^4)
For steel Young's Modulus E = 200GPa
Solving for I,
$I = 6.27 \times 10^{-6}\ m^4 = 627 \times 10^4\ mm^4$
SC 140 H-Beam (BIS) has an I value equal to 1470x104 mm^4. [15]
Mass of SC 140 is 33.3Kg/m
Max deflection with self weight included,

$$\delta_{max} = (PL^3)/48EI + (5wL^4)/384EI \quad (4)$$

where, w = Self weight (N)
This gives δ_{max} = 0.00991m = 9.91mm
This value is less than Allowable deflection, hence this beam is selected.
The comparison of traditional plant and mechanized plant is presented in Table 3.

Table 3: Comparison of mechanized model with existing model.

S No	Description	Existing plant	Mechanized plant
1	Installation cost	Rs 2,20,000	Rs 3,20,000
2	Modification cost only	-	Rs 1,00,000
3	Working area covered	5.5 x 5.5 m^2	7.5 x 3.0 m^2
4	Preparation time of compact (hay)	2 hours in main boiler drum	1.5 hour in compaction container
5	Time required for removal of processed hay	3 hours	10 min
6	Distillation time	4.5 hrs	4.5 hrs
7	Cycle time	9.5 hrs	5 hrs
8	Peppermint hay weight in 1 tank	400Kg	400Kg
9	No. of labourers	4	2
10	Labour cost	Rs 1600/ day	Rs 800/ day
11	No. of boilers used in one day	02	05
12	Earning per day of the plant@Rs1000/ cycle	Rs 2000/day	Rs 5000/day

Referring Table 3, it is clear that the earning per day is increased by Rs 3000 per day and labour cost is halved compared to traditional approach. The time required for removal of used leave is highly decreased. The chain pulley arrangement utilizes 9-10 minutes for removal of used/ processed hay.
It is depicted from above discussion that the incurred cost of setting up this can soon be covered with increased profit of the plant. It is also depicted that hand winch can be used ergonomically for moving the compact from compaction container to the vessel and vice versa.

However, this study is theoretical and assumptions are made at various stages. These assumptions may give very ideal results and more accurate results can only be obtained after successful trial in the real world. Uncertainties are involved, especially the time taken by various operations, may divert from these ideal values obtained here but not by much. But the reduction in effort that the mechanism offers is for sure.

4. Conclusion

Indian farmers largely depend on cultivation of crops. Cultivation of peppermint plant is common in northern part of India. The traditional methods being adopted in rural India is not at all safe and has several other problems as discussed in section 1.

The 2 dimensional mechanized system, presented in this paper, can easily replace the already existing traditional setup, with very little changes and modification of the existing system.

For Indian context, the traditionally used hydro-distillation technique for oil extraction from peppermint plant is found suitable due to its low cost, easier operation and high yield. There is no need of electricity. The drawbacks present in the currently used methods in Indian farms and the problems that arose are discussed and solved. The presented model eradicates the problems related to the material handling, long cycle times and will reduce the waiting time to greater extent. The proposed 2 dimensional model satisfies the need of farmers.

The 2 dimensional mechanized system has overall effective suitability for Indian villages. It will be easy to install as it a modification of the existing one. The setup cost required is moderate. This new model will increase the profit of the plant. No power supply is needed. No special technical knowledge is required for operation.

The major conclusions are:

1. Provides safety to workers.
2. Makes it easier for removal of processed hay. The time required for removal of used hay is largely decreased. The chain pulley arrangement takes about 9-10 minutes for removal of processed hay.
3. Compacts can be prepared while distillation is going on, hence saving large amount of time.
4. More distillation cycles can be completed in same time
5. Works without electricity and special technical knowledge.
6. The earning per day is increased by Rs 3000 per day and labour cost is halved compared to traditional approach.

A real world working setup will give more accurate results as the study presented in this paper is theoretical. Assumptions have been made at various stages for calculations and the results may be more on the idealistic side.

The 2 dimensional mechanized techniques for easy, safe and fast material handling, can bring a drastic change in the field of agriculture. The efficient method and highly economical setup will increase the productivity of the plant.

References

[1] Gavahian M, Farhoosh R, Farahnaky A, Javidnia K and Shahidi F 2015 Comparison of extraction parameters and extracted essential oils from Mentha piperita L. using hydrodistillation and steamdistillation *Int Food Res J* **22**(1) pp. 283-8.

[2] Sugiura T, Uchida S and Namiki N 2012 Taste-Masking Effect of Physical and Organoleptic Methods on Peppermint-scented Orally Disintegrating Tablet of Famotidine Based on Suspension Spray-coating Method. Chem. Pharm. Bull **60**(3) pp. 315–324.

[3] Pittler MH, Ernst E 1998 Peppermint oil for irritable bowel syndrome: a critical review and metaanalysis Am J Gastroenterol. Jul **93**(7) pp. 1131-5.

[4] http://www.nhb.gov.in/Horticulture%20Crops/Mint/Mint1.htm [02 March 2021].

[5] https://www.thehindubusinessline.com/economy/agri-business/high-prices-boost-mint-cultivation/article26332498.ece [02 March 2021].

[6] Anas M and Abusad 2019 Mechanization of Peppermint Oil Extraction Plant of Rural India *Proc. iNaCoMM2019* (IIT Mandi, Mandi) paper 195.

[7] Anas M, Mujeeb A and Sharma A 2020 Sustainable Model for Mechanization of Peppermint Oil Extraction Plants for Rural India *IPRoMM-2020* (BITS-Pilani, Hyderabad) paper 132.

[8] Braimah S R, Andoh P Y and Tawiah P O 2016 Designing a Mechanical System That Will Be Used To Extract and Separate Lemon Grass Oil *Int j sci eng* **5**(3) pp. 01-11.

[9] Lucchesi M E, Chemat F, and Smadja J 2004 Solvent-free microwave extraction of essential oil from aromatic herbs: comparison with conventional hydro-distillation *J Chromatogr A* **1043**(2) pp. 323–7.

[10] Gavahian M, Farahnaky A, Majzoobi M, Javidnia K, Saharkhiz M J and Mesbahi G 2011 Ohmic-assisted hydrodistillation of essential oils from Zataria multiflora Boiss (Shirazi thyme) *Int J Food Sci Technol* **46**(12) pp. 2619–2627.

[11] Chandna P, Deswal S and Chandra A 2010 An anthropometric survey of industrial workers of the northern region of India *Int J Ind Syst Eng* **6**(1) pp. 110-128.

[12] Kharade H, Deshmukh M, Tiwari P and Gite L 2010 Optimum crankshaft height and crank length for hand operated rotary maize sheller *PKV Res J* **34** (2010) pp. 177-180.

[13] https://www.indiamart.com [14 June 2021].

IOP Conf. Series: Materials Science and Engineering 1206 (2021) 012015 doi:10.1088/1757-899X/1206/1/012015

[14] Lippert D and Spektor J Calculating proper rolling resistance: A safer move for material handling, (2012). https://www.plantengineering.com/articles/calculating-proper-rolling-resistance-a-safer-move-for-material-handling/ [22 February 2021].

[15] Gorenc E B 2003 *Crane Runway Girders - Limit States Design* (Australian Institute of Steel Construction) pp. 43-44.

[16] Dimensions For Hot Rolled Steel Beam, Column, Channel And Angle Sections (Third Revision), IS 808:1989

Free vibration analysis on axially graded beam resting on variable Pasternak foundation

Saurabh Kumar

Department of Mechanical Engineering, School of Engineering, University of Petroleum and Energy Studies (UPES), Dehradun, 248007, India

Email - saurabhks88@gmail.com

Abstract. Free vibration analysis is conducted on axially functionally graded Euler-Bernoulli beam resting on variable Pasternak foundation. The material properties of the beam and the stiffness of the foundation are considered to be varying linearly along the axial direction. Two types of boundary conditions namely; clamped and simply supported are used in the analysis. The problem is formulated using Rayleigh-Ritz method and governing equations are derived with the help of Hamilton's principle. The numerical results are generated for different material gradation parameter, foundation parameter and boundary conditions and the effect of these parameters on the free vibration behaviour of the beam is discussed.

Keywords - Functionally graded material, elastic foundation, free vibration, natural frequency

1. Introduction

Functionally graded material (FGM) structures have gained popularity due to their inherent advantages over traditional composite materials. Recently, researchers are investigating the behaviour of these structures in different conditions. Structures resting of elastic foundation is one such condition. Fallah and Aghdam [1] analytically solved the nonlinear dynamic problem of FGM beams using He's variational method. Kanani et al. [2] conducted nonlinear vibration analysis on FGM beams using Galerkin method where the solution was obtained by variable iteration method. Calim [3, 4] carried out vibration analysis on FGM beams using complementary function method to study the influence of different parameters on its dynamic behaviour. Mirzabeigi and Madoliat [5] performed nonlinear dynamic analysis on FGM beams using energy method and Hamilton's principle. The Hencky bar-chain model was used by Zhang et al. [6] to carry out buckling and free vibration of FGM beams resting on variable elastic foundation. Yas et al. [7] used differential quadrature method to investigate similar beams. Deng et al. [8] investigated double functionally graded beams using Hamilton's principle and Wittrick-William algorithm. Avcar and Mohammed [9] carried out free vibration analysis on FGM beams resting on two parameter elastic foundation using classical beam theory and separation of variables technique. Mohamed et al. [10] examined the nonlinear dynamic characteristics of buckled FGM beam using differential-integral quadrature method where the solution of the governing equation was obtained by Newton's method. Esen [11] conducted vibration analysis on FGM beams subjected to moving mass of variable velocity using a modified finite element method. It was concluded that, the vibration can be controlled by adjusting the size of foundation parameters. Lohar et al. [12] conducted nonlinear forced vibration analysis using energy method and Hamilton's principle. M. A. Kumbhalkar et. al. [13] conducted free and forced vibration numerical analysis of non-linear curved beam. Hadji and Bernard [14] used Hamilton's principle to obtain Navier type analytical solutions of bending and free vibration of FGM beam. Esen et al. [15] investigated symmetric and sigmoid FGM beams considering the effect of a moving mass using Hamilton's method and finite element method where Newmark method was used to obtain the solutions.

From the above literature it is evident that investigating the dynamic aspects of FGM beams on elastic foundation is building up ample interest among the researchers. As such, in this paper, axially functionally graded material beams resting on variable two parameter Pasternak elastic foundation are investigated for free vibration.

2. Mathematical Formulation

A functionally graded beam with axial material gradation resting on two parameter elastic foundation with its geometrical dimensions is shown in Figure 1.

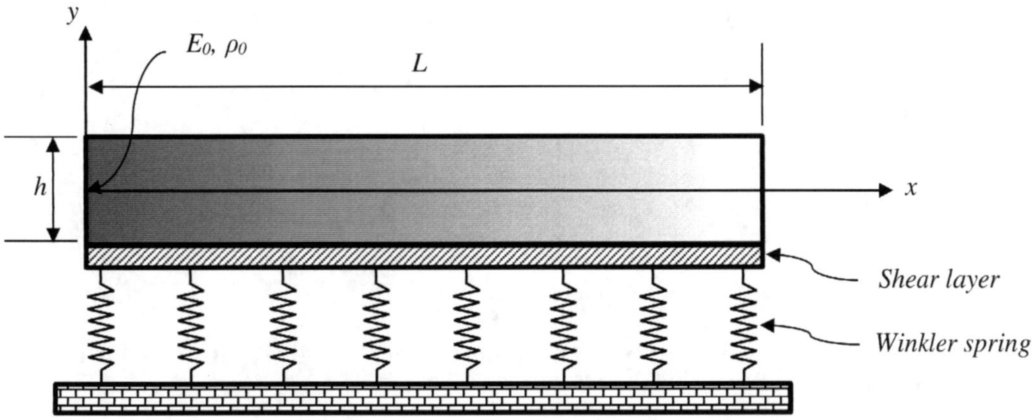

Figure 1. An axially functionally graded beam resting on two parameter elastic foundation.

The material properties (elastic modulus and material density) are considered to be varying along the x direction in the beam as per the following expression,

$$E(x) = E_0\left(1 - \frac{x}{L}\right), \rho(x) = \rho_0\left(1 - \frac{x}{L}\right) \tag{1}$$

where, E_0 and ρ_0 are elastic modulus and material density at the left end of the beam respectively. The beam is supported on an elastic foundation which contains a shear layer and a Winkler layer (group of linear springs). The stiffness of the shear layer (K_p) is kept constant whereas the stiffness of the Winkler layer (K_w) is varying linearly along the x-axis as per the following expression,

$$K_w(x) = K_0\left(1 + s\left(\frac{x}{L}\right)\right) \tag{2}$$

where, K_0 is the stiffness of the Winkler layer at the left end and s is the foundation variation parameter.

The beam is modelled using Euler-Bernoulli beam theory, following which the strain displacement relations can be written as,

$$\varepsilon_x = u_{,x} - z(v_{,xx}) \tag{3}$$

The total strain energy of the beam (U) is the summation of the strain energy stored in the beam, strain energy stored in the shear layer and the strain energy stored in the Winkler layer, and given as,

$$U = \frac{1}{2}A\int_0^L (u_{,x})^2 E(x)dx + \frac{1}{2}I\int_0^L (v_{,xx})^2 E(x)dx + \frac{1}{2}\int_0^L K_w(x)v^2 dx + \frac{1}{2}\int_0^L K_s(v_{,x})^2 dx \tag{4}$$

The kinetic energy of the system (T) can be written as,

$$T = \frac{1}{2}A\int_0^L (\dot{u}^2 + \dot{v}^2)dx \tag{5}$$

The displacement fields u (x-axis) and v (y-axis) for the beam are represented as linear combinations of unknown parameters (d_i) and orthogonal admissible functions (α_i) and (β_i) as per the Rayleigh-Ritz method. These displacement fields are expressed as follows,

$$u = \sum_{j=1}^{j=n} d_j \alpha_j e^{i\omega t}, \ v = \sum_{j=1}^{j=n} d_j \beta_j e^{i\omega t} \tag{6}$$

In the above expressions ω is the natural frequency of the system. The governing equation for the free vibration problem is derived from Hamilton's principle which states that,

$$\delta\left(\int_{t_1}^{t_2}(T-U)dt\right) = 0 \tag{7}$$

Solving the above equation gives the characteristic equation of free vibration for the problem in the following form,

$$[K] - \omega^2[M] = 0 \tag{8}$$

where, $[K]$ is the stiffness matrix and $[M]$ is the mass matrix of the beam. The natural frequency (ω) can be obtained by solving the above equation.

3. Results and Discussion

Numerical results for the present problem were generated using Matlab software following the methodology presented above. The results are presented for two boundary conditions namely clamped and simply supported in non-dimensional form. The non-dimensional parameters are as follows,

$$\overline{\omega} = \omega L^2\sqrt{\frac{\rho_0 A}{E_0 I}}\ , \overline{K}_w = \frac{K_0 L^4}{E_0 I}, \overline{K}_p = \frac{K_p L^2}{E_0 I} \tag{9}$$

Validation study is carried out by comparing the present results with those established in literature and furnished in Table 1.

Table 1. Validation of present results with literature

Boundary Condition	$\overline{K}_w$	$\sqrt{\overline{\omega}}$ $(s = 0.2)$	$\sqrt{\overline{\omega}}$ Ref. [16]
CC	10	4.7424	4.7511
	10^2	4.9219	4.9296
	10^3	6.1132	6.1172
SS	10	3.2118	3.2117
	10^2	3.6999	3.6999
	10^3	5.6185	5.6185

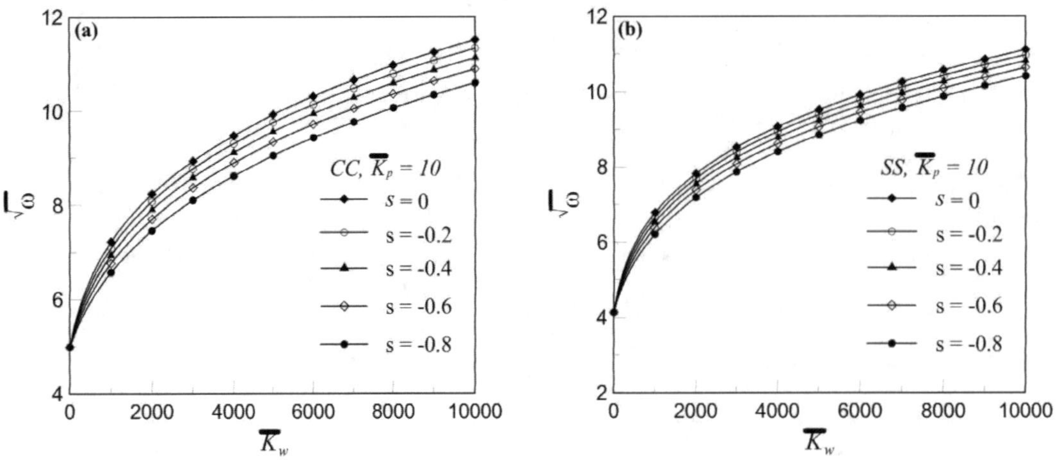

Figure 2. Effect of foundation stiffness ($\overline{K}_w$) on natural frequencies of FGM beam for different values of foundation variation parameter (s).

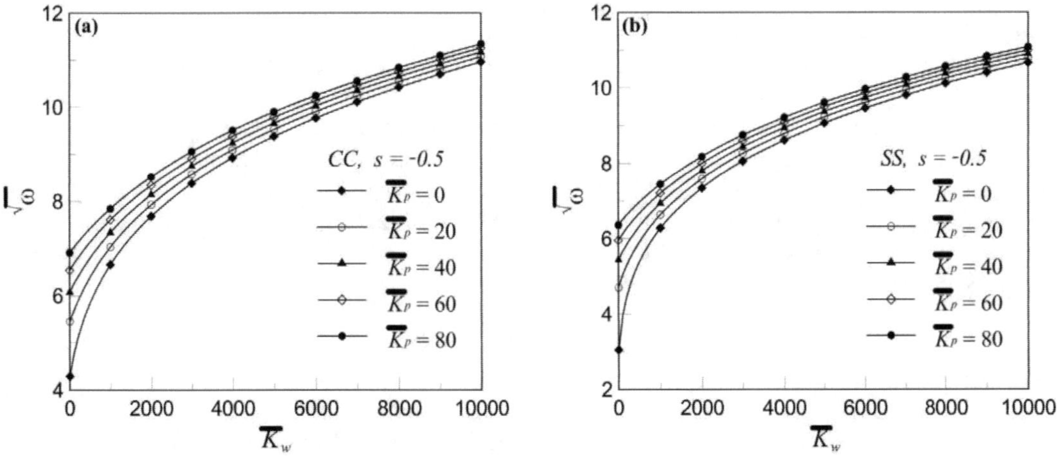

Figure 3. Effect of foundation stiffness ($\overline{K}_w$) on natural frequencies of FGM beam for different values of foundation stiffness ($\overline{K}_p$).

Following the validation study, new results are presented to study the effect of different parameters on the natural frequency of FGM beam. The effect of foundation stiffness ($\overline{K}_w$) on the natural frequencies of FGM beam for different values of foundation variation parameter is shown in figure 2. The value of $\overline{K}_w$ is varied from 0 to 10000 whereas $\overline{K}_p$ is kept constant at 10. The graphs are presented for 5 values of the foundation variation parameter (0, -0.2, -0.4, -0.6, -0.8). Similar results are also presented for different values of $\overline{K}_p$ in figure 3, where the value of s is kept constant at -0.5. The results are furnished for two boundary conditions and it can be seen from the figures that $\overline{K}_w$ significantly affect the natural frequency of the beam.

Figure 4 shows the effect of foundation variation parameter on natural frequencies of FGM beam. Here, the values of $\overline{K}_w$ and $\overline{K}_p$ are kept constant at 1000 and 10 respectively and s is varied from -1 to 1. It is evident from the figure that as the value of s increases the natural frequency also increases. Figure 5 shows the behaviour of FGM beam for the two selected boundary conditions. It can be seen that the natural frequency of SS beam is lower compared to CC beam owing to the less stiffness but they follow similar pattern for foundation stiffness variation.

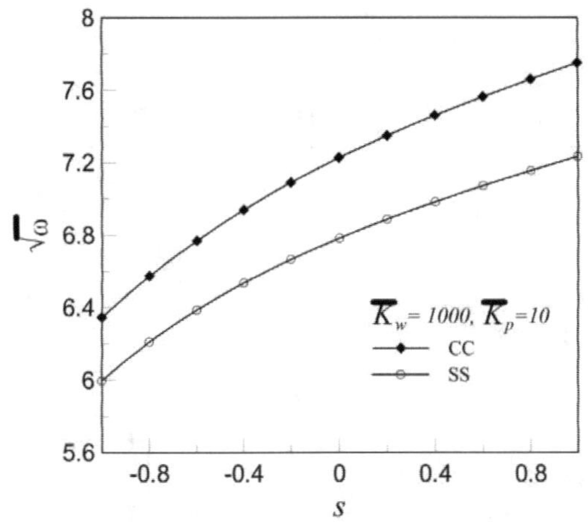

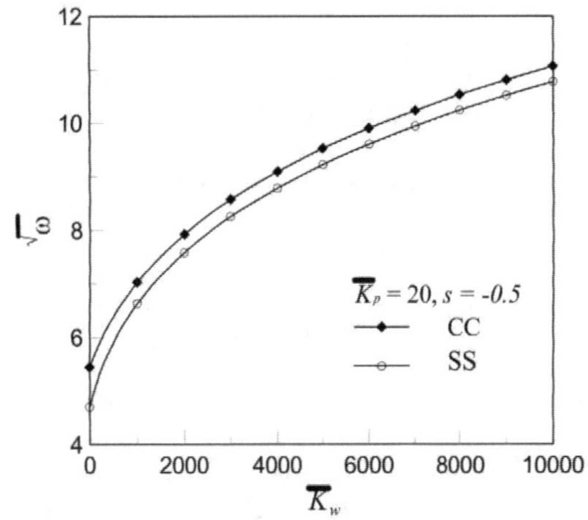

Figure 4. Effect of foundation variation parameter (s) on natural frequencies of FGM beam.

Figure 5. Effect of boundary conditions on natural frequencies of FGM beam.

4. Conclusion

Present study was carried out to investigate the dynamic behaviour of FGM beam resting on variable elastic foundation. An FGM beam with axial material variation is selected for this purpose along with a two parameter Pasternak foundation with variable foundation stiffness. Euler-Bernoulli beam theory was used for the formulation in conjunction with Rayleigh-Ritz method and Hamilton's principle was used to generate the governing equations. FGM beams with clamped-clamped and simply-supported-simply supported end conditions are investigated. The investigation revealed that the foundation stiffness significantly affects the natural frequency of the beam as it adds to the stiffness of the beam. Also, it was found out that the variation of stiffness greatly influences the dynamic behaviour of the beam.

5. References

[1] Fallah A and Aghdam M M 2011 Nonlinear free vibration and post-buckling analysis of functionally graded beams on nonlinear elastic foundation *Euro. J. Mech. A. Solids* **30(4)** pp 571-83.

[2] Kanani A S, Niknam H, Ohadi A R and Aghdam M M 2014 Effect of nonlinear elastic foundation on large amplitude free and forced vibration of functionally graded beam *Compos. Struct.* **115** pp 60-8.

[3] Calim F F 2016 Free and forced vibration analysis of axially functionally graded Timoshenko beams on two-parameter viscoelastic foundation *Compos. B. Eng.* **103** pp 98-112.

[4] Calim F F 2020 Vibration analysis of functionally graded Timoshenko beams on Winkler–Pasternak elastic foundation *Iran J. Sci. Technol. Trans. Civ. Eng.* **44** pp 901–20.

[5] Mirzabeigy A, Madoliat R 2016 Large amplitude free vibration of axially loaded beams resting on variable elastic foundation *Alex. Eng. J.* **55(2)** pp 1107-14.

[6] Zhang H, Wang C M, Ruocco E, Challamel N 2016 Hencky bar-chain model for buckling and vibration analyses of non-uniform beams on variable elastic foundation *Eng. Struct.* **126** pp 252-63.

[7] Yas M H, Kamarian S and Pourasghar A 2017 Free vibration analysis of functionally graded beams resting on variable elastic foundations using a generalized power-law distribution and GDQ method *Ann. Solid Struct. Mech.* **9(1-2)** pp 1-11.

[8] Deng H, Cheng W and Zhao S 2017 Vibration and buckling analysis of double-functionally graded Timoshenko beam system on Winkler-Pasternak elastic foundation *Compos. Struct.* **160** pp 152-68.

[9] Avcar M and Mohammed W K M 2018 Free vibration of functionally graded beams resting on Winkler-Pasternak foundation *Arabian J. Geosci.* **11(10)** pp 1-8.

[10] Mohamed N, Eltaher M A, Mohamed S A and Seddek L F 2018 Numerical analysis of nonlinear free and forced vibrations of buckled curved beams resting on nonlinear elastic foundations *Int. J. Non. Linear. Mech.* **101** pp 157-73.

[11] Esen I 2019 Dynamic response of a functionally graded Timoshenko beam on two-parameter elastic foundations due to a variable velocity moving mass *Int. J. Mech. Sci.* **153** pp 21-35.

[12] Lohar H, Mitra A and Sahoo S 2019 Nonlinear response of axially functionally graded Timoshenko beams on elastic foundation under harmonic excitation. *Curved Layer. Struct.* **6(1)** pp 90-104.

[13] Manoj A. Kumbhalkar; D.V. Bhope; A.V. Vanalkar, "Evaluation of Frequency Excitation of Helical Suspension Spring Using Finite Element Analysis", International Journal of Computer Aided Engineering and Technology, Vol. 9, No. 4, pp 420-433, 2017.

[14] Hadji L, Bernard F 2020 Bending and free vibration analysis of functionally graded beams on elastic foundations with analytical validation *Adv. Mat. Res.* **9(1)** pp 63-98.

IOP Conf. Series: Materials Science and Engineering 1206 (2021) 012016 doi:10.1088/1757-899X/1206/1/012016

[15] Esen I, Eltaher M A, Abdelrahman A A 2021 Vibration response of symmetric and sigmoid functionally graded beam rested on elastic foundation under moving point mass *Mech. Based. Des. Struct. Mach.* **26** pp 1-25.
[16] Kacar A, Tan H T, Kaya M O 2011 Free vibration analysis of beams on variable winkler elastic foundation by using the differential transform method *Math. Comput. Appl.* **16(3)** pp 773-83.

Application of Jaw Type Flexible Clutches in Human Powered Machine: A Literature Review

Nitin Sawarkar[1, *], K S Zakiuddin[2], Roshan Umate[3], Rajkumar Chagde[4]

[1, 2]Department of Mechanical Engineering, Priyadarshini College of Engineering, Nagpur, Maharashtra, 440019.
[3] Jawaharlal Nehru Medical College, Datta Meghe Institute of Medical Sciences, Nagpur, Maharashtra
[4] Department of Mechanical Engineering, Yeshwantrao Chavan College of Engineering, Nagpur
[*]Corresponding Author E-mail: nitinsawarkar304@gmail.com

Abstract. This article presents the topical progresses in the arena of an approach to articulate an experimental databased model for jaw type flexible clutches. The developing countries are facing very acute problems of the energy crisis. It has become essential to use human energy for energizing the manufacturing process machine. This kind of machine necessitates a flexible clutch for torque transmission. Due to the intricate impact phenomenon, flexible clutches cannot be deliberate without experimental data. Therefore, the experimentation methodology and experimental setup to evaluate the performance of torque transmitted by the clutch have been augment. It is obliging for selection of the optimum clutch-type for various applications in low powered processing machines. This study of analysing suitable material can be useful for the comparative study of Jaw type flexible clutches and their transmission ability. Mathematical model and Artificial Neural Network (ANN) development are essential, to evaluate the belongings of altered components and to make forecasts about behaviour.

Keywords. Jaw Type Flexible Clutches Torque, Human Energy, and Artificial Neural Network.

1. Introduction

In today's industrial environment, equipment system designers are exacting higher speed, torque, transmission efficiency, greater flexibility for clutches. As the technology advances, the need for flexible clutches is becoming more and more acute. Clutches used in engineering machineries to connect or disconnect the cause of influence [1]. It used in machineries for transmission of speed and torque. The researchers have executed the analysis of the various type of clutch, but a comparative study of Jaw clutches and their transmission ability not yet been studied. Therefore, this research mainly focused on Jaw Type Flexible Clutch (JTFC) and requisites torque output. The benefit of JTFC are having optimistic engagement and can transfer great torque with no-slip, when united [2]. Which impart in no heat generation during operation. The precise machining of teeth on two splits to ensure perfect contact stretches an adequate operation of the JTFC. Friction clutches are moderately greater as compared to JTFC for the same power transmission [3]. The spiral and square JTFC comprise of sloping and square teeth respectively, which lock into mating recesses in pebbledash members.
The Experiments carried out on the setup for various values of load torques and observations fed to the computer by using the instrumentation system. The experimentation performed by changing

IOP Conf. Series: Materials Science and Engineering 1206 (2021) 012017 doi:10.1088/1757-899X/1206/1/012017

independent physical quantities and responses gathered, to generate design data by performing experimentation involving independent parametric variation. Due to the nonlinear structure and intricate phenomenon, Artificial Neural Network (ANN) development is necessary [4, 27]. The significance of this grid decided by relating the output from the network with output from the mathematical model and experimental response as per the observed data.

2. Material And Methods

The material selection is significant for any potential concerns that definite materials may present. This will make the process of engineering design analysis easier in the culmination. Tirasuntarakul [2] developed a micro triggering device, which used a single motor to drive multi degrees of freedom. The torque transmission is vital in the progression. The magnetic jaw clutch is designate to create high conceivable torque and bi-directional rotation of the module. To control the operation of the magnetic jaw clutch used a controller design. Experimental setup is required to determine the accurate results of output torque. Overrunning clutches can also be used in various application if requires one direction torque transmission.

Yusuke [3] developed, "a single degree of freedom combined drive mechanism using regenerative servo clutches". This is consists of a regenerative brake and a differential gear mechanism. It can control the transferred torque by regulatory the regenerative brake torque. The main factor of the dynamic condition is the output torque imposed. Whereas Jaw type flexible clutches provides both directional torque transmission. It is compact in size and low cost as compared with an exploited clutch.

2.1. Human Powered Flywheel Motor

Human Powered Process Machines for several rural-based production activities developed by Chandak [4] as shown(Fig.1). The evolved machine system comprised of an energy entity, clutch, torque escalation gear-pair, and a process unit.

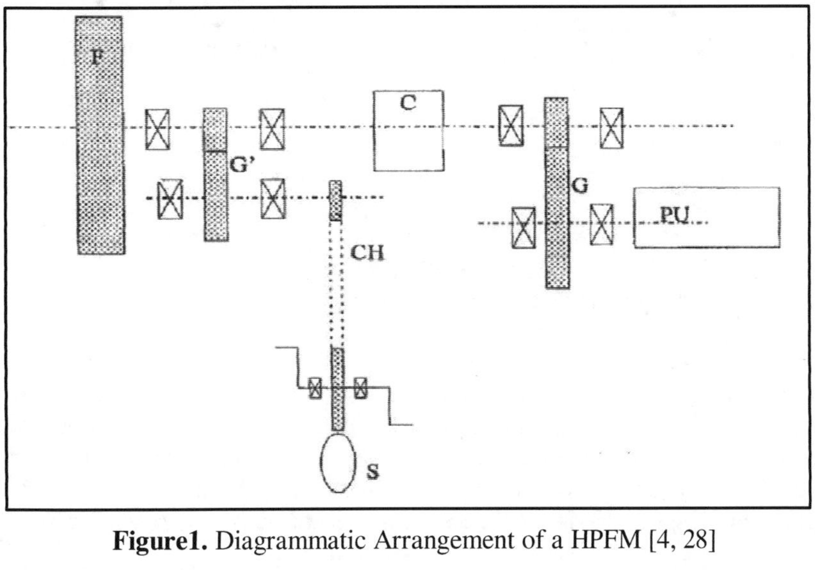

Figure1. Diagrammatic Arrangement of a HPFM [4, 28]

In this comparison between neural prediction and empirical prediction with experimental findings has carried out. According to him, clutch plays an important role in Human Powered Flywheel Motor and various low powered machines. No-slip occurs during power transmission therefore, no heat generation during the operation. A comparative study of these types of clutches can provide inheritance while selecting a clutch for essential torque output.

2.2. Selection of Dependent and In-dependent Variable

According to Mishra and Srinivasan [5], "clutch-to-clutch shifts are existing in automatic transmissions, motivating the need for formal and robust methods for controlling these shifts". The off-going clutch controlled to imitate a one-way clutch, which ensures reduced overall variation in the output shaft torque and smooth coordination of the two clutches, during the torque phase. Thereafter the oncoming clutch is control to ensure subsequent driveline oscillations, engagement at lock-up, and ensuing in a reduction of shock. The shift control intentions encountered by the proposed valuation and control strategy in the presence of significant model uncertainty. Thus, validating the practical effectiveness and robustness.

The Dependent and independent variables play avital role while designing a mathematical model for the clutch. The various independent and dependent variables identified for the designing and formulation of a mathematical model for jaw type flexible clutch listed in table 1.

Table 1. The Process Variables with Symbols and Dimensions

Sr. No	Description	Variable Type	Symbol	Dimensions
1	Load Torque (T_L)	Independent	N-m	ML^2T^{-2}
2	Speed of Shaft (n)	Independent	rpm	T^{-1}
3	Length of Jaw (L)	Independent	m	L
4	Width of Jaw (W)	Independent	m	L
5	Depth of Jaw (D)	Independent	m	L
6	Diameter of Hub (d)	Independent	m	L
7	Modulus of Elasticity (E)	Independent	N/m^2	$ML^{-1}T^{-2}$
8	Radius of Jaw measured from the center (r)	Independent	m	L
9	Acceleration due to Gravity (g)	Independent	m/sec	LT^{-2}
10	Bearing Friction Constant (b_F)	Independent	-	ML^2T^{-1}
11	No of Jaw (N)	Independent	-	$M^0L^0T^0$
12	Period of Operation	Independent	sec	T
12	Input Power (P_i)	Dependent	watts	ML^2T^{-3}
13	Driving Torque (T_d)	Dependent	N-m	ML^2T^{-2}
14	Torque Transmission Efficiency (η)	Dependent	-	$M^0L^0T^0$

To find the optimum shape constraints that deliver the mark stiffness curve, the shape optimization is important. After solving the optimization problem with the differential development method, optimum shape parameters of the cushion disc had found for N. Kaya et.al. [6] and Martin Steinberger et.al. [7], two case studies. Continuous variable transmissions play an important role in mechanical systems. Oday I. Abdullah et.al [8] states that in the process of power transmission, clutch is a necessary factor. It is essential to investigate the stresses of the clutch to avoid failure. The effort presented deformations during the steady-state period of the rigid clutch disc and the mathematical modelling of computing the stresses. In this research, response equated to the reference model and new model intended. The numerical results showed that by adjusting design parameters, the stresses of the clutch could be control. Zakiuddin et.al. [9], worked on The Pedal Operated Energized Flywheel Motor. "It has established functional feasibility and economic viability of pedal-powered process machines of 3 to 7 h.p. capacity", evaluated in their study.

According to the investigation, an experimental setup is required to determine the accurate result due to complexity in design. Rothabart [10] and Schenck [11] proposed a classical plan of experimentation, by which the experimentation mainly carried out. In such a plan all but one of the independent terms are to be maintain at their planned fixed level values and said the independent term is to be varied over its wide range as decided by the test points, taken one at a time. Jaw clutches also called a positive clutch because of the positive engagement between two parts of the clutch. It is useful for driving in both directions. The design of the Jaw clutch should base on strength consideration. Cast Iron is mainly use as a good quality material for this type of clutches [11].

2.3. Assortment of Artificial Neural Network Paradigm for Experimentation

According to ANN study paradigm[12-15], in the field of mechanical engineering, ANN has extensive application. From the origin of the mechanical engineering arena to the manufacturing, machine design, and operation, ANN technology is ubiquitous. Due to the hasty evolution of technology, mechanical engineering is extensively considered as a basic chastisement in the routine life. However, the technology based on mechanical engineering has several disadvantages such as the inconsistent system. The AI can promptly exchange the input and it can process it in a timely manner such that the output is not pretentious. It can also make meritorious verdicts for this defect. It is recognized that selecting a proper ANN parameter plays a significant role in verdict. The artificial neural network designated parameters are, coatings in the network, neurons, training approaches, learning procedure, transfer functions used, and performance function.

2.4. Experimentation System

Modelling of performances of such systems over Artificial Neural Networks comprises the routine of the above paradigm applicably. All the network factors have their own choices of parametric statistics [18, 20]. It is significant to differ a single parametric figure of associated network parameter for respective set of experiment, keeping all other figures of ANN parameters constant, to perceive the effect of parametric value, the subsequent parameter on ANN prediction [15-17].

3. Experimental Model to Investigate Jaw Type Flexible Clutches

For accurate engagement, two halves on the shafts is essential to mounting appropriately [23]. It requires greater exactitude in machining compared with toothed disks for trouble free meshing [24]. The engagement causes noise and fierce shock due to metal contact. JTFC clutch is consists of two robust disks with jaws that can apt with each other. The right part is fastening to the input shaft and other mounted over load shaft. Motion could be transmitting by direct intervention between the projections on parts of the clutch.

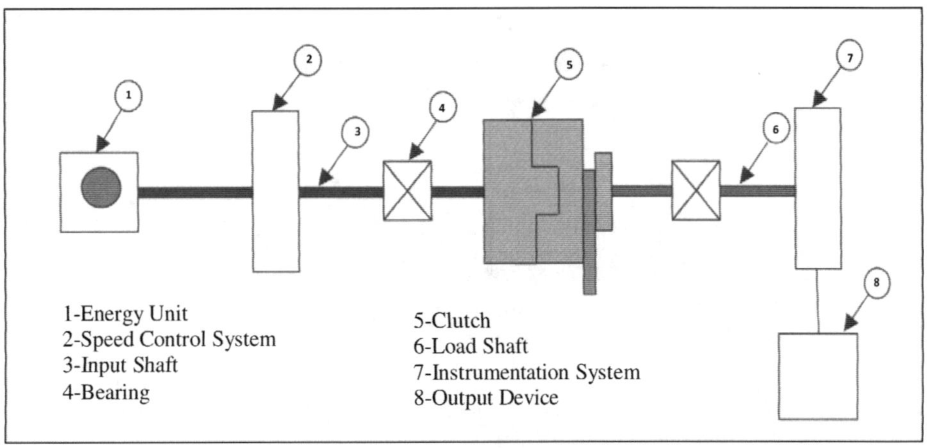

Figure 2. Schematic Representation of Experimental Setup

The Energy Unit consists of Electric Motor and Speed Control Unit on the left side of the setup as shown (fig. 2). The Input Shaft (IS) had connected with the Energy Unit. The Load Shaft (LS) has mounted on another side of the system. Both Input and Load Shaft has supported with the Bearings. Input Shaft and Load Shaft connected through Jaw Type Flexible Clutch (JTFC). The clutch consists of Two Parts. The first part attached to the driving shaft (Input shaft) and another attached to the driven

Shaft (load shaft) which consists of a jaw integral with the outer periphery. The Hand Lever mounted on the one part of the clutch to engage and disengage the Input and Load Shaft. The Speed in rpm selected as a first fixed independent term. As per the experimental plan, the load torque will be varied over its range [29]. Similarly, the experimentation performed for different levels of speed. This Process repeated for testing various Jaw Type Flexible Clutches [30].

4. Result and Discussion

"The clutch failed when the perilous load was reached and the beam buckled" [1]. Occurrence of the clutch dynamics is having non-linearity so the model formulation done using an artificial neural network to curtail the imprecision [25]. The output of this network evaluated by comparing it with the mathematical model intended data and the experimental data. ANN model can work very meritoriously for problems approximation. Requires more iterations, exercise and simulation time with the higher neurons size while lower size of neurons gives a deprived performance of the network in the hidden layer. This leads to result in overfitting thereby governing the forecast of ANN. Consequently, in the hidden layer optimum value of neuron size selection is protuberant. "For Human Powered Flywheel Motor with three independent variables, the neuron size of 50 originate to be optimal" [4]. According to Oday I. Abdullah [8], von mises stresses induced in the clutch disc which is ranges from 163.6 to 421.6 n/mm^2 for 0.040 to 0.187 mm displacement respectively. "The corresponding values of reliability and coefficient of determination for Human Powered Flywheel Motor of the proposed ANN model is found to be 96.68% and 0.95 respectively"[4]. The stiffness of the clutch plays an important role in part life. The stiffness ranges from 0.0023 to 0.124 mm has been analized by N. Kaya[6]. By taking these all factors into consideration final model has been formulize for analysis.

The evaluations have been made for two jaw clutch assembly with three material. The comparative study of structural steel, grey cast iron and stainless steel is as following.

Table 2. Results of Two Jaw Type Flexible Clutch Assembly with three type of materials.

Type	Total Deformation	Stress
Results (Structural Steel)		
Minimum	0. mm	-249.93 MPa
Maximum	0.60127 mm	245.69 MPa

Type	Total Deformation	Stress
Results (Grey Cast Iron)		
Minimum	0. mm	-246.83 MPa
Maximum	0.34774 mm	242.85 MPa

Type	Total Deformation	Stress
Results (Stainless steel)		
Minimum	0. mm	-249.93 MPa
Maximum	0.60127 mm	245.69 MPa

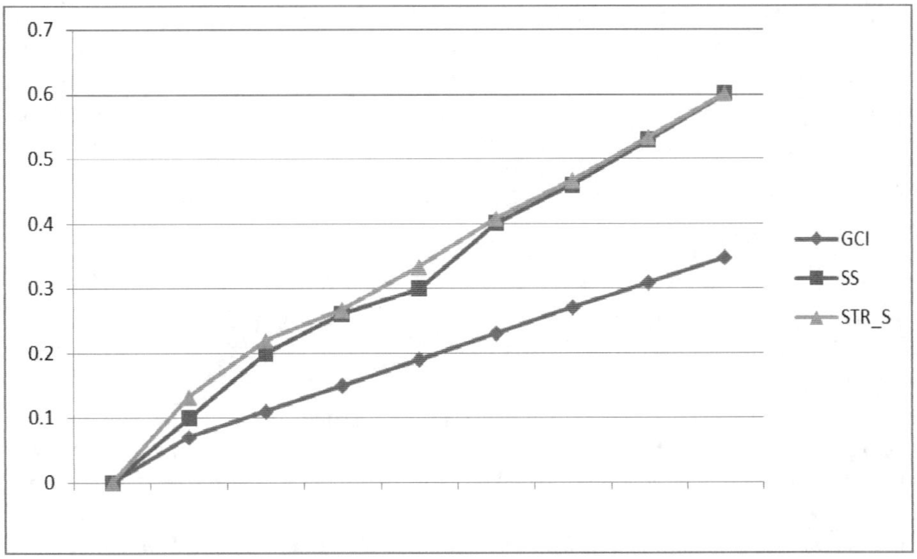

Figure 3. Comparative Study of Two Jaw clutch Assembly with 3 Materials

According to the Fig.3, Grey Cast Iron (GCI) is having less deformation. As per shear stress result, there is no much variation. But GCI is having better results than other two. With the experimental data of a JTFC, the model analysed qualitatively for variables like driving torque and torque transmission efficiency. During experimentation, an attempt has made for the quantitative analysis of the model for all response variables. Model represents the degree of interaction of numerous independent variables. The torque transmission efficiency varies with load torque but varies contrariwise with the speed of the clutch shaft. The load torque greatly varies by the independent variable like input power and driving torque. This experimental setup will also be useful for testing any type of clutch by some trifling variations.

5. Conclusion

The various type of Clutches used for various applications in the industry [2]. The analysis of the various type of clutch has been executed by the researchers [29, 30] but a comparative study of Jaw clutches and their transmission ability not yet been studied. In a conventional loom mechanical clutches used on main shaft of the machine. Until the motor attains its required torque and speed, the drive to the loom is disengaged. Similarly, Feed roller drive on a card, Yarn under winding in ring spinning, Lap roller drive on sliver doubling machine, etc. As far as the Mechanical type of Flexible Clutches is concerned, they have some obvious shortcomings such as a) While transmitting torque, the clutch should not absorb more energy, b) Impact phenomenon is very complex and hence cannot studied without experimental data [5].

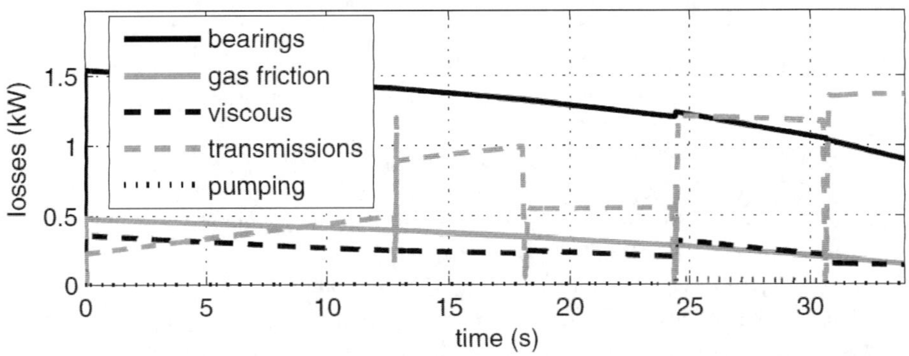

Figure 4. Analysis of clutch types by martin to overcome shortcomings [2]

It is quite possible to overcome the shortcoming. Thus, Jaw Type Flexible Clutches considered as the better option among the other [26, 27]. Experimentally it established the functional viability of various possibilities for these clutches. Therefore, it decided to generate design data to analyse the Jaw type of flexible clutches by performing experimentation involving independent parametric variation. Mathematical model development is essential, to study the properties of different components, to explain a system and to make predictions about performance. Formulation of design-databased model for jaw type flexible clutches has augmented. The Models if one optimizes then for minimization of responses, the best set of designed variables would accomplish.

References

[1] Gregory M. Roach, Larry L. Howell, Evaluation and Comparison of Alternative Compliant Overrunning Clutch Designs, Journal of Mechanical Design, Transaction of ASME, Provo, Utah, Vol.124, pp.485-491, 2002.

[2] Narongsak Tirasuntarakul, Arbtip Dheeravongkit, Torque Analysis of a Micro Actuating Mechanism Driving Multi Degree of Freedom via Single Motor,3rd International Conference on Control, Automation and Robotics, IEEE Transaction, 978-1-5090-6088-7117, pp.529-532, 2017.

[3] Yusuke Sugahara, Kensuke Kikui, Mitsuru Endo, Jun Okamoto, Daisuke Matsuura, and Yukio Takeda, A Human-Powered Joint Drive Mechanism Using Regenerative Clutches, RSJ International Conference on Intelligent Robots and Systems (IROS), IEEE Transaction, 978-1-5386-2682-5, 6337-6342Vancouver, BC, Canada, 2017.

[4] Pawan Chandak, Arati Lende, Jayant Modak, Modeling of Human Power Flywheel Motor through Artificial Neural Network-A Novel Approach, Elsevier, Procedia Computer Science, 125, pp.77-84, 2018.

[5] Kirti D. Mishra, K.Srinvasan, Robust Control and Estimation of Clutch-to-Clutch Shifts, Elsevier, Control Engineering Practice, Vol.65, pp.100-114,2017.

[6] N.Kaya, S.Kartal, T.Cakmak, F.Karpat, A.Karaduman, Shape Optimization of Clutch Cushion Disc Using Differential Evolution Method, Proceedings of the ASME, International Mechanical Engineering Congress and Exposition, Houston, Texas, pp.1-9, 2015.

[7] Martin Steinberger, Christian Milwisch, Walter Rosinger, Martin Horn, A clutch-based transmission for mechanical flywheel applications, Proceedingsofthe19th World Congress, The International Federation of Automatic Control Cape Town, South Africa. August 24-29, 2014.

[8] Oday I. Abdullah, Josef Schlattmann, Emir Pireci, Optimization of Shape and Design Parameters of the rigid Clutch Disc Using FEM, FME Transactions, Vol.41, No.4, pp.317-324, 2013.

[9] K.S. Zakiuddin, J.P. Modak, Application of pedal operated flywheel motor as an energy source for fodder chopper, International Journal of Agricultural Engineering, Vol. 3 No. 2, pp.251 -256, 2010.

[10] Mechanical Design and System Handbook, Second Edition, by Harold A. Rothabart, Published by Mac-Graw-Hill Book Company.

[11] Hilbert Schenck, Jr., Theories of Engineering Experimentation, McGraw Hill Book Company, New York, USA,1961.

[12] S. N. Shvanandam, S Sumathi, and S. N. Deepa, Introduction to Neural Network using Matlab 6.0, 2017, McGraw Hill publisher.

[13] Stamtios V. Kartaplopoulos, Understanding Neural Networks, and Fuzzy Logics, 1996, Wiley-IEEE Press, ISBN: 9780470546826.

[14] Neural Network Toolbox TM 7 User's GuideR2010a.

[15] P. A. Chandak, A. R. Lende, J. P. Modak, A literature review on fundamentals & methodologyofdevelopmentofmathematicalmodelthroughartificialneural network, International Journal of Computer Application, Issue 4, Volume 1, 2014.

[16] S. M. Sheikh, K. S. Zakiuddin, History of Human Powered Oil Expeller: A Literature Review, Springer Nature Switzerl and A G, B. Zhangand M. Ceccarelli(Eds.): Explorations in the History and Heritage of Machines and Mechanisms, HMMS 37, pp. 77–88,2019.

[17] M. Sohail Pervez, K. S. Zakiuddin, Literature Review on the Developments of Rice Milling

Machines, Springer Nature Switzerland AG, B. Zhang and M.Ceccarelli (Eds.): Explorations in the History and Heritage of Machines and Mechanisms, HMMS 37, pp. 89–100, 2019.
[18] Siraj Mohamad Ali Sheikh, Zakiuddin Syed Kazi, Design & Development of Human Energized Oil Expeller Machine, International Journal of Recent Trendsin Engineering & Research, Volume 02, Issue 06, pp.342-345, 2016.
[19] J.J.Bos, J.Mangus, N.Bellamy, A Comparison of Clutch Types for Naval Propulsion, ASMETurboExpo2018,TurbomachineryTechnicalConferenceandExhibitionGT 2018, Oslo, Norway, pp.1-8, 2018.
[20] V.N.Bhaiswar, J.P.Modak, M.P.Singh, Dynamic Analysis of Lobe Coupling by FormulationofanApproximateGeneralizedExperimentalDataBasedModel, International Journal of Scientific and Research Publications, Vol.01, Issue 01, pp.1- 5, 2011.
[21] Yonggang Liu, Datong Qin, Hong Jiang, Yi Zhang, A Systematic Model for Dynamics and Control of Dual Clutch Transmissions, Journal of Mechanical Design, Transactions of ASME, Vol.131, pp.061012, 1-10, 2009.
[22] Jignesh Patel, Kaushal Ajmera, Raghav Thanki, Rohit Maitar, Design and Theoretical Analysis of Single Plate Clutch by Varying Friction Lining Materials, International Journal of Advance Engineering and Research Development, Vol.2, Issue 11, pp.68- 72, 2015.
[23] Shakebuddin M, Khandelwal K.K, Langde A.M, Hussain H., Experimental Data Based Model for Process of Manufacturing Utensils Using Human Powered Flywheel Motor as an Energy Source, IOSR Journal of Mechanical and Civil Engineering (IOSR-JMCE) e-ISSN: 2278-1684, p- ISSN: 2320-334X, 14-16, 2014.
[24] M. Sohail Pervez, K. S. Zakiuddin. "Chapter 8Literature Review on the Developments of RiceMilling Machines", Springer Science and Business Media LLC, 2019.
[25] Harrsh Kumar Dubey, M.P. Singh, J.P. Modak, Rahul Makade. "A review on the advancement of human powered flywheel motor (HPFM) in India and its application for rural empowerment", Materials Today: Proceedings, 2020
[26] Yusuke Sugahara, Kensuke Kikui, MitsuruEndo, Jun Okamoto, Daisuke Matsuura, Yukio Takeda. "A human-powered joint drive mechanism using regenerative clutches", 2017IEEE/RSJ International Conference on Intelligent Robots and Systems (IROS), 2017.
[27] "Computing in Engineering and Technology", Springer Science and Business Media LLC,2020.
[28] "Advances in Mechanism and Machine Science", Springer Science and Business Media LLC, 2019
[29] Pranav R Morchapure, KS Zakiuddin, Nitin S Sawarkar, "Investigation of Flexible Clutches Over a Wide Range", Int. Journal of Innovations in Engg. and Science, Volume 5, No.10, pg.15-19, 2020. DOI: 10.46335/IJIES.2020.5.10.4
[30] N S Sawarkar, K S Zakiuddin, O Noman, P Palsodkar, "Analysis of Three Jaw Type Flexible Clutch", Zeichen Journal, Vol.6, Issue 11, pg.180-188, 2020.

Investigation of substitute jar materials for Laboratory grade ball milling machine to process electrode materials for energy storage devices

Sourabh Shinde[1*], Taukir Momin[2], Vispi Karkaria[3], Parshuram Karandikar[4]
[1]Dept. of Mechanical Engg., College of Engineering, Pune, India
[2]Dept. of Mechanical Engg., College of Engineering, Pune, India
[3]Dept. of Mechanical Engg., Northwestern University, IL, USA
[4]Dept. of Electronics and Telecommunication Engg. Army Institute of Technology, Pune, India

E-mail: shindesr19.mech@coep.ac.in

Abstract

Copious forms of energy are available in nature, but electrical energy is the convenient form of energy. As a result of this, it is expected that the need for electrical energy will increase considerably by the end of this decade. Thus, the storage of electrical energy is now becoming of paramount importance. Nevertheless, ultra-capacitors are currently a central area of research for energy storage devices due to their high-power density rating, short charging time and long cycling time. The capacitance of an ultra-capacitor is majorly a result of the processing of its electrode materials. Ball milling is one of the most profitable and cost-effective processes of electrode material processing. However, in most of the ball milling research, the focal point is on materials used for balls in ball milling. It is also observed that the material used to produce ball mill jars is of equal momentousness. So, this research aims to examine various materials as jars for a ball milling machine.

Keywords. Ultra-Capacitors, Capacitance, Ball Milling, Power density, Energy, Jars.

1. Introduction

To improve the performance of electric vehicles, it is critical to investigate the fundamental gaps in understanding the atomic and molecular level processes that govern their performance limits, operation, and failure [01-03].

There are two types of energy storage devices: Primary energy storage devices and Secondary energy storage devices: The chemical reaction in primary energy storage devices is irretrievable, and hence the entire device is required to be disposed of when the energy is discharged. Alternately, secondary energy storage devices are those which can be charged again and again as and when needed. Ultra-Capacitors is a new technology for secondary energy storage devices. Ultra-Capacitors can be an excellent option for tomorrow's energy storage devices. Ultra-Capacitor is an energy storage technology that offers high power density, high reliability, almost instant charging and discharging and very long lifetimes. Ultracapacitors have been in evolution for decades, but the technology has developed swiftly in recent years. This development has been driven by advances in nanomaterials [01-06]. Ultra-Capacitors are dealing with increased concerns around fuel efficiency and emissions in the automotive and transportation segments. Ultracapacitors are now delivering note-worthy economic benefits across a wide range of markets, including automotive, transportation, grid and renewables, and industrial applications. Three key components of an ultra-capacitor are electrodes, a dielectric separator, and electrolyte. The electrodes are commonly of three types: Metal oxides, polymers, and carbon. For obtaining high energy and power densities, the ions between the cathode and anode of ultra-capacitors should commute rapidly. To have high performance characteristics, the ultra-capacitor electrodes should meet the following standards [01]:

- Mechanically sturdy and chemically balanced electrode microstructure.
- Electron-ion transport path that is effective.
- Effective Electrochemical-active association between the electrode and the electrolyte.

Proper processing of ultra-capacitor electrode materials is very important. The finer the electrode material, the higher its capacitance rating. The increase in charge carrying capacity of the ultra-capacitor can only be observed when the surface area of the material is less than 3500 m^2/g, pore volumes of 3 cm^3/g and the pore sizes in the range of 1-2 nm [01]. There are various processes to grind electrode materials effectively. Ball milling is one of the most effective yet cheap manufacturing processes for the production of very fine electrode materials [07-12]. Most of the research revolves around the fact that only the balls of the ball milling machine play a vital role in the effectiveness of the ball milling process. However, it has been discovered that not only the ball mill ball material, but also the material used to make ball mill jars, has a significant impact on the effectiveness of the ball milling process. The common materials used for ball milling jars are stainless steel or high-grade steel. But, for small-scale applications, aluminium is also preferred as a material for ball milling jars. The problem with these materials is that the mass of these materials is excessively high, which results in an increased load on the motor of the ball milling jar. Moreover, the excess vibrations and noise are unavoidable during the operation. So, there is a need for substitute materials for manufacturing ball milling jars which can overcome all these drawbacks. As a result, three materials, viz., Polypropylene, Nylon, and Aluminum, are examined in this research as ball milling jar materials for small-scale applications. Special ball mill jars with steel layering are also suggested for extreme large-scale applications in this research.

The flow of the research paper is organised as follows,
Section 2 deals with the properties of electrode materials for ultra-capacitors. Section 3 illustrates ball milling of electrode materials for ultra-capacitors, section 4 examines various materials used to make ball milling jars, and section 5 deals with experimentation procedures for the research, following with conclusions and acknowledgement in section 6 and section 7 respectively.

2. Properties of electrode materials for ultra-capacitors

Activated carbon, metal oxides, and polymers are among the various types of electrode materials used for ultra-capacitors [13]. Activated carbon is one of the most readily available materials used as an electrode. Activated carbon is one of the most acceptable materials for making electrodes because of its good conductivity, high porosity, ease of availability, and thermal stability [13]. Pseudo capacitors are mostly manufactured out of metal oxides. A pseudo capacitor is part of an electrochemical capacitor, and forms together with an electric double-layer capacitor (EDLC) to create an ultra-capacitor. Nickle, Manganese, Magnesium, and cobalt oxides are more commonly used in the ultra-capacitor industry as electrode materials. The conductivity of these metal oxides depends on the crystallinity of the metal oxide. The higher the crystallinity of the material, the greater its conductivity [13]. Conducting polymers are also excellent as an electrode material in ultra-capacitors. Polymers for battery electrodes are made with the appropriate monomers and connectivity to meet these requirements. Some other conducting polymers used as electrodes in ultra-capacitors are polyaniline, polythiophene, and polypyrrole [13].

A mixture of manganese dioxide and activated carbon is used as an electrode material for research purposes. The mixture is made in the ratio of 1:1 by weight. As manganese dioxide is cheap and has an appropriate crystallinity, If the crystallinity of manganese oxide is very high, positive ion exchange will be limited. But this will lead to a lower surface area, and an increase in conductivity [14]. If there is low crystallinity, conductivity will be reduced, so there should be an adjustment between electrical conductivity and the surface area of the material. Activated carbon, on the other hand, is the most acceptable material to make electrodes for ultra-capacitors because of its excellent conductivity, small pore size, and large surface area due to its carbon microstructure [13].

Ideally, there are two commonly used electrode fabricating processes: Powder technology and film technology. Powder technology refers to a process in which electrode material granules are crushed into nano-sized powder. This powder is then adhered over the conducting core. The powder technology process is very easy and is extensively used for laboratory experiments. This powder of electrode materials can be produced through a variety of operations like crushing, ball milling, powder metallurgy, etc. But film technology is a complex process as compared to powder technology. Film technology involves chemical decomposition of the electrode material over the conducting core. A variety of chemical reactions happen to get electrode material over the conducting core. Powder technology is employed in the research to fabricate electrodes from manganese dioxide and activated carbon. Manganese dioxide is grinded in a special type of ball milling machine known as 'Laboratory grade ball milling machine'. This type of ball milling machine provides magnificent advantages for laboratory experiments.

3. Ball milling of electrode materials

A ball mill is a type of grinder used to crush materials into fine powder for use in pyrotechnics, mineral dressing processes, paints, ceramics, selective laser sintering, and electrode manufacturing. The ball mill works the principle of impact and enfeebled. The ball mill is one of the most effective yet ignored fields in the electrode manufacturing industry. A ball mill consists of a hollow jar where material to be crushed or ground is fed into the machine. This jar has several spherical balls made of various materials like steel, zirconia or even ceramics and rubber in some cases. These balls apply impact on the material to grind, and after a long period of time, very fine, nano-sized grinded materials are obtained. Ball milling machines are broadly classified into various types: Vertical Axis ball milling machines, Horizontal axis ball milling machines, Industrial ball milling machines, Planetary (Multi-jar) ball milling machines. The ball milling machine used in this research is a special type of ball milling machine which is designed as a combination of a horizontal axis and a planetary ball milling machine, termed as 'Laboratory grade ball milling machine'.

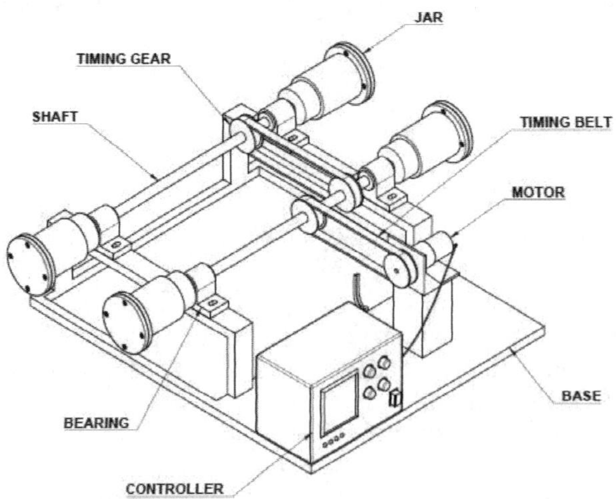

Figure 1. Laboratory grade ball milling machine [01]

Fig 1. shows a diagrammatic representation of a laboratory grade ball milling machine which is designed and manufactured by us. Four jars are accompanied in this case. The motor used is a 300 rpm, 100-Watt planetary DC gear motor. The controller enables the operator to control the speed of the motor and allows the operator to set a specific time for ball milling. Some of the many advantages of this type of ball milling machine are: It can crush materials in small and large quantities, which means that it is possible to build a ball mill machine of any capacity as needed. Moreover, the cost of installation is less as compared to other robust ball milling machines. Furthermore, it is suitable for grinding materials in small as well as large quantities, implying that a ball mill machine of any capacity can be manufactured as required. Also, the machine produces minimum noise during operation. Applications of this type of ball milling machine are:

- Electrode material processing for ultra-capacitors
- Electrode material processing for batteries and other energy storage devices
- Powder metallurgy
- Experimental use in laboratories, etc.

As a result of all these factors, the 'Laboratory grade ball milling machine' is used to crush manganese dioxide for ultra-capacitors. The jars of the ball milling machine are rotated at approximately 80 rpm and the cycle time is kept at 20 hrs [03]. The appropriate speed of ball milling is very important as more speed will cause high centrifugal forces and hence, material will adhere to the inner walls of the jar during ball milling, and hence the effectiveness of the process will be lost. But on the other hand, if the speed of ball milling is very low, very few centrifugal forces will be generated, so the material in the ball mill will experience less impact and enfeeble. A lot of research has already been done on various electrode materials for ultra-capacitors. But the processing of this electrode material is also an important aspect. Even though ball milling is a type of electrode processing technique, it is a neglected field of research. Hence, this research mainly focuses on the ball milling process for electrode material processing.

4. Materials for Ball milling jars

A wide range of materials are used to manufacture the jars of a ball milling machine. Austenitic stainless steel is the most used jar material for ball milling machines. Austenitic stainless steel has a face centred cubic structure. This structure is achieved by adding austenite stabilising elements like manganese, nickel, and nitrogen. Austenitic stainless steel has a superior ultimate tensile strength of 510 MPa. Aluminium is another metal used as a jar material for small-scale ball milling machines. Aluminium has a good ultimate tensile strength rating of 180-280 MPa. But aluminium is much lighter as compared to steel. Austenitic stainless steel has a mass density of 7800 kg/m^3 and, on the other hand, aluminium has a mass density of 2720 kg/m^3. Aluminium is approximately 30% lighter than austenitic stainless steel. But still, aluminium can not be used for many small-scale applications because it cannot absorb vibrations and noise. The leakage through ball milling jars made of aluminium is uncontrollable. Even more, 2720 kg/m^3 of mass density is even though it is less than aluminium, it is not tolerable for many applications of laboratory use. So, there is a need for lighter materials which can also absorb vibrations and noise. Composites can be used as a substitute for jar materials in ball milling machines to overcome all these problems. Composites, also known as Fibre-Reinforced Polymer (FRP) composites, are generated from a polymer matrix that is reinforced with a man-made or natural fibre (like glass, aramid, or carbon) or other reinforcing material. Nylon and Polypropylene are used to make ball milling jars in this research for small scale applications. Nylon is the most widely used synthetic substance; its applications are spread across a wide range of activities, from daily household activities to various industrial applications. Nylon is a silk-like thermoplastic, usually made from petroleum, that can be melt-processed into fibres, shapes, or films. Nylon polymers can be mixed with a huge variety of additives to achieve many different variations in their properties. Nylon has an ultimate tensile strength of 70 MPa and has a mass density of 1140 kg/m^3. Alternately, polypropylene is a thermoplastic polymer used in a wide variety of applications. It is generated through chain-growth polymerization from the monomer propylene. Polypropylene belongs to the group of polyolefins and is partly crystalline and non-polar. Its properties are like polyethylene, but it is somewhat harder and more heat resistant. Polypropylene has the lowest mass density of 855 kg/m^3 and a good ultimate tensile strength of 40 MPa. Table 1. shows the properties of the materials linked with the material cost for each selected material. These properties and the cost of material are taken by performing a market survey while purchasing material for jars. These jars made of Nylon and Polypropylene are compatible with only small to medium scale applications as the strength of these jars is not enough to sustain extreme large forces during the grinding or crushing at a considerably larger scale in a ball milling machine.

Table 1. Material properties and material cost

Material	Ultimate Tensile Strength	Density	Material Cost
Austenitic Stainless-Steel	510 MPa	7800 kg/m^3	2.04 \$/kg
Aluminum	280 MPa	2730 kg/m^3	1.98 \$/kg
Nylon	70 MPa	1140 kg/m^3	2.20 \$/kg
Polypropylene	40 MPa	855 kg/m^3	2.60 \$/kg

So, for large scale applications, there is a need for some strong materials which can not only withstand extremely high forces but also be cost-effective. Some of the carbon reinforced polymers are very strong, having an ultimate tensile strength in the range of 1000-1200 MPa. But the cost of these materials is exceedingly high, in the range of $25-30/kg. Moreover, these materials are very difficult to machine and only a few machining processes can be carried out on them, like forming, etc. This research also examines the use of a thin mild-steel layer at the inner periphery of the jars, with the rest of the jar being made of nylon or polypropylene. Through trial and error, the thickness of this mild-steel layer was found to be ideally more than 1.8 mm. For research purposes, the jars are specially designed to be compatible with laboratory grade ball milling machines. Jars are manufactured on the CNC lathe machine. Special care is taken to machine these materials as the temperature withstanding ability of these materials is not as high as steel and aluminium. So, the jars are turned at not more than 200-400 rpm.

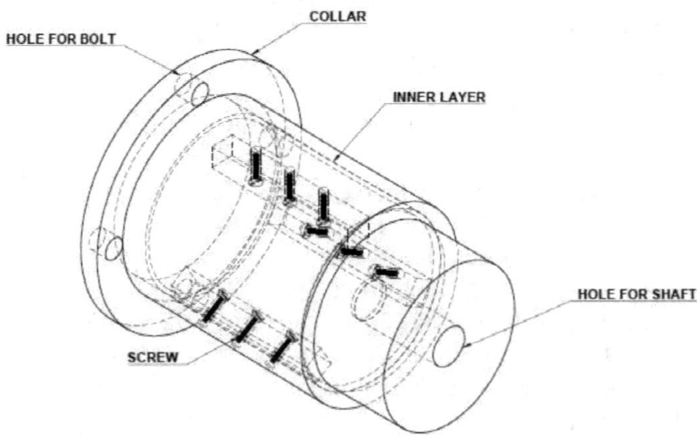

Figure 2. Ball milling jars for Laboratory grade ball milling machine

Fig 2. shows the design and labelled diagram of the ball milling jars used in research which is designed and manufactured by us. A lid is also manufactured to prevent the fall of the electrode materials from the jars during manufacturing. For special types of jars, the mild steel layer is press fitted into the previously turned Nylon and Polypropylene jars. The overall manufacturing cost of all these jars is shown in Table-2.

Table 2. Overall manufacturing cost for selected jars

Type of Jar	Overall manufacturing cost
Aluminum Jars	20.22 $

Nylon Jars	28.78 $
Polypropylene Jars	29.63 $
Nylon jars with mild-steel inner layer	39.91 $
Polypropylene jars with mild-steel inner layer	40.76 $

So, the cost of the jars will be increased by 15-20% for small-scale applications, but this will lead to the use of a considerably smaller motor and save electricity.

5. Experimentation

Ultra-capacitors are made from a mixture of crushed manganese dioxide and carbon in the ratio of 1:1 by weight. Fig. 3 shown here is the ultra-capacitor is made by us and the same is used to carry out the experimentation. First, manganese dioxide is crushed in a laboratory grade ball milling machine into a nano-sized powder. 50% of the wastage factor was also considered. Five samples of crushed manganese dioxide each from different types of jars, viz. Aluminium jars, Nylon jars, Polypropylene jars, Nylon jars with a mild-steel inner layer, and Polypropylene jars with a mild-steel inner layer. A slurry is formed from the mixture of crushed manganese dioxide and activated carbon with the use of an IPA solution. Slurry is then loaded over a steel mesh and then is allowed to dry in the proportion of 20 mg/cm^2. Polyethylene is used as a dielectric separator between the two electrodes. Following this process, five stacked type ultra-capacitors are made from the manganese dioxide crushed in each of five jars. The adhering agent used to paste the polyethylene sheets together is cyanoacrylate. These ultra-capacitors are now pressed tightly to ensure proper adhesion of the dielectric to prevent leakage. Potassium sulphate of molarity 0.65 is used as an electrolyte [15-20].

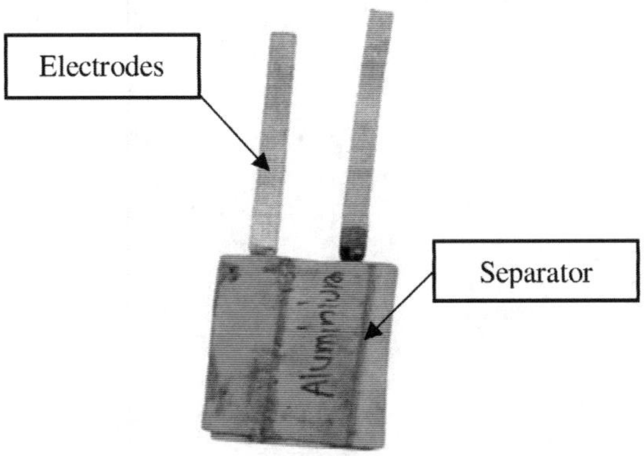

Figure 3. Stacked typed ultra-capacitor

The performance analysis of the electrodes of ultra-capacitors manufactured from each of the five jars can also be done using an electron beam microscope. The magnifying capacity of an electron beam microscope can be as large as 400-1900 times. These microscopes can be used to measure the size of ball milled product particles. But these microscopes are very delicate and expensive to work with. Nevertheless, an indirect approach is used to evaluate the performance ratings of all the ultra-capacitors manufactured from electrodes crushed in five different jars. In this indirect approach, ultra-capacitors are evaluated to determine the effectiveness of these ball milling jars made of five different materials. The capacitance rating of these ultra-capacitors is calculated using charge discharge tests. In this research, indirect approach turns out to be more beneficial than direct approach as indirect approach directly affiliates the application-based investigation of proposed ball mill machine jars and furthermore, indirect approach is cost effective and easy to apply. The performance of these ultra-capacitors manufactured from electrodes processed in Aluminium jars, Nylon jars, Polypropylene jars, Nylon jars with mild-steel inner layer, and Polypropylene jars with mild-steel inner layer was evaluated and noted. From fig. 5, and fig. 6, it can be observed that the specific capacitance values of ultra-capacitors made from electrodes crushed in jars made of Nylon and Propylene show better ratings as compared to ultra-capacitors made from the electrodes crushed in aluminium jars. Jars with a mild-steel inner layer show less capacitance rating because they are meant for large-scale ball milling. The equivalent resistances of Aluminium and Polypropylene jars are almost similar, with a difference of just 1.9 Ohms. Ideally, the equivalent series resistance of an ultra-capacitor should be as low as possible [21-23].

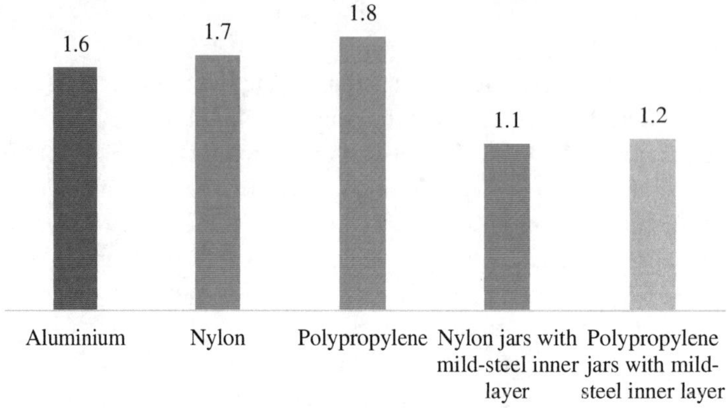

Figure 4. Specific Capacitance of ultra-capacitors with selected jar materials (Farad/gm)

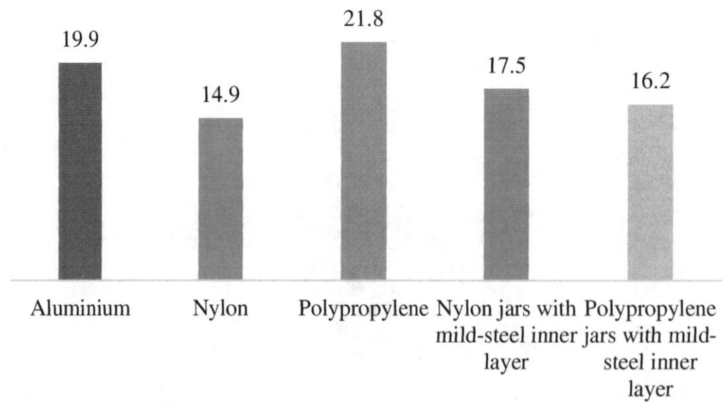

Figure 5. Equivalent series resistance of ultra-capacitors with selected jar materials (Ohms)

IOP Conf. Series: Materials Science and Engineering 1206 (2021) 012018 doi:10.1088/1757-899X/1206/1/012018

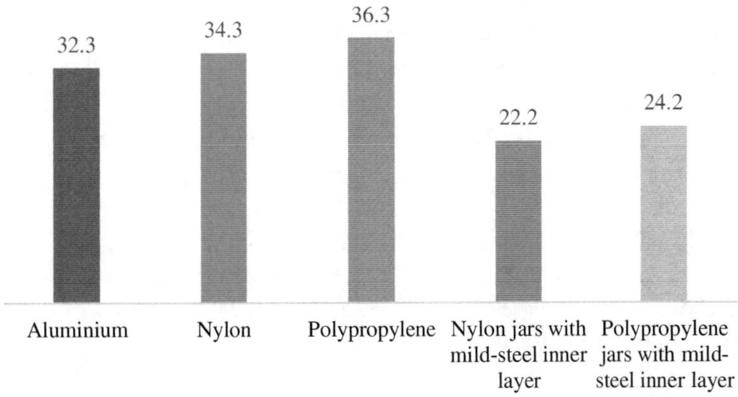

Figure 6. Energy density of ultra-capacitors with selected jar materials (Joule/gm)

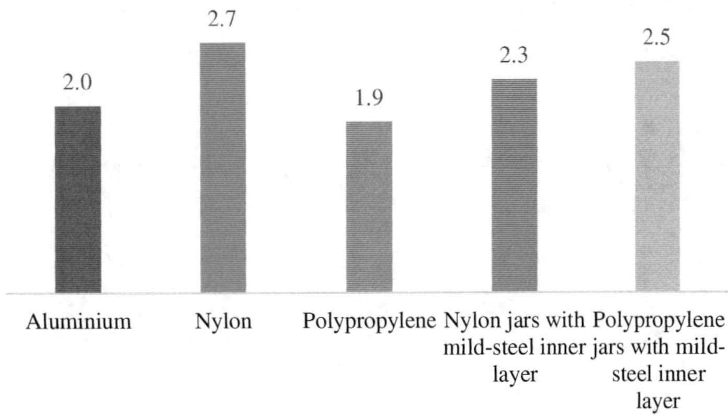

Figure 7. Power density of ultra-capacitors with selected jar materials (Watt/gm)

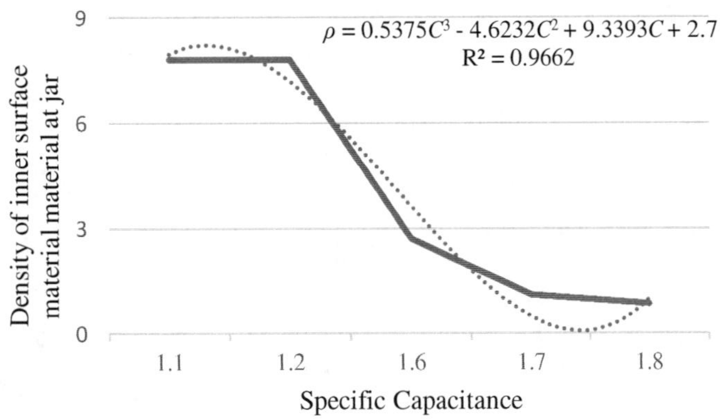

Figure 8. Correlation between the density of surface material at inner surface of ball milling jars (g/cm3) and specific capacitance of ultra-capacitors (Farads) manufactured from electrodes in this jar

From fig. 7 and fig. 8, it is evident that polypropylene is an excellent substitute for jar material for small-scale laboratory grade ball milling machines. But if the applications extend to a larger scale, i.e., for industrial applications, polypropylene will no longer be strong enough to withstand extremely high impact forces. Even aluminium will fail to withstand these forces for industrial scale applications. The only option in that case would be the use of steel. But steel has a very high density of 7.8 gm/cm^3. Therefore, the load on the motor will be increased. Due to this, an innovative idea of using a mild-steel inner layer for Nylon and Polypropylene is proposed in this research. The effectiveness of this jar material is very much equal to the jars made entirely of steel. Fig. 8 shows a curve depicting the correlation between the materials used for jars and the performance rating of the ultra-capacitors manufactured from the electrodes processed in these jars. An equation is derived to predict the specific capacitance of the ultra-capacitors based on the density of the jar material in which the electrodes of the ultra-capacitors were crushed. The equation derived is:

$$\rho = 0.5375C^3 - 4.6232C^2 + 9.3393C + 2.7 \quad (1)$$

Where, ρ – Density of material at the inner surface of ball milling jar (gm/cm^3).

C – Specific Capacitance of the ultra-capacitors manufactured from the electrodes crushed in these jars (Farad/gm).

The equation was derived by us using a small regression model for the purpose of curve fitting using (Fig. 8), considering all the jar materials taken into consideration. The efficiency of the equation for a given set of data was closely equal to 96.6%. Special care must be taken that specific capacitance values are always positive and hence, negative values after solving the equation must be neglected.

6. Conclusion

The research examines the performance of the ultra-capacitors based on the jar inner surface materials in which the electrode materials of ultra-capacitors are crushed. An indirect approach is portrayed in the research paper, where ball milled material is used to make electrodes for ultra-capacitors, and then the performance of these ultra-capacitors is compared and results are evaluated for five different types of jar materials: Aluminium, Nylon, Polypropylene, Nylon with mild-steel inner layer, and Polypropylene with a mild-steel inner layer. A laboratory-grade ball milling machine is used for experimental purposes. The ball milling was done at 80 rpm and with 20 hours of cycle time [03]. After the experiments, it was observed that the ultra-capacitors made from manganese dioxide and activated carbon grinded in polypropylene jars using stainless steel ball mill balls showed a better capacitance rating for small scale applications. But, for large-scale applications, Polypropylene jars with a mild-steel inner layer are recommended. These jars show performance ratings very similar to steel jars. Using the recommended jars, a few more advantages are observed:

- The overall weight of the ball milling jars is reduced by 69% with the use of Polypropylene and by 41.5% with the use of Nylon as jar materials for small scale applications as a substitute for Aluminium jars.
- Polypropylene and Nylon jars meant for low scale applications produce 25% less noise and vibrations.
- The power required for ball milling is reduced by 18%, so the load on the motor is also minimized.
- Cleaning of the jars can be easier. As the jar colour can be used in contrast with the colour of the material to be grind. This is because Polypropylene and Nylon are available in various colors.
- Aluminium jars are more prone to leakage. But Polypropylene and Nylon are composite materials. So, jars made from composite materials get compressed when the bolts on the lid are applied, minimising the leakage by 70%.

- Research also proposes composite jars with a mild-steel inner layer for large scale industrial applications.

The correlation curve between the 'density of surface material at the inner surface of ball milling jars and the specific capacitance of ultra-capacitors manufactured from electrodes in this jar' can be used as a tool to predict the effectiveness of a particular type of jar material. This can be an effective equation to decide excellent jar materials and thereby also evaluate the capacitance rating for ultra-capacitors whose electrodes are processed in this jar of material under consideration.

7. Acknowledgement

The authors of this paper highly acknowledge the support of the management of the Army Institute of Technology, Pune, and the College of Engineering, Pune for providing laboratory facilities during the experimentation. Last but not least, the authors also express their gratitude towards the workshop staff of the Army Institute of Technology for helping and guiding them wherever needed.

8. References

1. L S Godse, Vispi Karkaria, Dr P.B Karandikar, Dr N.R Kulkarni: Innovative Methods of Ball Milling to Grind Activated Carbon as an Electrode Material for Enhancing the Performance of Ultracapacitor. *International Conference on Energy, Communication, Data Analytics and Soft Computing*, pp. 2034-2041. IEEE, (2017).
2. He Liu, Xinbing Cheng, Yan Chong, Hong Yuan, Jia-Qi Huang, Qiang Zhang: Advanced electrode processing of lithium-ion batteries: A review of powder technology in battery fabrication. Particuology, **57**, pp. 56-71 (2021).
3. L.S. Godse, Dr. P.B. Karandikar, Dr. M.Y. Khaladkar: Study of Carbon Materials and Effect of Its Ball Milling, on Capacitance of Supercapacitor. *4th International Conference on Advances in Energy Research, Energy Procedia*, **54**, pp. 302 – 309, (2014).
4. Adnan Alamili, Yiqin Xue, Faith Anayi: An experimental and analytical study of the ultra-capacitor storage unit used in regenerative braking systems. *Applied Energy Symposium and Forum, Renewable Energy Integration with Mini/Microgrids, Energy Procedia* **159**, pp. 376-381, (2019).
5. Dan Li, Ziye Kang, Hao Sun, Ying Wang, Haiming Xie, Jia Liu, Jiefang Zhu: A bifunctional MnxCo3-xO4-decorated separator for efficient Li-LiI-O2 batteries: A novel strategy to promote redox coupling and inhibit redox shuttling. *Chemical Engineering Journal*, **428**, January (2022).
6. R Nandhini, P.A. Mini, B. Avinash, S.V. Nair, K.R.V. Subramanian: Supercapacitor electrodes using nanoscale activated carbon from graphite by ball milling. *Materials Letters*, **87**, pp. 165-168, (2012).
7. Gelines Moreno-Fernandez, Nicolas Boulanger, Andreas Nordenstrom, Artem Iakunkov, Alexandr Talyzin, Daniel Carriazo, Roman Mysyk: Ball-milling-enhanced capacitive charge storage of activated graphene in aqueous, organic, and ionic liquid electrolytes. *Electrochimica Acta*, **370**, (2021).
8. Karine R. R., Claire Mayer-Laigle, Bruno Piriou, Xavier Rouau: Comparative comminution efficiencies of rotary, stirred and vibrating ball-mills for the production of ultrafine biomass powders. *Energy*, **227**, (2021).
9. Sara I. Ahmad, Hicham Hamoudi, Janarthanan Ponraj, Khaled M. Youssef: In-situ growth of single-crystal plasmonic aluminumelithium-graphene nanosheets with a hexagonal platelet-like morphology using ball-milling. *Carbon,* **178**, pp. 657-665, (2021).
10. Yiru Zou, Chao Wang, Hanxiang Chen, Haiyan Ji, Qian Zhu, Wenshu Yang, Linlin Chen, Zhigang Chen, Wenshuai Zhu: Scalable and facile synthesis of V2O5 nanoparticles via ball milling for improved aerobic oxidative desulfurization. *ScienceDirect,* **6**, issue 2, pp. 169-175, (2021).

11. Trimurti L. Lambat, Sami H. Mahmood, Deeb Taher, Subhash Banerjee: Sulfamic acid catalyzed oxonium-ene reactions under ball milling conditions: Straightforward access to highly functionalized Oxabicyclo [3.2.1] octenes. *Current Research in Green and Sustainable Chemistry*, **4**, (2021).
12. Gaoce Han, Jize Yan, Zhen Guo, David Greenwood, James Marco, Yifei Yu: A review on various optical fibre sensing methods for batteries. *Renewable and Sustainable Energy Reviews*, **150**, (2021).
13. Priyanka Sharma, Vinod Kumar: Study of electrode and electrolyte material of supercapacitor. *Materials Today: Proceedings*, **33**, pp. 1573-1578, (2020).
14. Zhaowang Dong, Yang Xia, Xueyi Guo, Jinlong Zhao, Linfeng Jiang, Qinghua Tian, Yong Liu: Direct reduction of upgraded titania slag by magnesium for making low oxygen containing titanium alloy hydride powder. *Powder Technology*, **368**, pp. 160-169, (2020).
15. Shreyas Thombare: Dual Battery Fast Charging System for Electric Vehicles. Global Transitions Proceedings, August (2021).
16. Xiaolin Xu, Jinrui Nana, Ju Wang, Zepeng Gao: Estimate of super capacitor's dynamic capacity. *Energy Procedia*, **105**, pp. 2194 – 2200, (2017).
17. Haoxiang Zhang, Wenbin Niu, Shufen Zhang: Extremely stretchable, sticky and conductive double-network ionic hydrogel for ultra-stretchable and compressible supercapacitors. *Chemical Engineering Journal*, **387**, (2020).
18. Dacheng Zhang, Xiong Zhang, Xianzhong Sun, Haitao Zhang, Changhui Wang, Yanwei Ma: High performance supercapacitor electrodes based on deoxygenated graphite oxide by ball milling. *Electrochimica Acta*, **109**, pp. 874-880, (2013).
19. Zhuwei Lv, Yuqiang Yan, Chenchen Yuan, Bo Huang, Can Yang, Jiang Ma, Junqiang Wang, Lishan Huo, Zhiqiang Cui, Xunli Wang, Weihua Wang, Baolong Shen: Making Fe-Si-B amorphous powders as an effective catalyst for dye degradation by high-energy ultrasonic vibration. *Materials and Design*, **194**, (2020).
20. Ping Wang, Geng Zhang, Meng-Yu Li, Ya-Xia Yin, Jin-Yi Li, Ge Li, Wen-Peng Wang, Wen Peng, Fei-Fei Cao, Yu-Guo Guo: Porous carbon for high-energy density symmetrical supercapacitor and lithium-ion hybrid electrochemical capacitors. *Chemical Engineering Journal*, **375**, (2019).
21. Henry Kahimbi, Seok Bok Hong, MinHo Yang, Bong Gill Choi: Simultaneous synthesis of NiO/reduced graphene oxide composites by ball milling using bulk Ni and graphite oxide for supercapacitor applications. *Journal of Electroanalytical Chemistry*, **786**, pp. 14-19, (2017).
22. Jing Zhao, Botao Zhu, Guijin Yang, Yujun Fu, Yanna Lin, Jinyun Li: Vacuum annealed MnO2 ultra-thin nanosheets with oxygen defects for high performance supercapacitors. *Journal of Physics and Chemistry of Solids*, **150**, (2021).
23. Aranizadeh, A. Zaboli, O. Asgari Gashteroodkhani, B. Vahidi: Wind turbine and ultra-capacitor harvested energy increasing in microgrid using wind speed forecasting. *Engineering SciSSence and Technology, an International Journal*, **22**, Issue 5, pp. 1161-1167, (2019).

Finite Element Model for Free Vibration Analyses of FG-CNT Reinforced Composite Beams using Refined Shear Deformation Theories

Surojit Biswas[1] **and Priyankar Datta**[1,2]

[1] Mechanical Engineering Department, Jadavpur University, Kolkata-700032, India

[2] Corresponding author: priyankardatta.mech@jadavpuruniversity.in; priyomech@gmail.com

Abstract. The present article deals with the free vibration of functionally graded carbon nanotube reinforced composite (FG-CNTRC) beams employing various refined deformation theories and validates the accuracy and feasibility of these proposed theories. The theories involved are the first order shear deformation theory (FSDT) and other refined theories involving additional higher order terms. Carbon nanotubes (CNTs) are assumed to be oriented along the axis of the beam. Uniform and three types of different functionally graded (FG) distributions of CNTs through the thickness of the beam are considered. The rule of mixture is used to describe the effective material properties of the beams. The governing equations are derived using Hamilton's principle and solved using the finite element method (FEM). A FEM code is compiled in MATLAB considering a C^0 finite element. The influences of different key parameters such as CNT volume fraction, distribution type of CNTs, boundary conditions and slenderness ratio on the natural frequencies of FG-CNTRC beams are investigated. It can be concluded that the above-mentioned parameters have significant influence on the free vibration of the beam and the accuracy of the proposed refined theories is good.

Keywords - carbon nanotube, finite element method, functionally graded carbon nanotube reinforced composite, shear deformation theories

1. Introduction

The use of composite materials for weight-dependent applications in spacecrafts, marine and automotive has escalated exponentially over the last few decades. Shortly after CNTs were observed by Iijima [1], researchers have been continuously investigating its properties and also trying to improve its method of synthesis. CNTs have gained popularity as reinforcement in composites due to its extraordinarily high strength, stiffness and aspect ratio with lower density. The superior electrical [2], mechanical [3-4], optical [5] and thermal [6-7] properties of CNTs have drawn the attention of researchers over past three decades. Yu et al. [3] conducted tensile test on a set of 19 multiwalled carbon nanotubes (MWCNTs). They loaded each MWCNTs individually and measured the tensile strength which varied from 11 to 63 GPa to identify the elastic modulus (E) of the outermost layer from the individual stress-strain curves and the value ranged from 270 to 950 GPa. Salvetat et al. [4] experimentally determined that CNTs have extremely high strength, flexibility and resilience and the result is consistent with the theoretical predictions. Wong et al. [8] used a cantilever beam model in which silicon carbide (SiC) nanorods (NRs) and MWCNTs were bent using an atomic force microscope (AFM) tip. The MWCNTs found to be two times stiffer than SiC NRs and the toughness is found to be 5 to 10 times compared to SiC NRs. CNTs have thermal conductivity as high as diamond crystals [6]. Experimental observations found thermal conductance and thermal conductivity of a single-walled carbon nanotube (SWCNT) of length 2.6 μm and diameter 1.7 nm to be around 2.4 nW/K and 3500 $Wm^{-1}K^{-1}$ at room temperature [7]. An isolated CNT at room temperature is taken by Berber et al. [9] to determine the thermal conductivity of the same from molecular dynamics (MD) simulation. The simulation results indicated thermal conductivity of 6600 $Wm^{-1}K^{-1}$ which is exceptionally high like that of a high-purity diamond. Overney et al. [10] used a cantilever beam model to determine the elastic modulus of SWCNT and reported it to be 1,500 GPa which is 10 times greater than that of a similar beam made up of Iridium. Lu [11] used an empirical model to calculate moduli of SWCNTs and MWCNTs. Young's modulus, shear modulus and bulk modulus are found to be approximately 1 TPa, 0.45 TPa and 0.74 TPa, respectively. It is also observed that elastic properties are independent of size and shape of CNTs. The results confirmed that CNTs are even stronger than diamond. Hata et al. [12] successfully synthesized SWCNTs of 2.5 mm height using chemical vapor deposition (CVD). Using MD simulations, Arash et al. [13] demonstrated how Young's modulus of Poly methyl methacrylate (PMMA) increased 16 times when reinforced with CNTs compared to pure PMMA. Qian et al. [14] conducted a tensile test on polystyrene matrices to measure the influence of CNT reinforcement. Addition of 1 wt% MWCNTs improves elastic modulus by 36%–42% whereas the break stress improves around ~25%. Wuite and Adali [15] examined stresses developed in nanocomposite reinforced beams for multi scale model. Their results indicated that a small amount of CNT addition can lead to noteworthy increase in beam stiffness. CNTs can be reinforced along with other reinforcements like clay particles in hybrid composites to enhance its mechanical properties [16]. Due to overwhelming properties of CNTRC, many scholars have investigated its static and dynamic response. Kanu et al. [17] compared various applications, fabrication methods, cost and time of manufacturing of CNTRC. Jain et al. [18] presented various case studies to capture the widespread applications of CNTs in multiple areas like electronics, electrical, biomedical, agricultural, pollution control, etc.

Scientific community has shown huge interest to explore functionally graded materials (FGMs) since it was first proposed by Japanese scientists in the mid-1980s [19,20]. FGMs are a class of advanced materials, consisting of two (or more) different materials, fabricated in a specific way to have gradation of composition with dimensions [19]. This helps to achieve continuous variation of mechanical properties in a preferred direction. Kanu et al. [21] elaborately discussed about the various manufacturing processes of FGMs and its response during fracture. Recent developments in numerical and analytical methods to compute stress, vibration and buckling of FGMs have been reviewed. Drawing inspiration from the idea of FGMs, CNTs are embedded in an FG pattern which has resulted in a new class of composites i.e., FG-CNTRC. Beams are indispensable structural elements due to its wide application both in large and micro structures. They are commonly used in mechanical, civil, aerospace and marine structures. Various deformation theories are available to analyse dynamic behaviour of such structural elements. Understanding the feasibility and accuracy of various theories is of interest to both researchers and engineers. Nuttawit and Variddhi [22] analytically found solutions for vibration of CNTRC beams with elastic foundation and simply supported boundary. Different patterns of reinforcements are considered and the results are obtained employing multiple deformation theories. The result showed that the frequencies are inversely proportional to thickness ratios (L/h). Also, natural frequency is highest for the FG-X beam, eventually decreases in the order of UD, FG-V and O beams. The natural frequencies don't vary significantly with various deformation theories. Rokni et al. [23] gave an exponential CNT distribution along the length of the FG-CNTRC micro-beam which maximize the natural frequency for a pre-set weight percent (wt.%) of MWCNTs. Penna et al [24] considered free vibrations of FG-CNTRC nano-beams with geometric imperfection and nonlinearity under initial pretension force. Equation of motions is derived with the help of stress-driven nonlocal integral model (SDM) for different boundary conditions. Yas and Samadi [25] investigated the effects of CNT volume fraction and pattern, foundation stiffness, aspect ratios and edge conditions on natural frequency of FG-CNTRC Timoshenko beam. For laminated FG-CNTRC beams, Vo-Duy [26] studied free vibration using FSDT. The results revealed that 0° orientation of CNT for all layers produces the largest frequency. With addition of layers, the frequency decreases for FG-X distribution but increases for FG-O and FG-V. However, the lamination sequence has negligible effect on the frequency. Thermal environment is considered by Shen and Xiang [27] to carry out numerical studies on vibration of FG-CNTRC beams on Pasternak foundations with large deformations. Mohseni and Shakouri [28] employed Timoshenko beam theory to evaluate natural frequency of FG-CNTRC beams with variable thickness. Rokni et al. [29-30] developed FEM of a micro-beam to identify optimized CNT distribution for each vibration mode under different boundary conditions. A comparative study of nonlinear free vibration of FG-CNTRC beam employing FSDT and third-order shear deformation theory (TOSDT) was performed by Lin and Xiang [31]. Mohammadimehr et al. [32] scrutinized viscoelastically damped FG-CNTRC beams using MCST. The results revealed that micro composite beams reinforced with SWCNTs have higher natural frequency than that of MWCNTs. Kitipornchai et al. [33] presented free vibration of FG porous beams with graphene platelets (GPLs) as reinforcements. Formulations are derived using Timoshenko beam theory and the fundamental frequency is evaluated with the help of Ritz method. It is found that free vibrations can be most effectively improved by the symmetric and non-uniform distribution of both internal pores and GPLs. Momeni and Dehkordi [34] carried out frequency analysis on sandwich beam with flexible core and FG-CNT face sheets. Parametric studies of CNT volume fraction, types of CNT distribution, boundary conditions and geometry have been performed. Dabbagh et al. [35] introduced a new refined higher-order beam theory to study vibration of laminated hybrid nanocomposite beams. The beam is composed of polymer matrix reinforced with carbon fibres and CNTs. They presented both analytical and numerical solutions to the problem. Also, several recent research work can be found reporting the vibration control of both FG [36,37] structures and CNT reinforced [38-41] structures.

FG-CNTRC is considered to be a promising material because it can sustain at high pressures and in harsh conditions, resistant to wear and tear and relieves stress concentrations. Knowledge about the natural frequency of any structural element is a pre-requisite for design. Over the years FEM has been established as a robust numerical modelling tool for structural analysis. Also, for composite and FG structures several shear deformation theories have been developed to accurately predict the natural frequency and deformation of structures. Further, researchers have refined the available theories to improve the accuracy of computation, especially for thick beams, plates and shells. However, the literature review disclosed that most of the refined theories have the mathematical form which requires C^1 finite element modelling and eventually this increases the computation cost when implemented on large and complicated structures. In this article, the authors have presented some deformation theories by refining the FSDT which is suitable for structures with high aspect ratio and also requires less computational cost as the FE modelling can be done using C^0 finite elements. Various articles on free vibrations of different FG-CNTRC structures have been reviewed due to completeness of the topic. However, the present work is restricted to study the free vibration of FG-CNTRC beams only.

2. Mechanical Properties of FG-CNTRC Beam

The UD and FG distributions (pattern 'X', 'V' and 'O') of CNTs along the thickness of the beam (z-direction) are shown in the figure 1. We have considered a prismatic beam of length L and cross-section $b \times h$ defined along the x-, y- and z-axes, respectively as depicted in the figure 2. Density of CNTs remains unchanged while the volume fraction changes along the thickness of the beam. Irrespective of the type of distributions, the total amount of CNT volume fraction (V_{tcnt}) is same and distributed across the thickness based on the following equations [25].

$$\text{UD: } V_{cnt} = V_{tcnt} \tag{1}$$

$$\text{FG-X: } V_{cnt} = 4\frac{|z|}{h} V_{tcnt} \tag{2}$$

$$\text{FG-V: } V_{cnt} = (1 + \frac{2z}{h}) V_{tcnt} \tag{3}$$

$$\text{FG-O: } V_{cnt} = (2 - 4\frac{|z|}{h}) V_{tcnt} \tag{4}$$

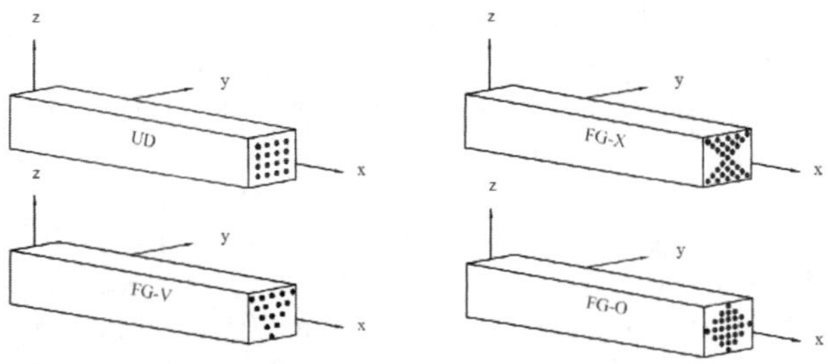

Figure 1. Configurations of the FG-CNTRC beams

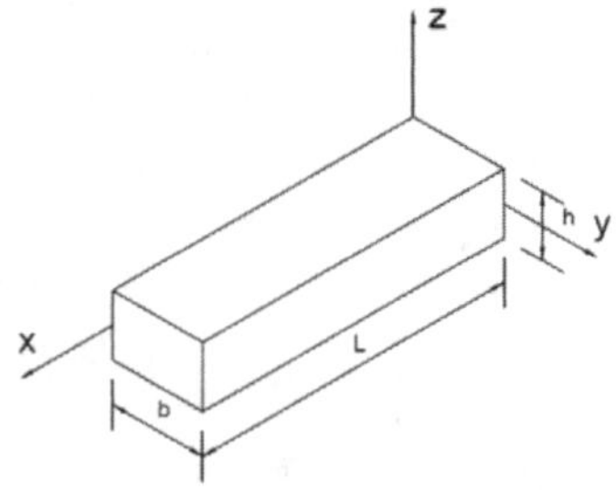

Figure 2. Reference coordinate system of the beam

Polymer matrix is reinforced with SWCNTs along the length direction such that there is no discontinuity at the fiber-matrix interface. In the present discussion, the isotropic matrix considered is PMMA having the following material properties: ν^m = 0.3, ρ^m =1190 kg/m^3, E^m = 2.5 GPa. It is reinforced with armchair (10,10) SWCNTs having material properties ν_{12}^{cnt}= 0.19, G_{12}^{cnt} = 17.2 GPa, E_{11}^{cnt} = 600 GPa, E_{22}^{cnt} = 10 GPa, ρ^{cnt} = 1400 kg/m^3. Effective material properties are obtained by using the rule of mixture. Effective Young's modulus (E_{11} and E_{22}), shear modulus (G_{12}), Poisson's ratios (ν_{12} *and* ν_{21}) and density (ρ) can be obtained as given below [25]:

$$E_{11} = \eta_1 V_{cnt} E_{11}^{cnt} + V_m E^m \tag{5}$$

$$\frac{\eta_2}{E_{22}} = \frac{V_{cnt}}{E_{22}^{cnt}} + \frac{V_m}{E^m} \tag{6}$$

$$\frac{\eta_3}{G_{12}} = \frac{V_{cnt}}{G_{12}^{cnt}} + \frac{V_m}{G^m} \tag{7}$$

$$V_m = 1 - V_{cnt} \tag{8}$$

$$\nu_{12} = V_{cnt}\nu_{12}^{cnt} + V_m \nu^m \tag{9}$$

$$\nu_{21} = \nu_{12}\frac{E_{22}}{E_{11}} \tag{10}$$

$$\rho = V_{cnt}\rho^{cnt} + V_m \rho^m \tag{11}$$

where E_{11}^{cnt}, E_{22}^{cnt} represent the Young's modulus of CNTs along the axial and the transverse direction, respectively; G_{12}^{cnt} is the shear modulus of CNTs. E^m and G^m denote Young's modulus and shear modulus of the matrix, respectively. ν_{12}^{cnt} and ν^m represent Poisson's ratio of CNT and matrix, respectively, and ρ^{cnt} and ρ^m are the respective mass densities. V_m denotes volume fraction of the matrix. CNT efficiency parameters η_j (j= 1,2,3) mentioned in equations (5) to (7) incorporates the size effect on the material properties and are calculated from the results of MD simulations. Value for the CNT efficiency parameters for some particular CNT volume fractions are known from the available literature. η_1= 1.2833, η_2 = 1.0556 for V_{tcnt} = 0.12; η_1= 1.3414, η_2 = 1.7101 for V_{tcnt} = 0.17; η_1 = 1.3238, η_2 = 1.7380 for V_{tcnt} = 0.28. Also, it has been assumed that η_3 = η_2 [25].

3. Theory and Formulation

The refined theories considered here includes ESDT, TSDT and HSDT. These theories have been developed by extrapolating the displacement field of FSDT. Unlike FSDT, these theories exclude shear correction factor (k) for computing the shear stresses. Under the assumption of these refined theories, displacement field for the FG-CNTRC beams can be expressed as

$$u(x,z,t) = u_0(x,t) + z\theta_x(x,t) + f(z)\phi_x(x,t) \quad (12)$$
$$v(x,z,t) = z\theta_y(x,t) + f(z)\phi_y(x,t) \quad (13)$$
$$w(x,t) = w_0(x,t) \quad (14)$$

where u_0 and w_0 denote change in position of any point on the mid-plane (z=0) along the x and z axes, respectively. θ_x and ϕ_x indicates the rotations of a transverse normal about the y axis, whereas, θ_y and ϕ_y are the rotations of the same about the x axis. The expressions for $f(z)$ in the equations (12-14) are assumed to be as follows:

$$f(z) = \begin{cases} 0, \text{ FSDT} \\ z\,exp[-2\left(\frac{z}{h}\right)^2], \text{ ESDT} \\ \frac{h}{\pi} sin\frac{\pi z}{h}, \text{ TSDT} \\ z\,cos\,h\left(\frac{1}{2}\right) - h\,sin\,h\left(\frac{z}{h}\right), \text{ HSDT} \end{cases} \quad (15)$$

We have assumed k =5/6 for FSDT and k =1 for the refined theories. The reference coordinate system is located on the mid-plane (z=0) in such a way that L and b units along x and y axes, respectively define the extent of the beam. Top and the bottom surfaces have z coordinates $\frac{h}{2}$and $-\frac{h}{2}$, respectively. For the ease of computation, translational and rotational component of the generalized displacement variables are arranged into two vectors $\{d_t\}$and $\{d_r\}$, respectively.

$$\{d_t\}= [u_0 \quad w_0]^T \quad \text{and} \quad \{d_r\}= [\theta_x \quad \theta_y \quad \phi_x \quad \phi_y]^T \quad (16)$$

Generalized displacement vectors $\{d_t\}$ and $\{d_r\}$ can be interpolated from the generalized nodal displacement vectors $\{d_t^e\}$ and $\{d_r^e\}$with the help of shape functions $[N_t]$ and $[N_r]$ as follows:

$$\{d_t\}= [N_t]\{d_t^e\} \quad \text{and} \quad \{d_r\}= [N_r]\{d_r^e\} \quad (17)$$

in which

$$\{d_t\} = [u_{01} \quad w_{01} \quad u_{02} \quad w_{02} \quad u_{03} \quad w_{03}]^T$$
$$\{d_r\}= [\theta_{x1} \quad \theta_{y1} \quad \phi_{x1} \quad \phi_{y1} \quad \theta_{x2} \quad \theta_{y2} \quad \phi_{x2} \quad \phi_{y2} \quad \theta_{x3} \quad \theta_{y3} \quad \phi_{x3} \quad \phi_{y3}]^T \quad (18)$$

$[N_t]$ and $[N_r]$ have been explicitly expressed in Appendix A. The state of strain at any point within the beam have been grouped into two vectors $\{\varepsilon_b\}$ and $\{\varepsilon_s\}$, which denote strains due to bending and strains due to shear, respectively. ε_{xx} denotes the normal strains along the x- direction, γ_{xy} is the in-plane shear strain and γ_{xz} and γ_{yz} are the transverse shear strains.

$$\{\varepsilon_b\} = [\varepsilon_{xx}\,\gamma_{xy}]^T \quad \text{and} \quad \{\varepsilon_s\} = [\gamma_{xz}\,\gamma_{yz}]^T \quad (19)$$

Using the displacement field from equations (12-14) and assuming linear strain–displacement relations, strains can be expressed as

$$\{\varepsilon_b\} = [B_{tb}]\{d_t^e\} + [Z_1][B_{rb}]\{d_r^e\}$$
$$\{\varepsilon_s\} = [B_{ts}]\{d_t^e\} + [Z_2][B_{rs}]\{d_r^e\} \quad (20)$$

Elaborated form of $[B_{tb}]$, $[B_{rb}]$, $[B_{ts}]$, $[B_{rs}]$, $[Z_1]$ and $[Z_2]$ mentioned in equation (20) have been shown in the Appendix A. The state of stress at any point can be expressed with the help of following two stress vectors:

$$\{\sigma_b\} = [\sigma_{xx} \quad \sigma_{xy}]^T \quad \text{and} \quad \{\sigma_s\} = [\sigma_{xz} \quad \sigma_{yz}]^T \quad (21)$$

where σ_{xx} is the normal stress along the x- direction, σ_{xy} is the in-plane while σ_{xz} and σ_{yz} are the out of plane shear stresses. Let $[C_b]$ and $[C_s]$ be bending and transverse shear elastic coefficients, respectively. The constitutive relations are given by

$$\{\sigma_b\} = [C_b]\{\varepsilon_b\} \quad \text{and} \quad \{\sigma_s\} = [C_s]\{\varepsilon_s\} \quad (22)$$

where $[C_b]$ and $[C_s]$ have their usual form.
The governing equation is derived using the dynamic version of the principle of virtual work, which is given by

$$\int_0^T (\delta U + \delta V - \delta K) = 0 \tag{23}$$

where U, V *and* K denote the strain energy, work done by the applied force and the kinetic energy, respectively. Using Virtual work principle on strain energy,

$$\delta U = b \int_{-h/2}^{h/2} \int_0^{l_e} \delta\{\epsilon_b\}^T\{\sigma_b\} + \delta\{\epsilon_s\}^T\{\sigma_s\}\, dx\, dz \tag{24}$$

Substituting the value from equation (20) and (22) into equation (24), we get

$$= b \int_{-h/2}^{h/2} \int_0^{l_e} [\, \delta\{d_t^e\}^T([K_{tb}^e] + [K_{ts}^e])\{d_t^e\} + \delta\{d_t^e\}^T([K_{trb}^e] + [K_{trs}^e])\{d_r^e\} +$$
$$\delta\{d_r^e\}^T([K_{trb}^e]^T + [K_{trs}^e]^T)\{d_t^e\} + \delta\{d_r^e\}^T([K_{rrb}^e] + [K_{rrs}^e])\{d_r^e\}\,]\, dx\, dz \tag{25}$$

The elemental stiffness matrix appearing in equation (25) are given by

$$[K_{tb}^e]=[B_{tb}]^T[C_b][B_{tb}]\,;\; [K_{ts}^e]=[B_{ts}]^T[C_s][B_{ts}];$$
$$[K_{trb}^e]=[B_{tb}]^T[C_b][Z_1][B_{rb}]\,;\; [K_{trs}^e]=[B_{ts}]^T[C_s][Z_2][B_{rs}]\,;$$
$$[K_{rrb}^e]=[B_{rb}]^T[Z_1]^T[C_b][Z_1][B_{rb}]\,;\; [K_{rrs}^e]=[B_{rs}]^T[Z_2]^T[C_s]\,[Z_2][B_{rs}] \tag{26}$$

For loading along z-direction, the elemental load vector $\{F^e\}$ is computed using virtual work done by the applied force and is given by

$$\{F^e\} = \int_0^{l_e} [N_t]^T q_0\, [0\ \ 1]^T\ \ dx \tag{27}$$

whereas the elemental mass matrix $[M^e]$ is computed using virtual kinetic energy and is expressed by

$$[M^e] = b \int_{-h/2}^{h/2} \int_0^{l_e} \rho\, [N_t]^T[N_t] dx\, dz \tag{28}$$

Substituting the value from equations (25), (27) and (28) into equation (23), we get the elemental governing equation of motion as follows:

$$[M^e]\{\ddot{d}_t^e\} + [K_{tt}^e]\{d_t^e\} + [K_{tr}^e]\{d_r^e\} = \{F^e\}$$
$$[K_{tr}^e]^T\{d_t^e\} + [K_{rr}^e]\{d_r^e\} = \{0\} \tag{29}$$

where, $[K_{tt}^e] = [K_{tb}^e] + [K_{ts}^e]$; $[K_{tr}^e] = [K_{trb}^e] + [K_{trs}^e]$; $[K_{rr}^e] = [K_{rrb}^e] + [K_{rrs}^e]$.

After assembling all elemental matrices, we get the global equation of motion

$$[M]\{\ddot{X}_t\} + [K_{tt}]\{X_t\} + [K_{tr}]\{X_r\} = \{F\}$$
$$[K_{tr}]^T\{X_t\} + [K_{rr}]\{X_r\} = \{0\} \tag{30}$$

where $[M]$ indicates the global mass matrix; $[K_{tt}]$, $[K_{tr}]$ and $[K_{rr}]$ are the global stiffness matrices; $\{X_t\}$ and $\{X_r\}$are the global generalized nodal displacement vectors corresponding to $\{d_t^e\}$ and $\{d_r^e\}$, respectively and $\{F\}$is the global nodal load vector. The stiffness matrix can be obtained by setting the time derivative terms to zero in the above equation. For free vibration, the eigen values of $[M]^{-1}[K]$ depicts square of circular natural frequency, ω^2.

4. Numerical Results and Discussion

A MATLAB code is developed for the present numerical analyses. Four different deformation theories i.e., FSDT, TSDT, ESDT and HSDT have been used to deal with the free vibration of FG-CNTRC beams. Several numerical results have been generated to validate the accuracy and feasibility of these proposed refined theories. There is not much variation in the results obtained using different theories. Effects of key parameters like CNT volume fraction, its distribution, *L/h* ratio and boundary conditions have been presented elaborately. Results obtained from the FEM indicated that these parameters strongly influence the free vibrations. The results are in well agreement with the published results. Results are obtained in terms of non-dimensional frequency parameters $\bar{\Omega} = \omega L\sqrt{\rho^m[1-(\nu^m)^2]/E^m}$.

4.1. Validation of Present Theories

Before studying the various trends in the vibration, we must establish the validity of the present FE model so that the results obtained can be presented accurately. First three non-dimensional natural frequency parameters ($\bar{\Omega}$) obtained using the present theories are compared for various boundary conditions with the available results for *L/h*=15. The results are tabulated in table 1 and table 2. It can be inferred that the results obtained using these theories are in well agreement with the results in the published articles. Shi et al. [42] gave an exact solution whereas Yas and Samadi [25] assumed Timoshenko beam theory. Deformation theories considered in the present article and the reference articles are different.

Table 1. First three non-dimensional natural frequencies of (S-S) FG-CNTRC beam (*L/h*=15).

V_{CNT}	CNT Distribution Pattern	Present FSDT			Present TSDT			Present ESDT			Present HSDT			Shi [41]		
		$\bar{\Omega}_1$	$\bar{\Omega}_2$	$\bar{\Omega}_3$	$\bar{\Omega}_1$	$\bar{\Omega}_2$	$\bar{\Omega}_3$	$\bar{\Omega}_1$	$\bar{\Omega}_2$	$\bar{\Omega}_3$	$\bar{\Omega}_1$	$\bar{\Omega}_2$	$\bar{\Omega}_3$	$\bar{\Omega}_1$	$\bar{\Omega}_2$	$\bar{\Omega}_3$
0.28	UD	1.4002	4.0559	6.8192	1.4014	4.0823	6.9286	1.4018	4.0856	6.9378	1.4012	4.0797	6.9202	1.3981	4.0500	6.8086
	FG-X	1.6104	4.3969	7.1993	1.5819	4.2909	7.0863	1.5821	4.2937	7.0959	1.5823	4.2900	7.0789	1.6086	4.3928	7.1913
	FG-V	1.2165	3.7799	6.6057	1.2136	3.7745	6.6431	1.2138	3.7769	6.6511	1.2136	3.7730	6.6363	1.3639	3.7703	6.6307

Table 2. First three non-dimensional natural frequencies of (C-C) FG-CNTRC beam (L/h=15).

V_{CNT}	CNT Distribution Pattern	Present FSDT			Present TSDT			Present ESDT			Present HSDT			Yas [25]		
		$\bar{\Omega}_1$	$\bar{\Omega}_2$	$\bar{\Omega}_3$	$\bar{\Omega}_1$	$\bar{\Omega}_2$	$\bar{\Omega}_3$	$\bar{\Omega}_1$	$\bar{\Omega}_2$	$\bar{\Omega}_3$	$\bar{\Omega}_1$	$\bar{\Omega}_2$	$\bar{\Omega}_3$	$\bar{\Omega}_1$	$\bar{\Omega}_2$	$\bar{\Omega}_3$
0.12	UD	1.4828	3.0982	4.9512	1.5015	3.1675	5.0943	1.5027	3.1713	5.1013	1.5004	3.1641	5.0876	1.5085	3.1353	4.9979
	FG-X	1.5779	3.2293	5.1085	1.6048	3.3164	5.2788	1.6065	3.3208	5.2863	1.6033	3.3123	5.2712	1.6000	3.2629	5.1514
	FG-V	1.3792	2.9597	4.7863	1.3928	3.0183	4.9155	1.3938	3.0216	4.9219	1.3920	3.0153	4.9092	1.4068	2.9997	4.8363
	FG-O	1.2884	2.8328	4.6309	1.2607	2.7712	4.5450	1.2604	2.7720	4.5485	1.2615	2.7720	4.5438	1.3180	2.8762	4.6840
0.17	UD	1.8783	3.9641	6.3631	1.8989	4.0440	6.5318	1.9003	4.0485	6.5404	1.8977	4.0400	6.5236	1.9144	4.0187	6.4348
	FG-X	2.0175	4.1600	6.6066	2.0331	4.2306	6.7595	2.0350	4.2358	6.7690	2.0316	4.2259	6.7505	2.0498	4.2111	6.6753
	FG-V	1.7344	3.7732	6.1387	1.7474	3.8348	6.2815	1.7485	3.8387	6.2893	1.7466	3.8314	6.2741	1.7721	3.8312	6.2139
	FG-O	1.6104	3.5946	5.9187	1.5855	3.5379	5.8418	1.5854	3.5391	5.8462	1.5862	3.5384	5.8401	1.6500	3.6565	5.9970
0.28	UD	2.1211	4.3866	6.9766	2.1520	4.4965	7.1988	2.1540	4.5022	7.2090	2.1504	4.4912	7.1886	2.1618	4.4556	7.0745
	FG-X	2.2772	4.6338	7.3057	2.2472	4.6238	7.3439	2.2490	4.6299	7.3555	2.2461	4.6188	7.3333	2.3169	4.7051	7.4093
	FG-V	2.0064	4.2662	6.8715	2.0174	4.3319	7.0282	2.0189	4.3371	7.0384	2.0162	4.3272	7.0184	2.0504	4.3414	6.9783
	FG-O	1.8823	4.0959	6.6638	1.8809	4.1155	6.7299	1.8820	4.1200	6.7397	1.8803	4.1122	6.7216	1.9284	4.1740	6.7728

4.2. Effect of CNT Distribution

Several numerical results are presented in table 3 to table 6. First three non-dimensional frequencies are obtained for UD, FG-X, FG-V and FG-O beams for various CNT volume fractions and boundary conditions. It is observed that FG-X produces the maximum frequency, whereas minimum frequency is obtained with FG-O. This trend is obtained for all vibration modes, irrespective of CNT volume fraction and boundary conditions. This happens as the pattern of CNT along the thickness direction also distributes the stiffness in a graded manner. The FG-X distribution has higher CNT concentration near the top and bottom surface. Thus, the stiffness is low near mid plane and high near the top and bottom surfaces which increases resistance against bending and the natural frequency also goes up. Since the FG-O pattern is reverse of FG-X, it gives lowest natural frequency. It can be concluded that symmetric distributions of CNTs are more effective in increasing or decreasing the frequency compared to uniform and unsymmetric distributions.

Table 3. First three non-dimensional natural frequencies of FG-CNTRC beams (L/h= 15) using FSDT.

V_{CNT}	CNT Distribution Pattern	Clamped-Free			Clamped-Clamped			Simply-Supported		
		$\bar{\Omega}_1$	$\bar{\Omega}_2$	$\bar{\Omega}_3$	$\bar{\Omega}_1$	$\bar{\Omega}_2$	$\bar{\Omega}_3$	$\bar{\Omega}_1$	$\bar{\Omega}_2$	$\bar{\Omega}_3$
0.12	UD	0.3633	1.6677	3.6146	1.4828	3.0982	4.9512	0.9458	2.8207	4.8133
	FG-X	0.4278	1.8194	3.8316	1.5779	3.2293	5.1085	1.0858	3.0353	5.0198
	FG-V	0.3081	1.5142	3.3866	1.3792	2.9597	4.7863	0.8178	2.5880	4.5748
	FG-O	0.2705	1.3926	3.1955	1.2884	2.8328	4.6309	0.7274	2.4001	4.3645
0.17	UD	0.4430	2.0927	4.5912	1.8783	3.9641	6.3631	1.1635	3.5563	6.1518
	FG-X	0.5248	2.3028	4.9042	2.0175	4.1600	6.6066	1.3466	3.8642	6.4674
	FG-V	0.3734	1.8858	4.2798	1.7344	3.7732	6.1387	0.9983	3.2365	5.8144
	FG-O	0.3272	1.7249	4.0192	1.6104	3.5946	5.9187	0.8850	2.9847	5.5195
0.28	UD	0.5437	2.4123	5.1587	2.1211	4.3866	6.9766	1.4002	4.0559	6.8192
	FG-X	0.6411	2.6506	5.5277	2.2772	4.6338	7.3057	1.6104	4.3969	7.1993
	FG-V	0.4610	2.2189	4.9117	2.0064	4.2662	6.8715	1.2165	3.7799	6.6057
	FG-O	0.4054	2.0485	4.6495	1.8823	4.0959	6.6638	1.0847	3.5198	6.3240

Table 4. First three non-dimensional natural frequencies of FG-CNTRC beams (L/h= 15) using TSDT.

V_{CNT}	CNT Distribution Pattern	Clamped-Free			Clamped-Clamped			Simply-Supported		
		$\bar{\Omega}_1$	$\bar{\Omega}_2$	$\bar{\Omega}_3$	$\bar{\Omega}_1$	$\bar{\Omega}_2$	$\bar{\Omega}_3$	$\bar{\Omega}_1$	$\bar{\Omega}_2$	$\bar{\Omega}_3$
0.12	UD	0.3637	1.6832	3.6722	1.5015	3.1675	5.0943	0.9465	2.8354	4.8780
	FG-X	0.4288	1.8432	3.9078	1.6048	3.3164	5.2788	1.0880	3.0615	5.1114
	FG-V	0.3083	1.5249	3.4333	1.3928	3.0183	4.9155	0.8179	2.5973	4.6252
	FG-O	0.2692	1.3681	3.1255	1.2607	2.7712	4.5450	0.7199	2.3401	4.2283
0.17	UD	0.4435	2.1094	4.6565	1.8989	4.0440	6.5318	1.1641	3.5715	6.2222
	FG-X	0.5247	2.3152	4.9564	2.0331	4.2306	6.7595	1.3435	3.8628	6.5135
	FG-V	0.3735	1.8954	4.3270	1.7474	3.8348	6.2815	0.9981	3.2427	5.8603
	FG-O	0.3262	1.7033	3.9549	1.5855	3.5379	5.8418	0.8787	2.9300	5.3884
0.28	UD	0.5445	2.4387	5.2517	2.1520	4.4965	7.1988	1.4014	4.0823	6.9286
	FG-X	0.6357	2.6177	5.4756	2.2472	4.6238	7.3439	1.5819	4.2909	7.0863
	FG-V	0.4608	2.2264	4.9580	2.0174	4.3319	7.0282	1.2136	3.7745	6.6431
	FG-O	0.4050	2.0459	4.6555	1.8809	4.1155	6.7299	1.0816	3.5005	6.3011

Table 5. First three non-dimensional natural frequencies of FG-CNTRC beams (L/h= 15) using ESDT.

V_{CNT}	CNT Distribution Pattern	Clamped-Free			Clamped-Clamped			Simply-Supported		
		$\bar{\Omega}_1$	$\bar{\Omega}_2$	$\bar{\Omega}_3$	$\bar{\Omega}_1$	$\bar{\Omega}_2$	$\bar{\Omega}_3$	$\bar{\Omega}_1$	$\bar{\Omega}_2$	$\bar{\Omega}_3$
0.12	UD	0.3638	1.6843	3.6758	1.5027	3.1713	5.1013	0.9467	2.8374	4.8839
	FG-X	0.4289	1.8449	3.9124	1.6065	3.3208	5.2863	1.0885	3.0646	5.1188
	FG-V	0.3083	1.5257	3.4364	1.3938	3.0216	4.9219	0.8181	2.5989	4.6303
	FG-O	0.2692	1.3678	3.1257	1.2604	2.7720	4.5485	0.7197	2.3391	4.2280
0.17	UD	0.4435	2.1106	4.6608	1.9003	4.0485	6.5404	1.1644	3.5738	6.2291
	FG-X	0.5248	2.3169	4.9617	2.0350	4.2358	6.7690	1.3440	3.8661	6.5220
	FG-V	0.3736	1.8963	4.3306	1.7485	3.8387	6.2893	0.9982	3.2443	5.8660
	FG-O	0.3261	1.7031	3.9555	1.5854	3.5391	5.8462	0.8786	2.9293	5.3888
0.28	UD	0.5446	2.4405	5.2573	2.1540	4.5022	7.2090	1.4018	4.0856	6.9378
	FG-X	0.6357	2.6194	5.4813	2.2490	4.6299	7.3555	1.5821	4.2937	7.0959
	FG-V	0.4608	2.2277	4.9628	2.0189	4.3371	7.0384	1.2138	3.7769	6.6511
	FG-O	0.4050	2.0468	4.6594	1.8820	4.1200	6.7397	1.0817	3.5019	6.3070

Table 6. First three non-dimensional natural frequencies of FG-CNTRC beams (*L/h*= 15) using HSDT.

V_{CNT}	CNT Distribution Pattern	Clamped-Free			Clamped-Clamped			Simply-Supported		
		$\bar{\Omega}_1$	$\bar{\Omega}_2$	$\bar{\Omega}_3$	$\bar{\Omega}_1$	$\bar{\Omega}_2$	$\bar{\Omega}_3$	$\bar{\Omega}_1$	$\bar{\Omega}_2$	$\bar{\Omega}_3$
0.12	UD	0.3637	1.6823	3.6690	1.5004	3.1641	5.0876	0.9463	2.8339	4.8728
	FG-X	0.4287	1.8418	3.9036	1.6033	3.3123	5.2712	1.0876	3.0588	5.1046
	FG-V	0.3083	1.5242	3.4306	1.3920	3.0153	4.9092	0.8178	2.5962	4.6209
	FG-O	0.2693	1.3689	3.1268	1.2615	2.7720	4.5438	0.7203	2.3423	4.2312
0.17	UD	0.4434	2.1084	4.6528	1.8977	4.0400	6.5236	1.1640	3.5698	6.2163
	FG-X	0.5247	2.3138	4.9518	2.0316	4.2259	6.7505	1.3433	3.8605	6.5062
	FG-V	0.3735	1.8947	4.3241	1.7466	3.8314	6.2741	0.9980	3.2416	5.8556
	FG-O	0.3262	1.7039	3.9559	1.5862	3.5384	5.8401	0.8790	2.9319	5.3910
0.28	UD	0.5445	2.4372	5.2466	2.1504	4.4912	7.1886	1.4012	4.0797	6.9202
	FG-X	0.6357	2.6168	5.4714	2.2461	4.6188	7.3333	1.5823	4.2900	7.0789
	FG-V	0.4608	2.2254	4.9538	2.0162	4.3272	7.0184	1.2136	3.7730	6.6363
	FG-O	0.4050	2.0455	4.6528	1.8803	4.1122	6.7216	1.0817	3.5002	6.2973

4.3. Effect of CNT Volume Fraction

It is evident from figure 3 and figure 4 that vibration frequency increase with increase in CNT volume fraction. This effect is more pronounced for higher modes of vibration. This trend can be observed irrespective of distribution and boundary conditions. The results indicate that addition of CNT fibres increases the stiffness of the beam significantly.

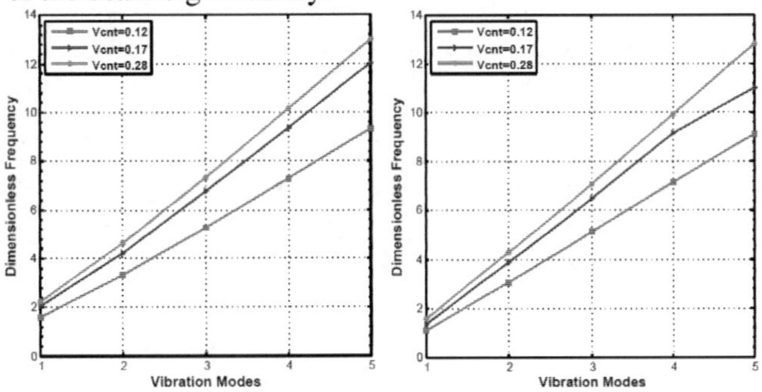

(a) (b)

Figure 3. First five non-dimensional natural frequencies of FG-CNTRC beams (*L/h*= 15) with various CNT volume fraction using TSDT. The CNT distribution is FG-X. Two different boundary conditions are considered: (a) C-C (b) S-S.

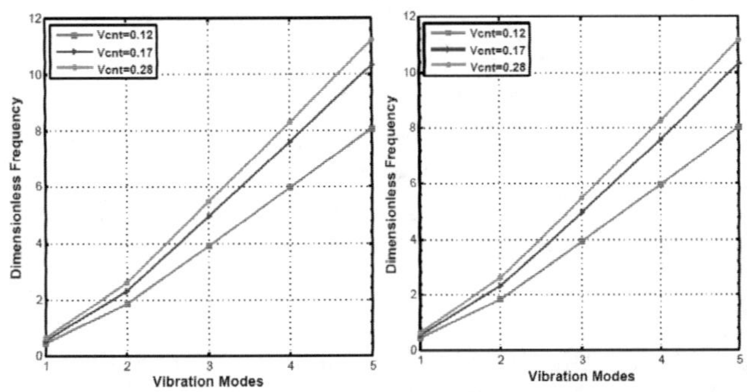

(a) (b)

Figure 4. First five non-dimensional natural frequencies of FG-CNTRC beams (*L/h*= 15) with various CNT volume fraction. The CNT distribution is FG-X and the boundary condition is C-F. Two deformation theories are considered: (a) ESDT (b) HSDT.

4.4. Effect of Boundary Conditions

Boundary conditions play a major role in free vibration. figure 5 (a) illustrates the variation of frequency due to boundary conditions using ESDT. The data showed that maximum frequency is obtained for C-C boundary conditions and the minimum frequency is obtained for C-F boundary conditions. However, for higher modes of vibration, there is no significant difference in frequency for C-C and S-S boundary conditions. As similar trends are observed with other deformation theories, they are not shown here. It can be inferred from figure 5 (b) that boundary conditions strongly influence the first mode of vibration both for thin and thick beams.

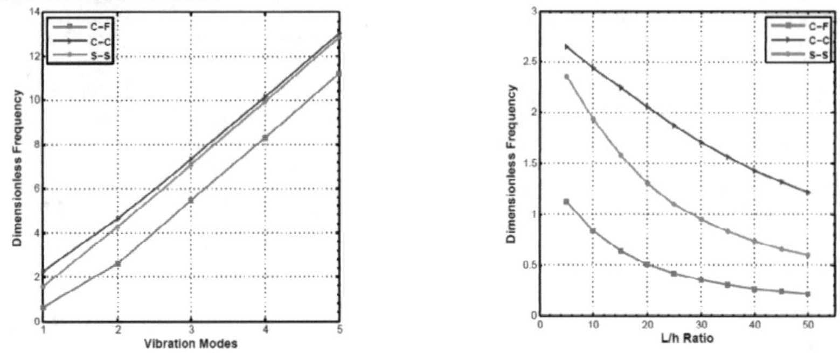

(a) (b)

Figure 5. (a) First five non-dimensional natural frequencies of FG-X beams (*L/h*= 15) with various boundary conditions using ESDT. (b) First non-dimensional natural frequencies of FG-CNTRC beams (FG-X) with various length-to-thickness ratios using TSDT. V_{tcnt} = 0.28.

4.5. Effect of Geometry

Figure 6 illustrates the influence of slenderness ratio on free vibration. It can be concluded that frequency is inversely proportional to L/h ratio. The stiffness of the beam reduces with increase in slenderness ratio. The trend observed is same irrespective of CNT distributions. However, for thin beams, variation of frequencies with CNT distributions is not very pronounced. To make it concise results using HSDT are presented only.

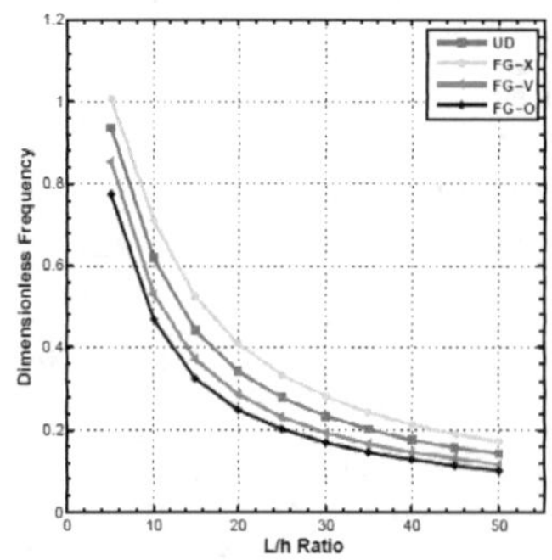

Figure 6. First non-dimensional natural frequencies of FG-CNTRC beams (C-F) with various length-to-thickness ratios using HSDT. V_{tcnt} = 0.17.

4.6. Accuracy of the Present Theories

From the above results, it is clear that these theories can be considered to be very accurate for solving engineering problems. The actual shear stresses profile along the thickness of the beam is better represented by a quadratic function or higher-order polynomial. Expanding the displacement fields of FSDT by adding few higher-order polynomial, variation of transverse shear strain and warpage of the cross section can be taken care of. Therefore, the accuracy of the results obtained using TSDT, ESDT and HSDT are generally higher than FSDT. Table 7 and 8 presents the non-dimensional natural frequencies for thick and thin beams, respectively. The results obtained using different deformation theories are fairly consistent. For thick beams, the frequency predicted by FSDT is smaller because

FSDT does not take into account warpage of the cross section. The higher order theories like TSDT, ESDT and HSDT better approximate the profile of the cross section of thick beams. However, the traction free boundary condition is not accounted for in the present refined theories. Hence, higher order theories predict higher results for thick beams compared to FSDT. Further study of this refined theories is necessary for deflection and distribution of transverse shear stress across the thickness of the beams to conclude their efficacy compared to FSDT. However, for thin beams, the difference in the frequency obtained using various theories is practically negligible. Hence, FSDT is the most suitable to analyse thin beams as the computational cost is minimum and the finite element formulation is comparatively easier.

Table 7. Comparison of natural frequencies obtained using different shear deformation theories for thick beam (L/h= 5).

V_{CNT}	CNT Distribution Pattern	Clamped-Free				Clamped-Clamped				Simply-Supported			
		FSDT	TSDT	ESDT	HSDT	FSDT	TSDT	ESDT	HSDT	FSDT	TSDT	ESDT	HSDT
0.12	UD	0.7303	0.7424	0.7431	0.7419	1.7810	1.8734	1.8764	1.8703	1.6032	1.6245	1.6265	1.6228
	FG-X	0.7805	0.7976	0.7985	0.7969	1.8019	1.9044	1.9071	1.9015	1.6721	1.7024	1.7049	1.7001
	FG-V	0.6755	0.6843	0.6848	0.6839	1.7634	1.8509	1.8537	1.8478	1.5240	1.5407	1.5423	1.5392
	FG-O	0.6285	0.6182	0.6183	0.6184	1.7408	1.7407	1.7444	1.7373	1.4533	1.4077	1.4077	1.4087
0.17	UD	0.9233	0.9366	0.9374	0.9360	2.3093	2.4214	2.4251	2.4173	2.0489	2.0721	2.0744	2.0701
	FG-X	0.9963	1.0093	1.0103	1.0085	2.3476	2.4593	2.4631	2.4552	2.1542	2.1693	2.1721	2.1669
	FG-V	0.8473	0.8560	0.8565	0.8555	2.2883	2.3912	2.3949	2.3872	1.9371	1.9522	1.9541	1.9506
	FG-O	0.7835	0.7744	0.7746	0.7746	2.2543	2.2603	2.2647	2.2564	1.8378	1.7939	1.7940	1.7947
0.28	UD	1.0468	1.0670	1.0680	1.0661	2.4860	2.6247	2.6287	2.6203	2.2713	2.3075	2.3106	2.3047
	FG-X	1.1278	1.1235	1.1246	1.1227	2.5613	2.6585	2.6634	2.6531	2.3981	2.3601	2.3633	2.3576
	FG-V	0.9844	0.9941	0.9949	0.9935	2.5115	2.6307	2.6351	2.6259	2.2005	2.2127	2.2154	2.2104
	FG-O	0.9200	0.9222	0.9228	0.9217	2.4827	2.5553	2.5610	2.5495	2.1060	2.0980	2.1000	2.0968

Table 8. Comparison of natural frequencies obtained using different shear deformation theories for thin beam (L/h= 50).

V_{CNT}	CNT Distribution Pattern	Clamped-Free				Clamped-Clamped				Simply-Supported			
		FSDT	TSDT	ESDT	HSDT	FSDT	TSDT	ESDT	HSDT	FSDT	TSDT	ESDT	HSDT
0.12	UD	0.1171	0.1171	0.1171	0.1171	0.6966	0.6974	0.6975	0.6974	0.3261	0.3261	0.3261	0.3261
	FG-X	0.1423	0.1423	0.1423	0.1423	0.8214	0.8232	0.8233	0.8230	0.3944	0.3945	0.3945	0.3945
	FG-V	0.0972	0.0972	0.0972	0.0972	0.5905	0.5910	0.5910	0.5910	0.2714	0.2714	0.2714	0.2714
	FG-O	0.0843	0.0843	0.0843	0.0843	0.5183	0.5161	0.5161	0.5162	0.2358	0.2355	0.2355	0.2355
0.17	UD	0.1415	0.1415	0.1415	0.1415	0.8493	0.8502	0.8503	0.8501	0.3944	0.3944	0.3944	0.3944
	FG-X	0.1722	0.1722	0.1722	0.1722	1.0072	1.0069	1.0071	1.0068	0.4782	0.4781	0.4781	0.4781
	FG-V	0.1170	0.1170	0.1170	0.1170	0.7156	0.7160	0.7160	0.7159	0.3268	0.3268	0.3268	0.3267
	FG-O	0.1014	0.1014	0.1014	0.1014	0.6269	0.6252	0.6252	0.6253	0.2837	0.2835	0.2835	0.2835
0.28	UD	0.1776	0.1776	0.1776	0.1776	1.0431	1.0446	1.0447	1.0445	0.4935	0.4935	0.4935	0.4935
	FG-X	0.2162	0.2159	0.2159	0.2159	1.2319	1.2212	1.2213	1.2214	0.5983	0.5967	0.5967	0.5967
	FG-V	0.1464	0.1464	0.1464	0.1464	0.8837	0.8833	0.8834	0.8833	0.4082	0.4081	0.4081	0.4081
	FG-O	0.1270	0.1270	0.1270	0.1270	0.7767	0.7761	0.7762	0.7762	0.3547	0.3546	0.3546	0.3546

5. Conclusions

In this article, three deformation theories have been developed from modifying the FSDT to numerically investigate free vibration of FG-CNTRC beams. An efficiency parameter is used to account for the multi scale effect while representing the effective material properties of the FG-CNTRC beams. The governing finite element equations are derived using virtual work principle. The effect of different parameters on free vibration of the beam are numerically computed. The results obtained are compared with the published articles in tabular form and the comparison of different refined shear deformation theories are graphically presented. The following conclusions can be drawn from the results presented here:

- Frequency parameters increases with increase in CNT volume fraction.
- Beams with FG-X CNT distributions exhibit maximum fundamental frequency, whereas FG-O distributions has lowest fundamental frequency.
- Boundary conditions significantly affect the vibration characteristics. C-C beams have the highest natural frequency which is followed by S-S and C-F, respectively.

- Frequency parameters decrease with increase in slenderness ratio.
- Refined theories i.e., TSDT, ESDT and HSDT predict fundamental frequency higher than FSDT for thick to moderately-thick beams.
- For thin beams, there is negligible difference in the natural frequency obtained using different theories.

Appendix A

$[\boldsymbol{N_t}]$ and $[\boldsymbol{N_r}]$ of equation (17) have the following expresssions

$$[N_t]=[N_1 I_2 \quad N_2 I_2 \quad N_3 I_2]\ ; \ [N_r]= [N_1 I_4 \quad N_2 I_4 \quad N_3 I_4] \qquad \text{(A1)}$$

Expressions $[B_{tb}]$, $[B_{rb}]$, $[B_{ts}]$, $[B_{rs}]$, $[Z_1]$ and $[Z_2]$ mentioned in equation (20) are as follows,

$$[B_{tb}]= \begin{bmatrix} \frac{\partial}{\partial x} & 0 \\ 0 & 0 \end{bmatrix} [N_t]; \ [B_{rb}]= \frac{\partial}{\partial x} I_4 \ [N_r]; \ [B_{ts}]= \begin{bmatrix} 0 & \frac{\partial}{\partial x} \\ 0 & 0 \end{bmatrix} [N_t]; \ [B_{rs}]= I_4 \ [N_r];$$

$$[Z_1] =[z I_2 \quad f(z) I_2] \text{ and } [Z_2] = \frac{\partial}{\partial z} [Z_1] \qquad \text{(A2)}$$

where I_j is a j x j identity matrix.

REFERENCES

[1] Iijima S, Helical microtubules of graphitic carbon. *Nature*. 1991 Nov;**354(6348)**:56-8
[2] Sandler J, Shaffer MS, Prasse T, Bauhofer W, Schulte K, Windle AH. Development of a dispersion process for carbon nanotubes in an epoxy matrix and the resulting electrical properties. *Polymer*. 1999 Oct 1;**40(21)**:5967-71.
[3] Yu MF, Lourie O, Dyer MJ, Moloni K, Kelly TF, Ruoff RS. Strength and breaking mechanism of multiwalled carbon nanotubes under tensile load. *Science*. 2000 Jan 28;**287(5453)**:637-40.
[4] Salvetat JP, Bonard JM, Thomson NH, Kulik AJ, Forro L, Benoit W, Zuppiroli L. Mechanical properties of carbon nanotubes. *Applied Physics A*. 1999 Sep 1;**69(3)**:255-60.
[5] Kataura H, Kumazawa Y, Maniwa Y, Umezu I, Suzuki S, Ohtsuka Y, Achiba Y. Optical properties of single-wall carbon nanotubes. *Synthetic metals*. 1999 Jun 1;**103(1-3)**:2555-8.
[6] Che J, Cagin T, Goddard III WA. Thermal conductivity of carbon nanotubes. *Nanotechnology*. 2000 Jun;**11(2)**:65.
[7] Pop E, Mann D, Wang Q, Goodson K, Dai H. Thermal conductance of an individual single-wall carbon nanotube above room temperature. *Nano letters*. 2006 Jan 11;**6(1)**:96-100.
[8] Wong EW, Sheehan PE, Lieber CM. Nanobeam mechanics: elasticity, strength, and toughness of nanorods and nanotubes. *science*. 1997 Sep 26;**277(5334)**:1971-5.
[9] Berber S, Kwon YK, Tománek D. Unusually high thermal conductivity of carbon nanotubes. *Physical review letters*. 2000 May 15;**84(20)**:4613.
[10] Overney G, Zhong W, Tomanek D. Structural rigidity and low frequency vibrational modes of long carbon tubules. *Zeitschrift für Physik D Atoms, Molecules and Clusters*. 1993 Mar 1;**27(1)**:93-6.
[11] Lu JP, Elastic properties of single and multilayered nanotubes. *Journal of physics and chemistry of solids*. 1997 Nov 1;**58(11)**:1649-52.
[12] Hata K, Futaba DN, Mizuno K, Namai T, Yumura M, Iijima S, Water-assisted highly efficient synthesis of impurity-free single-walled carbon nanotubes. *Science*. 2004 Nov 19;**306(5700)**:1362-4.
[13] Arash B, Wang Q, Varadan VK, Mechanical properties of carbon nanotube/polymer composites. *Scientific reports*. 2014 Oct 1;**4**:6479.
[14] Qian D, Dickey EC, Andrews R, Rantell T, Load transfer and deformation mechanisms in carbon nanotube-polystyrene composites. *Applied physics letters*. 2000 May 15;**76(20)**:2868-70.

[15] Wuite J, Adali S, Deflection and stress behaviour of nanocomposite reinforced beams using a multiscale analysis. *Composite Structures*. 2005 Dec 1;**71(3-4)**:388-96.

[16] Lal A, Kanu N. The nonlinear deflection response of CNT/nanoclay reinforced polymer hybrid composite plate under different loading conditions. *IOP Conference Series: Materials Science and Engineering*. 2020;**814**:012033.

[17] Kanu N, Bapat S, Deodhar H, Gupta E, Singh G, Vates U et al. An Insight into Processing and Properties of Smart Carbon Nanotubes Reinforced Nanocomposites. *Smart Science*. 2021:1-16.

[18] Jain N, Gupta E, Kanu N. Plethora of Carbon Nanotubes Applications in Various Fields – A State-of-the-Art-Review. *Smart Science*. 2021:1-24.

[19] Niino M, Maeda S. Recent development status of functionally gradient materials. *ISIJ International*. 1990 Sep 15;**30(9)**:699-703.

[20] Kawasaki A, Watanabe R. Concept and P/M fabrication of functionally gradient materials. *Ceramics international*. 1997 Jan 1;**23(1)**:73-83.

[21] Kanu N, Vates U, Singh G, Chavan S. Fracture problems, vibration, buckling, and bending analyses of functionally graded materials: A state-of-the-art review including smart FGMS. *Particulate Science and Technology*. 2018;**37(5)**:583-608.

[22] Wattanasakulpong N, Ungbhakorn V, Analytical solutions for bending, buckling and vibration responses of carbon nanotube-reinforced composite beams resting on elastic foundation. *Computational Materials Science*. 2013 Apr 1;**71**:201-8.

[23] Rokni H, Milani AS, Seethaler RJ, Size-dependent vibration behavior of functionally graded CNT-reinforced polymer microcantilevers: modeling and optimization. *European Journal of Mechanics-A/Solids*. 2015 Jan 1;**49**:26-34.

[24] Penna R, Feo L, Fortunato A, Luciano R, Nonlinear free vibrations analysis of geometrically imperfect FG nano-beams based on stress-driven nonlocal elasticity with initial pretension force. *Composite Structures*. 2020;**255**:112856.

[25] Yas MH, Samadi N, Free vibrations and buckling analysis of carbon nanotube-reinforced composite Timoshenko beams on elastic foundation. *International Journal of Pressure Vessels and Piping*. 2012 Oct 1;**98**:119-28.

[26] Vo-Duy T, Ho-Huu V, Nguyen-Thoi T, Free vibration analysis of laminated FG-CNT reinforced composite beams using finite element method. *Frontiers of Structural and Civil Engineering*. 2019 Apr 1;**13(2)**:324-36.

[27] Shen HS, Xiang Y, Nonlinear analysis of nanotube-reinforced composite beams resting on elastic foundations in thermal environments. *Engineering Structures*. 2013 Nov 1;**56**:698-708.

[28] Mohseni A, Shakouri M, Vibration and stability analysis of functionally graded CNT-reinforced composite beams with variable thickness on elastic foundation. *Proceedings of the Institution of Mechanical Engineers, Part L: Journal of Materials: Design and Applications*. 2019 Dec;**233(12)**:2478-89.

[29] Rokni H, Milani AS, Seethaler RJ, Maximum natural frequencies of polymer composite micro-beams by optimum distribution of carbon nanotubes. *Materials & Design*. 2011 Jun 1;**32(6)**:3389-98.

[30] Rokni H, Milani AS, Seethaler RJ, *2D* optimum distribution of carbon nanotubes to maximize fundamental natural frequency of polymer composite micro-beams. *Composites Part B: Engineering*. 2012 Mar 1;**43(2)**:779-85.

[31] Lin F, Xiang Y, Numerical analysis on nonlinear free vibration of carbon nanotube reinforced composite beams. *International Journal of Structural Stability and Dynamics*. 2014 Jan 12;**14(01)**:1350056.

[32] Mohammadimehr M, Monajemi AA, Afshari H, Free and forced vibration analysis of viscoelastic damped FG-CNT reinforced micro composite beams. *Microsystem Technologies*. 2017 Dec 27:1-5.

[33] Kitipornchai S, Chen D, Yang J, Free vibration and elastic buckling of functionally graded porous beams reinforced by graphene platelets. *Materials & Design*. 2017 Feb 15;**116**:656-65.

[34] Momeni M, Dehkordi MB, Frequency analysis of sandwich beam with FG carbon nanotubes face sheets and flexible core using high-order element. *Mechanics of Advanced Materials and Structures*. 2019 May 3;**26(9)**:805-15.

[35] Dabbagh A, Rastgoo A, Ebrahimi F, Finite element vibration analysis of multi-scale hybrid nanocomposite beams via a refined beam theory. *Thin-Walled Structures*. 2019 Jul 1;**140**:304-17.

[36] Bodaghi M, Damanpack AR, Aghdam MM, Shakeri M. Geometrically non-linear transient thermo-elastic response of FG beams integrated with a pair of FG piezoelectric sensors. *Composite Structures*. 2014 Jan 1;**107**:48-59.

[37] Datta P. Active vibration control of axially functionally graded cantilever beams by finite element method. *Materials Today: Proceedings*. 2021 Jan 1;**44**:2543-50.

[38] Nguyen-Quang K, Vo-Duy T, Dang-Trung H, Nguyen-Thoi T. An isogeometric approach for dynamic response of laminated FG-CNT reinforced composite plates integrated with piezoelectric layers. *Computer Methods in Applied Mechanics and Engineering*. 2018 Apr 15;**332**:25-46.

[39] Datta P, Ray MC. Effect of carbon nanotube waviness on smart damping of geometrically nonlinear vibrations of fuzzy-fiber reinforced composite plates. *Journal of Intelligent Material Systems and Structures*. 2019 Apr;**30(7)**:977-97.

[40] Manoj A. Kumbhalkar; D.V. Bhope; A.V. Vanalkar, "Evaluation of Frequency Excitation of Helical Suspension Spring Using Finite Element Analysis", International Journal of Computer Aided Engineering and Technology, Vol. 9, No. 4, pp 420-433, 2017.

[41] Song ZG, Zhang LW, Liew KM. Active vibration control of CNT-reinforced composite cylindrical shells via piezoelectric patches. *Composite Structures*. 2016 Dec 15;**158**:92-100.

[42] Shi Z, Yao X, Pang F, Wang Q, An exact solution for the free-vibration analysis of functionally graded carbon-nanotube-reinforced composite beams with arbitrary boundary conditions. *Scientific reports*. 2017 Oct 10;**7(1)**:1-8.

A review of wear resistance materials used in power-screw mechanism for aerospace applications

Ujjwal KS[1], Dr. Sharath Chandra N[2]*

[1*] Department of Mechanical Engineering, the National Institute of Engineering, Mysore.
[2] Assistant professor and Head CTAT, Department of Mechanical Engineering, the National Institute of Engineering, Mysore.

ujjwalkanambadi1804@gmail.com[1], sharath.chandra@nie.ac.in[2]

1. Abstract

In today's industrial requirement, good wear resistant materials with the same or better mechanical properties are the need of the hour. In the present paper, an actuator that is used to lift the load is under study. It uses a linear actuation mechanism involving a screw-like motion to lift the loads. It employs a stainless steel of high quality. Since there are various types of wearing involved in the system, choice of the materials becomes the prime factor for determining the maintenance cost and the cycle life of the actuator. There are many research advances made in this field which provides us with the best type of material for its particular type of job. The present paper discusses the materials that could be substituted in place of the existing material choice i.e., in place of AMS5659 and aluminium-bronze. The two materials stated above are known for its superior wear handling and excellent mechanical properties.

Keywords: AMS 5659, Aluminium bronze, wear resistance materials

2. Introduction

The necessity to find the highly wear resistant materials is of high importance to any industry as it affects the performance and quality of the products it produces. Wear of a material is defined as surface phenomenon and deals with the removal of material from one or both of the solid surfaces in sliding, rolling or impact motion relative to one another [1, 2]. The present paper deals with a power-screw mechanism that deals with two materials in contact with each other. A power-screw mechanism is used when the rotary motion of the screw is to be converted to a required linear motion overall [3]. It employs two parts, namely, the screw and the nut. The screw material under study is the AMS 5659 stainless steel and the nut material under study is the Aluminium bronze. Tables 1 and 2 shows the properties of the metals used. The major challenges faced during the usage of these materials is that due to the rise in temperature between the materials during the operations, the softer material being Aluminium bronze, melts over the surface of the screw material, AMS 5659. This adhesive wear causes problems in the mechanism over the long run. Due to the accumulation of the softer particles on the harder material, the service life of the mechanism reduces. Also, the maximum extension of the mechanism cannot be achieved due to the increase of friction between the two components. Hence, this reduces the overall usage of the mechanism. The main aim of this paper is to suggest a different range of materials that could be used for the mechanism that could replace either or both of the materials in the mechanism. The range of materials could be from metals, non-metals, metallic coatings, hybrid composites and polymers.

Table 1: Mechanical properties of AMS 5659 under different heating conditions [4]

Materials	Condition	Ultimate tensile strength	% Reduction in Area	Rockwell C Hardness
Alloy 15-5 PH Bar AMS 5659	Cond A	-	-	363 HB max
Alloy 15-5 PH Bar AMS 5659	H900	190	35	388-444
Alloy 15-5 PH Bar AMS 5659	H925	170	38	375-429
Alloy 15-5 PH Bar AMS 5659	H1025	155	45	331-401
Alloy 15-5 PH Bar AMS 5659	H1075	145	45	311-375
Alloy 15-5 PH Bar AMS 5659	H1100	140	45	302-363
Alloy 15-5 PH Bar AMS 5659	H1150	135	50	277-352

Table 2: Mechanical properties of Aluminium-bronze [5]

Properties	Values
Hardness, Rockwell C	27
Tensile Strength (MPa)	690
Yield Strength (MPa)	380
Compressive yield strength (MPa) (At permanent set of 1%)	880
Poisson's Ratio	0.34
Shear Modulus (GPa)	42.3

3. Materials used in the domain

As mentioned earlier, the materials selection could vary from metals to non-metals to polymers. The selection of materials is based on the cost of handling the material, the availability of the material, transportation costs and the mechanical properties of them. The wear resistance of a material is directly proportional to the hardness of the material. Keeping these factors in mind, the following is the list of materials that could be chosen in order to replace the above-mentioned parent materials.

3.1 Coatings

3.1.1 Diamond-like carbon coating: Diamond-like carbon (DLC) is defined as an amorphous carbon (a-C) having structural, mechanical, electrical, optical, chemical, and acoustic properties similar to those of diamond [6]. The name "diamond-like carbon" is currently used widely when at least one property is similar to diamond. Diamond-like carbon coating (DLC) are very popular as a result of their superior properties such as high hardness and low coefficient of friction. The major challenges faced in the diamond-like carbon coating is that it develops high internal compressive stress and has a low adhesion. The internal stresses increases with increasing sp_3 content in the (sp_3/sp_2) ratio.

3.1.2 Plasma Nitriding: Plasma nitriding produce high surface hardness, good wear resistance, increased fatigue strength and toughness. The major advantages of using the plasma nitriding process are that at a lower working temperature (about 300_0C), nitriding can be performed. Also, the treatment time taken for the plasma nitriding process is much faster when compared with the other nitriding processes [7]. Even though it has advantages, it has a certain drawbacks as well. The initial cost of setup for the plasma nitriding is high and also the nitriding of complex parts is difficult.

3.1.3 PVD Coating: PVD (or Physical Vapour Deposition) refers to a variety of thin film deposition techniques in which a solid material is vaporised in a vacuum environment and deposited on substrates as pure metals or alloy composition coating [8,9]. There are three methods of PVD coating which could be employed for a material. They are:
1. Evaporation PVD
2. Sputtering PVD
3. Cathodic Arc PVD

PVD Coatings have more advantages when compared to other types of deposition techniques as they do not use any chemicals in their process and they do not require extensive clean up [10]. The major disadvantage of using PVD coating is it has higher cost of operation. Also, the process requires a skilled operator and complex machinery. The rate of operation of PVD is also quite slow. However, PVD coating remains one of the most effective way of coating.

3.1.4 Salt bath Nitriding: In salt bath nitriding the nitrogen donating medium is a nitrogen containing salt such as cyanide salt. The salts used also donate carbon to the workpiece surface making salt bath a nitro carburizing process. The temperature used is typical of all nitro carburizing processes: 550 to 570 °C [11]. The advantages of salt nitriding is that it achieves higher diffusion in the same period of time compared to any other method. Also, the processing time for this process is very quick usually in the order of 4 hours to achieve. The drawbacks of this process are that because it uses highly toxic salts as a source of nitrogen to diffuse through the ferrous matrix, any substance containing Nitrogen (N) and Carbon (C) releases cyanide gas when subjected to heat conditions.

3.1.5 Silicon Carbide Coating: Silicon carbide (SiC) is one of the most important advanced ceramics in contemporary usage [12]. With an exceptional hardness of 25 GPa, and a low density of 3.21 g cm_{-3}, SiC ceramics see their most important commercial use as lightweight armour ceramics, with wear resistant linings another leading application [13,14]. The advantage of using this material is that it has a high temperature resistance and can be used as a refractory material. Also, due to its high hardness value, it could be used in multiple applications [15]. The thermal conductivity of this material is limited and the power loss in this material is considerably high.

3.2 Metals

3.2.1 HARDOX400 Stainless steel: HARDOX 400 stainless steel is an abrasion resistant stainless steel with a nominal hardness of 400 HBW. It is an all-around stainless steel. The very important property of HARDOX 400, which guarantees the long life of products, is a fact, that HARDOX 400 plates are full hardened. HARDOX 400 steel can be welded without initial warming up until the plates' thickness 40 mm. The impact strength and plasticity of HARDOX 400 are on the high level. The HARDOX 400 can be used in some cases as a construction plate carrying the load. The high hardness of HARDOX steel is acquired thanks to heat treatment in continuous process [17].

3.2.2 EN41B with active screen plasma nitriding: The process parameters of 500^0C temperature for 4 hours with the gas flow in the ratio of H_2:N_2= 1:1 were used for the treatment. The improvement in the surface was assessed by the micro hardness as well as SEM (Scanning electron microscope) and XRD (X-Ray Diffusion). The micro hardness value reveals that there is 4 times more hardness than the parent EN41B material [18].

3.2.3 Boron Carbo-nitride: Ternary compounds containing boron, carbon and nitrogen are potential candidates as super-hard materials. These compounds are essentially carbon alloys. The major advantages of using this metal is that it is known to have excellent erosion characteristics and high resistance to wear under high energy loads. However, one of the drawbacks of boron carbo-nitride is that it starts oxidising at temperatures of over 500_0C [19].

3.3 Hybrid Composites: Hybrid composites prepared from ceramics and polymers are a class of materials that combines the benefits of both phases to produce materials with enhanced properties not achieved by either phase alone. The advantages of using the hybrid composites are balanced strength and stiffness, balanced bending and membrane mechanical properties, balanced thermal distortion stability, reduced weight and/or cost and improved fatigue resistance [20, 21]. However, the challenges faced in this are high cost, high density, and high weight [22, 23]. In addition to this it have large impact on environment while preparing these fibres and their related composites.

3.4 Polymers: Polymers are replacing classical metal parts in terms of sliding parts because of their rapid development with better tribological, mechanical and corrosion properties. The successful applications of unfilled polymers in engineering purposes led to use of polymers as composites, where modifications to properties could be achieved by adding one or several reinforcing fillers [24]. The fillers provide high hardness, strength or elasticity, while the matrix, which is a polymeric material, adds a self- lubricating effect. This unique combination combines the best and suppresses the worst properties of the individuals. In the recent research of materials, self-healing polymers have been discovered [25]. These materials reorganize polymeric bonds and structures under external stresses, which makes them highly wear resistant. In bearing and wear applications, polymers provide extensive advantages over metals by allowing for lower power motors for moving parts due to lower frictional properties of polymer wear components compared to metals [26]. Also, the lower wear rates allow for the less maintenance related downtime. Even with the said advantages, the polymers have a certain set of disadvantages. It has low thermal resistance, high coefficient of thermal expansion. Under extensive usage, temperature is bound to increase which will change the geometric dimensions of the material, hence, rendering it as a liability. Also, polymers have a low strength and low stiffness (modulus of elasticity).

4. Comparison of properties

Coatings

Table 3: Comparison of properties of various coatings

	Hardness	Tensile Strength	Yield Strength
Diamond-like carbon Coating (DLC)	71-80 HRC	420 MPa	220 MPa
Plasma Nitriding	58-63 HRC	370-741 MPa	337-679 MPa
PVD Coating	67 HRC	570 MPa	243 MPa
Salt bath nitriding	50-70 HRC	535 MPa	240 MPa
Silicon Carbide coating	2480-2960 HK	240 ksi	21 GPa

From the above table, it can be seen that the hardness value of Silicon carbide coating is the highest owing to the fact that the material behaves as a ceramic and hence, the hardness value [6-15].

Metals

Table 4: Comparison of properties of various metals

	Hardness	Tensile Strength	Yield Strength
HARDOX400 SS	370-470 HB	1250 MPa	900 MPa
EN41B with active screen plasma nitriding	201-277 HB	700-925 MPa	480-525 MPa
Boron Carbonitride	1261 HV	240-315 MPa	150 MPa

The above table depicts the comparison of metallic elements that have a good wear resistance when compared with the parent materials. Hardox 400 stainless steel has very high values and is the best suitable metal for the application. The active screen plasma nitriding refers to the fact that the covering screen also participates in the nitriding process. The range of values for the EN41B material is due to the fact that there are two types of compounds formed during the nitriding: R and S type. The range of values is the combined values of both R and S types. The R and S types refers to the heating conditions of martensitic steel [16-19].

Non-metals

Table 5: Comparison of properties of various Non-metals

Material	Maximum service temperature (oF) (oxidising conditions)	Room temperature hardness (Knoop)	Melting point (oF)
Diamond	1500	7000	1800 (carbonizes)
TiB_2	1800	2700	5250
ZrO_2	4200	-	4850
WC	1200	1880	4770
BeO	4300	1220	4660
TiC	1800	2460	5660 (sublimates)
BC	1800	2800	4430
SiC	1750	2500	4160
Al_2O_3 (dense)	3500	2000	3720
Cr_3C_2	-	-	3600
Cr-Mo-Al_2O_3 cermet	2300	50 RC	3360 (approx.)

Diamond is the hardest known material among the non-metals and even in the metals. Owing to the fact that diamond is quite expensive, it cannot be used everywhere. A diamond tip could be given to tools which are involved in the manufacturing of very high important components that require high precision. Of the other non-metals, the ceramic natured compounds like TiC, SiC, and BC and so on have the hardness values exceeding 2000 in Knoops hardness. As mentioned earlier, higher the hardness value, higher is the wear resistance of that material. Hence, among the other non-metallic materials, BC and TiB_2 have the highest wear resistance [28].

5. Future Materials

Tribology, the science of friction, wear and lubrication, has often been regarded as standing on the side-lines of material sciences. It is important to appreciate that "wear resistance" is emphatically not a materials property. In considering a simple mechanical problem such as deflection of beams, it is clear that the behaviour does not depend only on the beam material but also the geometry of the beam.

Similarly, the rates of wear in a tribological system in which two solid bodies slide against each other depend not only on the properties of the materials involved, but also on geometry of contact, velocity of sliding, the ambient temperature and environmental conditions. There has been a rapid recent progress in the development of new materials. Among the "conventional" bulk metals, polymers and ceramics, most improvements in wear resistance have in the past been incremental and will probably remain so in the future. Nevertheless, with a better understanding of the factors that influence wear rate, significant further improvements will undoubtedly be possible, and maybe achieved with lesser problems that there has been in the past.

No review of future developments would be complete without the mentions of "smart materials" [29]. These are the materials that possess some ability to sense their environment and to respond in a controlled way to that stimulus. It can be argued that so far no truly smart materials have been developed and demonstrated. A smart material for a tribological application might respond to the stimulus of surface tractions or deformations by changing some aspects of its surface in order to improve its ability to resist wear or to reduce frictional traction. For example, the good resistance of Hadfield manganese steel to certain types of impact abrasion, for instance, is due to the stress-induced transformation of the austenitic structure to martensitic structure when the surface is deformed; the hardness increases locally from 200 to 600 HV [30]. Whether concepts such as these can provide methods for improving the tribological performance of the material and whether it might be possible locally to modify the surface stresses remains a stimulating question for future [31].

6. Conclusion

A lead screw, also known as a power screw or translation screw, is a screw shaft that rotates causing a nut to move linearly along the shaft. It's difficult to pinpoint when lead screws were first invented because you could say that the original grape and olive presses were a form of lead screw. Lead screws have also been used for many years in various types of vices, jacks, and machine slides. The power screw mechanism is used in places where a rotary motion is to be converted to a linear motion.

Various materials have been used in the power screw mechanism in order to evade the wearing of the materials and hence, limit the usage of the mechanism. A continuous research is being done with respect to finding the best suitable material for all the right applications. In the recent days, a platinum gold alloy has been developed that is claimed to have the hardness close to the value of the diamond. It is supposed to have a durability 100 times more than that of steel. The alloy consists of 90% platinum and 10% gold.

For the present case however, there cannot be a particular material that could replace the parent material as each class of material has its own properties. Hence, choosing a material from the class is more advisable than just choosing a single particular material.

Observing the properties of the above stated materials, the silicon carbide coating produces the highest wear resistance amongst all the other coatings. However, the salt bath nitriding produces a considerably high hardness value and is comparatively cheaper that the other processes. Hence, among the coatings, salt bath nitriding would be a good choice for the material coating.

Among the metals, replacement of the material is only possible. From the table 4, it is clearly visible that the HARDOX400 Stainless steel is the best choice of metal. It has the highest hardness value in the metals and also shows a good tensile strength. Also, the material cost is cheaper than that of other two metals. Platinum gold alloy is the hardest metal available but the major drawback is that the cost per gram of the material is very high. Hence, this metal is used only for high accuracy and precision jobs and cannot be used for all industrial purposes.

Polymeric compounds are being preferred owing to their high hardness value equal to their metallic counterparts but with more advantages. The hybrid composites are also considered to be a part of the polymeric compounds. The hybrid composites and polymers are considered to be the future of wear resistant materials. Of the Non-metallic compounds that has the highest hardness values from the table 5, SiC is found out to be the best choice of material. It has a very high hardness value and also the cost of the material is the lowest among all others.

It is to be noted that the comparison made above is with respect to the parent materials, AMS 5659 and Aluminium bronze. All the values are compared with respect to the two metals.

Reference

[1] "Uniform wear theory", *https://mechdiploma.com/explain-i-uniform-pressure-theory-iiuniform-wear-theory-clutches-and-bearing*

[2] J.B.K Das, P.L. Srinivasa Murthy, "Design of machine elements 2", *Sapna publications, PP 491.*

[3] Zhijian ling, Whenxiang Zhao, Peter Omand Rasmussen, Jinghua Ji, "Design and manufacture of linear actuator based on magnetic screw transmission", *IEEE transactions on Industrial electronics, Vol. 68, No. 2, 2021, PP 1095-1107.*

[4] "15-5 PH Stainless steel- AMS5659 data handbook", *https://www.ssacorp.com/documents/Data%20Sheet%2015-5PH-AMS5659.pdf*

[5] Jianlin Xu and Zhiping Wang, "Research into a new high-strength aluminium bronze alloy", *International journal of Materials and product technology, Vol.21, No.5, 2004, PP443-453.*

[6] Abdul Wasy Zia, Zhifeng Zhou and Lawrence Kwok-Yan Li, "Structural, mechanical and Tribological characteristics of diamond-like carbon coatings", *Diamond and related materials, Elsevier Publications, Volume 10, Issue 3-7, 2001, PP 171-189.*

[7] Wolfgang Tillmann, Nelson Filipe Lopes Dias, Dominic Stangier, "Influence of plasma Nitriding pretreatments on the tribo-mechanical properties of DLC coatings sputtered on AISI H11", *Surface and Coatings Technology, 2018,* https://doi.org/10.1016/j.surfcoat.2018.11.002, *S0257-8972(18)31212-X.*

[8] Kedhong Zhang, Jainxin Deng, Xuhong Guo, Lining Sun, Shuting Lei, "Study on adhesion and tribological behavior of PVD TiAlN Coatings with a multi-scale textured substrate surface", *International journal of refractory and hard materials, 2018, PP 292-305.*

[9] E.J. Bienk, H. Reitz, N.J. Mikkelsen, "Wear and friction properties of hard PVD coatings", *Surface coatings and technology, 1995, PP 475-480.*

[10] T. Leyendecker, O. Lemmer and S. Esser, "The development of the PVD coating TiAlN as a commercial coating for cutting tools", *Surface and Coatings technology, 1991, PP 175-178.*

[11] "Salt bath nitriding process and its safer alternative", *https://www.paulo.com/salt-bathnitriding- process-safer-alternative/*

[12] "SI-TUFF", https://haydale.com/wp-content/uploads/2019/10/HTI-SI-TUFFPerformance-SiC-Applications-Guide-Nov-2016.pdf

[13] S. Suresha, B.K. Shridhara, "Effect of silicon carbide particulates on wear resistance of graphitic aluminium matrix composites", *Materials and design, 2010, PP 4470-4477.*

[14] Nand Jee Kanu, Amogh Mangalam, Eva Gupta, Umesh Kumar Vates, Gyanendra Kumar Singh, Devendra Kumar Sinha, "Investigation on secondary deformation of ultrafine SiC particles reinforced LM25 metal matrix composites", *Materials Today: Proceedings, Vol. 47, Part 11, 2021, ISSN 2214-7853, PP 3054-3058, DOI: https://doi.org/10.1016/j.matpr.2021.05.640.*

[15] "4 wear resistant coatings for lightweight metals", *https://blog.keronite.com/4-wearresistant- coatings-for-lightweight-metal*

[16] "SSAB Boron", *https://www.ssab.com/products/brands/ssab-boron-steel*

[17] Henryk Buglacki, Monika Smajdor, "Mechanical properties of abrasion-resistant HARDOX400 Steel and their welded joints", *Advances in material science, 2003, Vol. 4, No. 2(4).*

[18] Nand kumar, B. Ganguli, Bidesh Roy, Bachu Deb, "Investigation of surface properties of EN8, EN24, EN41B low alloy steel treated by active screen plasma nitriding", *Indian institute of metals, 2021, No. 74(4), PP 799-810.*

[19] Osamu Takai, "Super hard materials", *Chapter 34, 2003, http://dx.doi.org/10.1016/B978-008044163-4/50034-6*

[20] Abhishek Chauhan, Umesh Kumar Vates, Nand Jee Kanu, Eva Gupta, Gyanendra Kumar Singh, Bhupendra Prakash Sharma, Srinivasa Rao Gorrepati, "Fabrication and characterization of novel nitinol particulate reinforced aluminium alloy metal matrix composites (NiTip/AA6061 MMCs)", Materials Today: Proceedings, Vol. 38, Part 5, 2021, PP 3027-3034, ISSN 2214-7853, DOI: https://doi.org/10.1016/j.matpr.2020.09.326.

[21] Dora Shiva Prasad, Chintada Shoba, "Hybrid composites- a better choice for high wear resistant materials", *Journals of materials research and technology, 2014, No. 3(2), PP 172-178.*

[22] Shankar Kadam and Sachin Chavan and Nand Jee Kanu, "An insight into advance self-healing composites", *Materials Research Express, 2021, Vol. 8, Number 5, DOI: 10.1088/2053-1591/abfba5*

[23] Nand Jee Kanu, Eva Gupta, Umesh Kumar Vates, Gyanendra Kumar Singh, "Self-healing composites: A state-of-the-art review", *Composites Part A: Applied Science and Manufacturing, Vol. 121, 2019, PP 474-486, ISSN 1359-835X, DOI: https://doi.org/10.1016/j.compositesa.2019.04.012*.
[24] "Wear of polymer composites", *Elsevier Publications, DOI: 10.1533/9781782421788.133*
[25] Klaus Friedrich, "Wear of reinforced polymers by different abrasive counterparts", *Polymer and composites group, Elsevier publications, Volume 1, 1986, PP 233-287.*
[26] Vahid Adibnia, Marziye Mirbhageri, Jimmy Faivre, Jordan Robert, Jeungjun Lee, Krzyst of Matyjaszewski, Dong Woog Lee, Xavier Banquy, "Bio-inspired polymers for wear resistance and lubrication", *Progress in polymer science, Elsevier publication, 2020, https://doi.org/10.1016/j.progpolymsci.2020.101298*
[27] Nand Jee Kanu, Saurabh Bapat, Harshad Deodhar, Eva Gupta, Gyanendra Kumar Singh, Umesh Kumar Vates, Girish C. Verma & Vivek Pandey, "An Insight into Processing and Properties of Smart Carbon Nanotubes Reinforced Nanocomposites", *2021, Smart science, DOI: 10.1080/23080477.2021.1972913*
[28] W.A. Glaeser, "Wear characteristics in non-metallic materials", *Wear journal, Elsevier publications, Volume 6, Issue 2, 1963, PP 93-105.*
[29] Nand Jee Kanu, Umesh Kumar Vates, Gyanendra Kumar Singh & Sachin Chavan, "Fracture problems, vibration, buckling, and bending analyses of functionally graded materials: A state-of-the-art review including smart FGMS, Particulate Science and Technology", *2019, Smart Science, 37:5, PP 583-608, DOI: 10.1080/02726351.2017.1410265*
[30] I.M. Hutchings, "Wear resistant materials: Into the next century", *Material science and engineering, 1994, PP 185-195.*
[31] Ayushi, Umesh Kumar Vates, Sanjay Mishra, Nand Jee Kanu, "Biomimetic 4D printed materials: A state-of-the-art review on concepts, opportunities, and challenges", *Materials Today: Proceedings, Vol. 47, Part 11, PP 3313-3319, ISSN 2214-7483, DOI: https://doi.org/10.1016/j.matpr.2021.07.148*.

Analysis of Electromagnetic Reflection Loss for Mesh Structure with Al6061 MMC for Aerospace Applications

Siva Chakra Avinash Bikkina*, P.V.Y. Jayasree.
Research scholar, Department of Electrical Electronics and Communication Engineering, GITAM Institute of Technology, GITAM Deemed to be University Visakhapatnam, Andhra Pradesh, India,530045.
Professor Department of Electrical Electronics and Communication Engineering, GITAM Institute of Technology, GITAM Deemed to be University Visakhapatnam, Andhra Pradesh, India,530045.

E-mail sbikkina@gitam.in, jpappu@gitam.edu

Abstract. One of the major problems facing by the aircraft was a lightning strike. To overcome this problem, fiber-reinforced materials have been used. The fiber-reinforced materials have less conductivity. These fiber-reinforced materials can't eliminate the lightning strike effect. For that purpose, the metal matrix composite materials significantly impacted the aircraft's internal circuits and physical components from the lightning strike effect. To meet industries dynamic and ever-increasing demands, Al6061 metal matrix composite reinforced with fly ash must be utilized to build the aircraft to offer HIRF. The material thickness should be kept low as possible then it can be used to cover the plane's surface. To prevent lightning strikes, it might be used to protect electronic components from a concentrated high-intensity radiated field, primarily in Aeroplan configuration. The electromagnetic characteristics of composites are measured using the X-band for normal incidence. The electromagnetic reflection properties of AL6061 reinforced with fly ash are studied in this study for mesh structure. Mat lab Software was used to calculate the maximum reflection loss of 33.88dB for 15% fly ash and 85 percent AL6061 at X-band.
Keywords. High intensity radiated fields (HIRF), AL6061, Reflection, Fly ash, lightning strike, and Fiber-reinforced plastic composites (FRPs)

1. Introduction

The lightning strike is one of the significant problems facing aerospace. Commercial flight gets hit by the lightning strike once every year. These lightning strike effects diminished the internal electronic components. Because the airplane skin is frequently composed of aluminum, an electrically conductive substance, the actual harm caused by a lightning strike is commonly limited to consuming lines on the skin and the following edges. During a lightning strike, the airplane's metallic exterior acts as a Faraday cage, protecting the aircraft from electromagnetic interference (EMI) [1][2]

Generally, with the presentation of composite materials in flight progressions, safeguarding airplanes from lightning strikes has turned into a genuine errand, as FRPs are less conductive than their metallic partners [3][4]. A couple of headways in lightning strike protection (LSP) for composite designs have been utilized in flight. The vital capacity of LSP is to give a constant conductive channel all through the plane's outside, especially in regions more helpless against lightning strikes, like the nose, wingtips, nacelles, radomes, and empennage limits. LSP is regularly developed of a lightweight metallic cross-area or foil made for most aluminum or copper, with traces of phosphor bronze and titanium remembered for the external overlay utilized that associates the outer surface to a metallic ground plane engine [5][6]. Another major problem facing by aerospace application was consumption of fuel to overcome these issues less weight materials has been used. The weight of the materials can be reduced by using mesh structure [7] [8]. Nevertheless, the imperfection of FRPs is its low electrical conductivity but having good mechanical properties, which keeps the electric field from going through it when lightning strikes the plane, achieving delamination and embrittlement [9] [10].

There is a lot of promise in aluminum composite complexes, and their fundamental characteristics may be enhanced by using other materials for reinforcing. Metals such as titanium, nickel, magnesium, copper, and aluminum have been utilized to create metal matrix composites [11]. However, aluminum is most commonly used due to its lightweight, strength, electrical, and reflective characteristics [12] [13] [14]. These Aluminum Metal Matrix Composite materials have the best mechanical properties [15]. Different physical models of lightning strike prevention exist based on their electrical characteristics. AL6061 was presented as an alternative to FRPs because it had suitable qualities such as low density, corrosion resistance, high strength, and good electrical properties [16].

Lightweight materials are commonly requested in the realm of electromagnetic protection. As a result, in actual designing applications, the thickness of the chosen materials should be kept as low as feasible under the constant protecting structure [17]. Because there are so many accidental E.M. waves that are fancifully coordinated, measuring the safeguarding productivity for the ordinary occurrence of E.M. waves on the cross-section sheet may apply to every single feasible application. The investigation is based on the transmission line model (T.L) shown in regular frequency [18]. Being lightweight is an essential inventive requirement for practical EMI protecting applications, especially in planes, cars, and family gadgets. Foamed composites with cell designs are great for the creation of lightweight EMI protecting materials [19]. The reinforced material is fly ash, which improves the electrical properties of AL6061 metal matrix composites [20]. Electromagnetic waves travel through fly ash; it enhances the microwave absorption property because fly ash contains porous carbon grain and Fe_2O_3 [21]. Zuoyong Dou and Gaohui Wu evaluated the Electromagnetic Shielding Effectiveness (EMSE) of an aluminium composite supported by fly debris and discovered that the electromagnetic protection was between 32 and 102.5 dB. [22]. Casey, Kendall F 1988proposed the wire mesh screen for the electromagnetics shielding for aluminium's metal material [23]. The main aim for going to the composite materials mesh structure was less weight than the metallic mesh [24].

This paper is classified into six sections; section 1 explains the introduction, section 2 describes the composite material preparation, section3 explains electrical parameters, section 4 presents the Electromagnetic Reflection calculations, section5 explains the results of reflection loss for the composite material, and section 6 conclusion.

2. Composite Material Preparation

The underlying method used for preparing materials was stir casting. The primary purpose of selecting stir casting was affordable to utilize. This strategy changes the composites into a fluid state at a temperature for every component [25]. The AL6061 melted at the $700c^0$ and fly ash reinforced with AL6061 at the $700c^0$and required structure can be implemented with molt.

3. Electrical Parameters

The electrical parameters that were considered for calculating the materials' shielding effectiveness were permittivity, permeability, and conductivity of the materials. The vector network analysis (VNA) can estimate these parameters. To calculate the required parameters for calculating the reflection loss of the materials, U.C. Hasar 2009 designed a two-port network analyzer [26]. These characteristics may be calculated by inserting the material with the necessary dimensions within the rectangular waveguide, the electrical parameters with AL6061, and the combination of different percentages of fly ash.

4. Electromagnetic Reflection Loss Calculation

The mesh wire's shielding effectiveness depends on the parameters of the three electrical properties and the materials' thickness. The fundamental cross-section structure is displayed in figure1. At the point when an electromagnetic wave goes through a shielding material, it is weakened by one of three systems: (1) Reflection Loss (RL), (2) Absorption Loss (AL), and (3) Multiple reflection Loss(ML) are three kinds of misfortunes. The complete EMI SE of safeguarding material might be communicated utilizing the formulae underneath. The going with the condition can infer the particular Reflection Loss characteristics. [17][23][27].

The reflection loss is denoted by.

$$R_{dB} = 20\log\frac{(Z_0 + Z_S)^2}{4.Z_O * Z_S} \quad (1)$$

Here free space is represented as Z_0 and sheet impedances represents as Z_S.

$$Z_S = Z_w * as + j\omega l_s \quad (2)$$

$$l_s = \frac{\mu_0 as}{2\pi}\ln\left(1 - e^{\frac{-2\pi r_w}{as}}\right)^{-1} \quad (3)$$

The radius of the wire mesh is represented as r_w and as is the mesh size. The internal impedance per unit length as Z_w.

$$Z_w = R'_w \frac{(\sqrt{j\omega\tau_w}.)I_0.(\sqrt{j\omega\tau_w})}{2I_1\sqrt{j\omega\tau_w}} \quad (4)$$

Diffusion time constant is τ_w , the modified Bessel function of the first kind of order n $I_n(.)$.and dc resistance per unit length of the mesh is R'_w.

$$R'_w = (\pi r_w^2 \sigma_w)^{-1} \quad (5)$$

$$\tau_w = \mu_w \sigma_w r_w^2 \quad (6)$$

The permeability and conductivity of the material employed in the mesh structure are represented by $\mu_w and\ \sigma_w$.

5. Results

The reflection loss of the X-band frequency is calculated. Figure 3 shows the variation in reflection loss for pure AL6061, which ranges from 27.04dB to 10.97dB. Figure 4 illustrates the reflection loss for 5% fly ash and 95% AL6061, ranging from 28.57 dB to 2.93 dB.

Table 1: Comparison of Maximum Reflection loss at X-band frequency for the combination of Al6061 with different percentages of fly ash.

S.no	Component name	Maximum Reflection loss (dB)
1	Pure AL6061	27.04
2	95%_AL6061 +5%_Fly ash	27.42
3	90%_AL6061+10%_Fly ash	33.78
4	85%_AL6061+15%_Fly ash	33.88

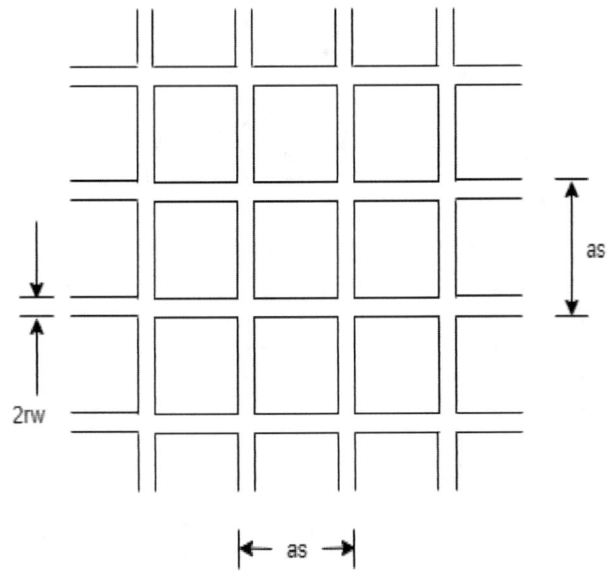

Figure1: Representation of mesh sheet.

Figure 5 shows that the reflection loss for a combination of 10% fly ash and 90% AL6061 ranged from 33.88 dB to 7.134 dB. Figure 6 illustrates the reflection loss for a variety of 15% fly ash and 85% AL6061, which ranged between 33.78 dB and 13.57 dB. As the fly ash reinforcement increases, the reflection loss increases. It is evident from the outcomes that reflection loss decreases as the frequency increases. Figure 7 shows the different combinations of AL6061 with fly ash are compared, and the maximum reflection loss is tabulated in table1.

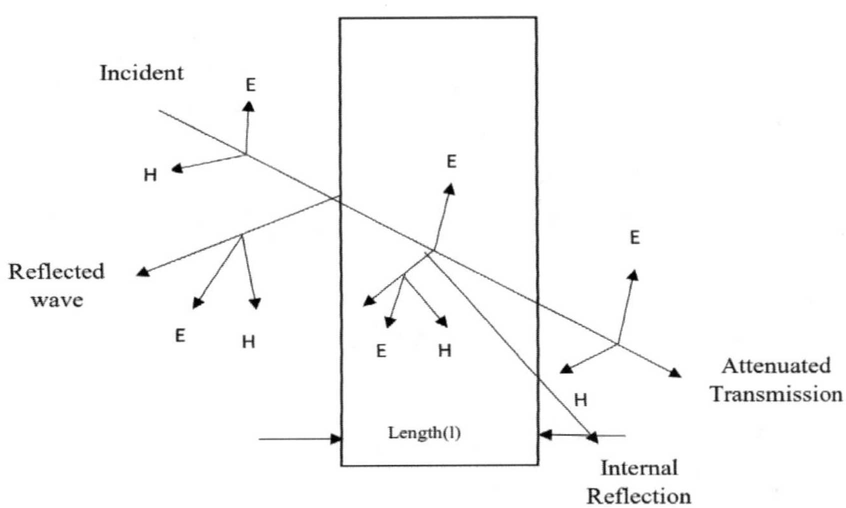

Figure 2: electromagnetic signal represented with reflection and transmission for normal incidences.

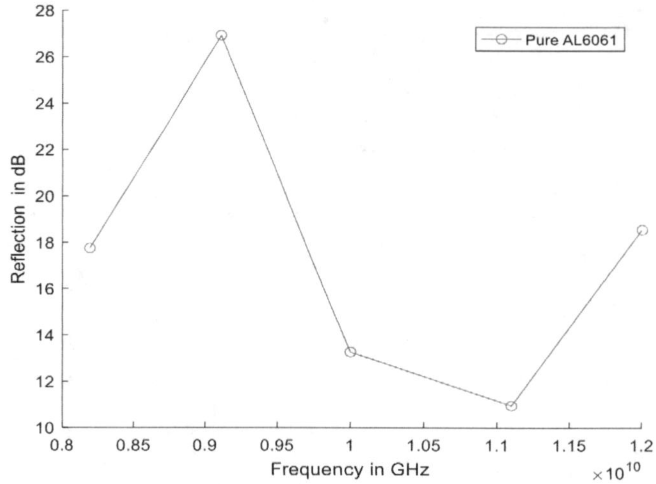

Figure 3: Reflection loss composite material for Pure AL6061.

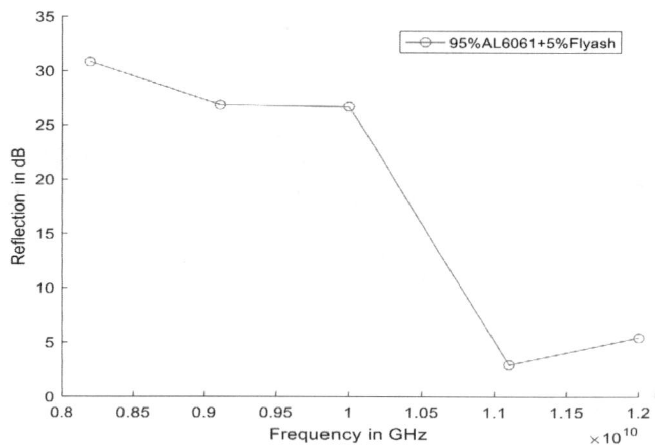

Figure 4: Reflection loss of composite materials 5% of Fly ash and 95% of AL6061.

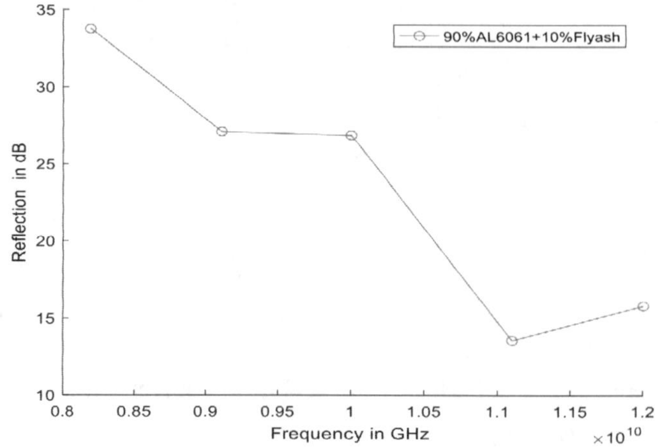

Figure 5: Reflection loss of composite materials 10% of Fly ash and 90 of AL6061.

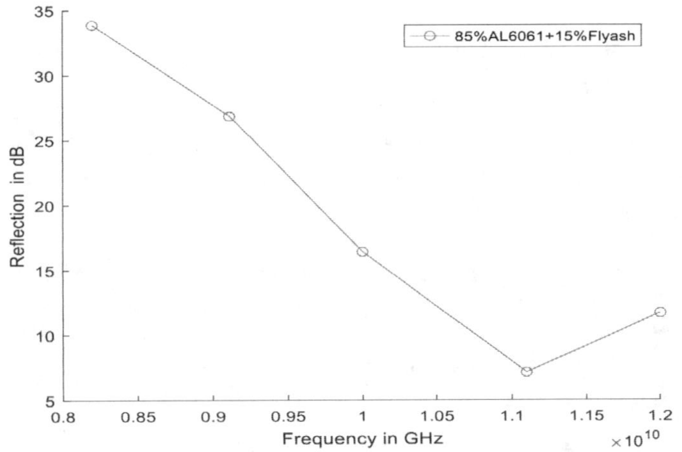

Figure 6: Reflection loss of composite materials 15% of Fly ash and 85% of AL6061.

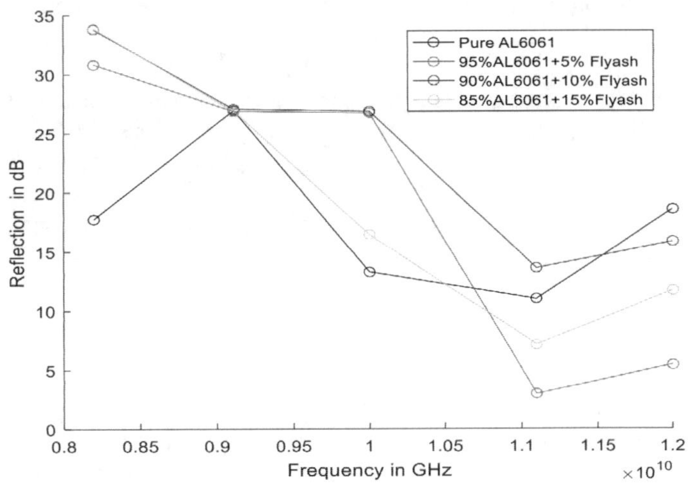

Figure 7: Comparison of different reinforcement of Fly ash with AL6061.

6. Conclusion

The aircraft industry today produces composite materials to defend against high intensity radiated fields. These composites have reduced weight and the best reduction of reflection. The AL6061 metal composite material supports fly ash at different rates in this present work (5 percent ,10 percent ,15 percent). The particular numbers represent the loss of reflectivity for a certain fly ash level. As the fly ash is introduced to the AL6061, reflection loss grows. Pure AL6061 reflects 27.04dB when 5% of fly ash is supported, 33.77dB 10%, 33.88dB is supported, 33.88dB is supported, 15%. The composite materials can be used to protect against radiation fields of high intensity.

7. References

[1]. Larsson, A., 2002. The interaction between a lightning flash and an aircraft in flight. *Comptes Rendus Physique*, *3*(10), pp.1423-1444.

[2]. Zhang, B., Soltani, S.A., Le, L.N. and Asmatulu, R., 2017. Fabrication and assessment of a thin, flexible surface coating made of pristine graphene for lightning strike protection. *Materials Science and Engineering: B*, *216*, pp.31-40.

[3]. Zantout, A.E. and Zhupanska, O.I., 2010. On the electrical resistance of carbon fiber polymer matrixcomposites. *Composites Part A: Applied Science and Manufacturing*, *41*(11), pp.1719-1727.

[4]. Tianchun, Z., Jin, W., Keyi, M. and Zhenyu, F., 2011. Simulation of lightning protection for composite civil aircraft. *Procedia Engineering*, *17*, pp.328-334.

[5]. Obaid, A.A. and Yarlagadda, S.H.R.I.D.H.A.R., 2008. Structural performance of the glass fiber–vinyl ester composites with interlaminar copper inserts. *Composites Part A: Applied Science and Manufacturing*, *39*(2), pp.195-203.

[6]. Black, S., 2013. Lightning strike protection strategies for composite aircraft. *High-Performance Composites*, *23*, pp.52-61.

[7]. Lal, A. and Kanu, N.J., 2020, March. The nonlinear deflection response of CNT/nanoclay reinforced polymer hybrid composite plate under different loading conditions. In *IOP Conference Series: Materials Science and Engineering* (Vol. 814, No. 1, p. 012033). IOP Publishing.

[8]. Jain, N., Gupta, E. and Kanu, N.J., 2021. Plethora of Carbon nanotubes applications in various fields–A state-of-the-art-review. *Smart Science*, pp.1-24.

[9]. Gagné, M. and Therriault, D., 2014. Lightning strike protection of composites. *Progress in Aerospace Sciences*, *64*, pp.1-16.

[10]. Kanu, N.J., Gupta, E., Vates, U.K. and Singh, G.K., 2019. Self-healing composites: A state-of-the-art review. *Composites Part A: Applied Science and Manufacturing*, *121*, pp.474-486.

[11]. Moona, G., Walia, R.S., Rastogi, V., and Sharma, R., 2018. Aluminum metal matrix composites: a retrospective investigation. *Indian Journal of Pure & Applied Physics (IJPAP)*, *56*(2), pp.164-175.

[12]. Gizatullin, Z.M., Gizatullin, R.M. and Nuriev, M.G., 2016. Technique of physical modeling of lightning strike effects on aircraft. *Russian Aeronautics (Iz VUZ)*, *59*(2), pp.157-160.\

[13]. Kanu, N.J., Mangalam, A., Gupta, E., Vates, U.K., Singh, G.K. and Sinha, D.K., 2021. Investigation on secondary deformation of ultrafine SiC particles reinforced LM25 metal matrix composites. *Materials Today: Proceedings*.

[14]. Chauhan, A., Vates, U.K., Kanu, N.J., Gupta, E., Singh, G.K., Sharma, B.P. and Gorrepati, S.R., 2021. Fabrication and characterization of novel nitinol particulate reinforced aluminium alloy metal matrix composites (NiTip/AA6061 MMCs). *Materials Today: Proceedings*, *38*, pp.3027-3034.

[15]. Kanu, N.J., Vates, U.K., Singh, G.K. and Chavan, S., 2019. Fracture problems, vibration, buckling, and bending analyses of functionally graded materials: A state-of-the-art review including smart FGMS. *Particulate Science and Technology*, *37*(5), pp.583-608.

[16]. Ramnath, B.V., Elanchezhian, C., Annamalai, R.M., Aravind, S., Atreya, T.S.A., Vignesh, V., and Subramanian, C., 2014. Aluminum metal matrix composites–a review. *Rev. Adv. Mater. Sci*, *38*(5), pp.55-60.

[17]. Liu, L., Chen, X., Wang, J., Qiao, L., Gao, S., Song, K., Zhao, C., Liu, X., Zhao, D. and Pan, F., 2019. Effects of Y and Zn additions on electrical conductivity and electromagnetic shielding effectiveness of Mg-Y-Zn alloys. *Journal of Materials Science & Technology*, *35*(6), pp.1074-1080

[18]. Han, F. and Zhang, L., 1996, August. Degeneration of shielding effectiveness of planar shields due to oblique incident plane waves. In *Proceedings of Symposium on Electromagnetic Compatibility* (pp. 82-86). IEEE.

[19]. Ma, J., Zhan, M. and Wang, K., 2015. Ultralightweight silver nanowires hybrid polyimide composite foams for high-performance electromagnetic interference shielding. *ACS applied materials & interfaces*, *7*(1), pp.563-576.

[20]. Swamy, A.R.K., Ramesha, A., Kumar, G.V. and Prakash, J.N., 2011. Effect of particulate reinforcements on the mechanical properties of Al6061-WC and Al6061-Gr MMCs. *Journal of minerals and materials characterization and engineering*, *10*(12), p.1141.

[21]. Selvam, J. David Raja, DS Robinson Smart, and I. Dinaharan. "Microstructure and some mechanical properties of fly ash particulate reinforced AA6061 aluminum alloy composites prepared by composting." *Materials & Design* 49 (2013): 28-34.

[22]. Dou, Zuoyong, et al. "Electromagnetic shielding effectiveness of aluminum alloy–fly ash composites." *Composites Part A: Applied Science and Manufacturing* 38.1 (2007): 186-191.

[23]. Casey, Kendall F. "Electromagnetic shielding behavior of wire-mesh screens." *IEEE transactions on Electromagnetic Compatibility* 30.3 (1988): 298-306.

[24]. Cheng, L., Zhang, T., Guo, M., Li, J., Wang, S., and Tang, H., 2015. Electromagnetic shielding effectiveness and mathematical model of stainless-steel composite fabric. *The Journal of The Textile Institute*, *106*(6), pp.577-586.

[25]. Manimaran, R., Jayakumar, I., Mohammad Giyahudeen, R. and Narayanan, L., 2018. Mechanical properties of fly ash composites—A review. *Energy Sources, Part A: Recovery, Utilization, and Environmental Effects*, *40*(8), pp.887-893.

[26]. Hasar, Ugur Cem. "Permittivity measurement of thin dielectric materials from reflection-only measurements using one-port vector network analyzers." *Progress in Electromagnetics Research* 95 (2009): 365-380.

[27]. Lovat, Giampiero, Paolo Burghignoli, and Salvatore Celozzi. "Shielding properties of a wire-medium screen." *IEEE transactions on electromagnetic compatibility* 50.1 (2008).

Statistical Quality Study of the Parts Produced in an Automobile Industry: A Daimler India Case Study

Tanmay Pati[1], Shruti Kabra[1], Utkarsh Chadha[1]

[1]Department of Manufacturing Engineering, School of Mechanical Engineering (SMEC), Vellore Institute of Technology (VIT), Vellore, Tamilnadu, India – 632014.

utkarshchadha1302@gmail.com

Abstract. Every manufacturing industry considers defect-free production in this competitive era. Though it is practically impossible to improve the rate of defect minimization, statistical tools can be used. In this case study, we take data from Daimler India, an automobile part manufacturing industry specialized in manufacturing OEM tools like chassis, oil filter nipple, and so on, and study how many defects are made during the day. The defect-causing process was eliminated after using statistical quality control tools such as np charts and cumulative sum charts, resulting in higher production quality.

Keywords. Control charts, np chart, cumulative sum charts, process quality control, non-conformities, control limits, parts acceptance.

1. Introduction

This research article is centered on the automotive business, an original equipment manufacturing (OEM) organization specializing in stamping metallic pieces, sub-assemblies, welded assemblies, chassis, and, more recently, precision parts. This research intends to investigate and analyze the statistical quality control applied to one-piece reference parts utilizing variable control charts for the parts that the plant produces and inspects. They use both manual and automated methods to inspect the quality of the finished product.

A procedure may be monitored, managed, and advanced the usage of statistical procedure manage. When random causes only influence a process, one speaks of a stable state of statistical control. When systematic influences influence a process, it is said to be statistically uncontrolled. The characteristic of a steady-state process is that they often obey random distributions because the statistical process is constantly subjected to regular variation. The data distribution will vary if a process is out of control. Statistical process control is a top-down tool mainly used as part of a quality improvement policy by the higher management of many industries. On paper, Statistical process control has the following characteristics:

- Statistical process control identifies a process by analyzing data from samples.
- The goal of statistical process control is to track a process's performance across time.
- Another purpose of Statistical process control is to bring a process back to its optimal state.

As an effective Statistical process control tool, the control chart analyses the stability of a process and identifies unstable elements. In today's high-tech era, some products have unusually short life cycles and can be manufactured at an early stage. Greater globalization and manufacturing capacity

boost the competitiveness of automotive factories. Quality in the automotive industry is determined by how satisfied customers are with the products and services. It is a method for increasing capability by reducing variability in the product, delivery, process, material, attitude, and equipment.
As previously stated, the industry is primarily an Original Equipment Manufacturer (OEM) specializing in, among other things, stamping metallic components, subassemblies, welded assemblies, and chassis. Many objects are rejected due to tolerance differences, and the quality of the objects is monitored using Statistical Quality Control tools. This paper aims to reduce the number of ways to detect errors during an inspection, improve production quality consistency, lay the groundwork for attainable specifications, identify bottlenecks and difficulty spots, and determine manufacturing process capability. [1]

2. Concepts & signs of progress

2.1. Concepts

2.1.1. NP control chart-based process control

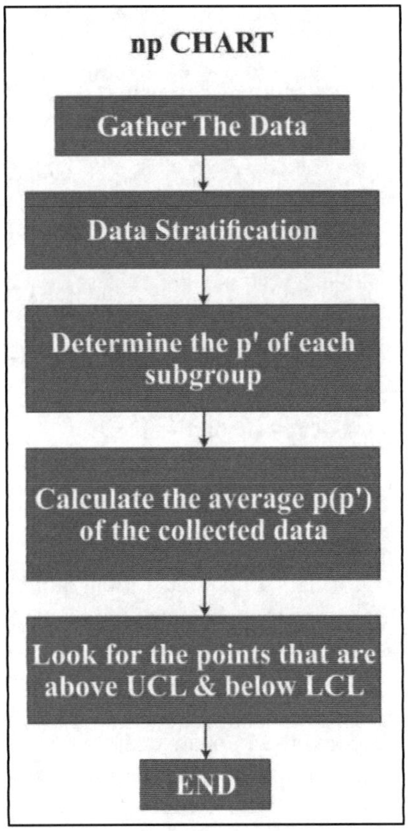

Figure 1. Using an np Chart.

A mechanism for monitoring and controlling the processes is proposed. For error inspection, the proposed method employs a control np chart and a binomial capacity analysis. An np chart is a type of attribute control chart frequently used with data collected in subsets of the same size that vary over time based on the number of harmful objects created. The process's attribute (or characteristic) is always expressed as a yes/no, or pass/fail situation (Figure 1). An np chart would be used to display the incomplete accident reports from a daily sample of five. NP charts are used to determine whether a process is stable and predictable and track the outcomes of various process improvement theories.

2.1.2. CUSUM control chart-based process control

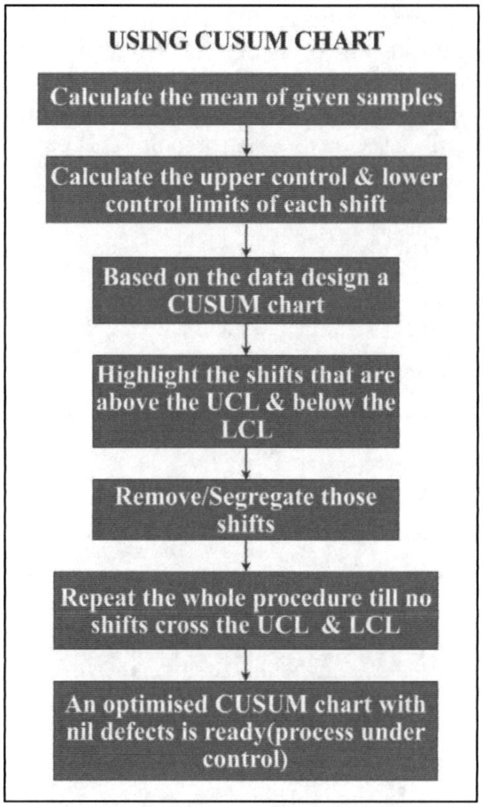

Figure 2. Using a CUSUM Chart.

The author conducted a thorough literature search and discovered that CUSUM-primarily based manipulate charts were used for fault detection and prediction in engineering applications. The proposed method successfully applied the incipient prognosis of fault states of gadgets used in the vehicle sector based on practical information about historical faults. The most important advantage of the tabular CUSUM chart is that it is far more sensitive to minor adjustments that suggest than different manipulate charts, particularly when comparing different manipulate charts (Figure 2).

2.2. Understanding the literature

In the research done in [2], researchers used a statistical analysis of the process data, the P-control chart, to identify abnormal process variability and take immediate action. The proposed statistical process control technique based on the P-control chart can control the process and improve the quality through data acquisition and statistical analysis, as shown in the application example. Using Statistical Process Control, the suggested P chart-based process control approach can assist businesses in improving the quality of their products. (Bo Yang, 2020)1.

The research in [3] is about the cumulative sum-based approach for monitoring conditions and detecting fault diagnosis of wind turbines (WTs) using Supervisory Control & Data Acquisition for the data work. The method is used to monitor the state of a WT in real-time utilizing temperature-related Supervisory Control & Data Acquisition data. (Dao, 2021).

The research in [4] develops two and three adaptive sample size control schemes based on the Max chart to simultaneously monitor the mean and standard deviation of the process since the main disadvantage of the Max chart is that it is a single chart that is variably controlled.A single statistical graph is required; the design and operation of adaptive sample size schemes for this graph proved simpler than the standard X and S graphs.The most dramatic improvement occurred when the two-state scheme was compared to the graphs of the non-adaptive X and S joints (Huang, 2014). This article uses an exponentially weighted average chart to illustrate integrated optimization techniques that solve maintenance and quality control techniques. (Mahmoud, 2020) [5]. The research in [6] discusses the perks, using np reduces the cost factor compared with other traditional charts. Towards the end, they concluded that adequate and accurate implementation of np charts in a real-time industrial scenario could prove to be beneficial in reducing cost and time to a great extent. (Ming Ha Lee, 2020).

The findings in [7] state that the quality of any product in an industry can be described and evaluated mainly based on the following:

1. Features
2. Reliability
3. Durability
4. Performance
5. Conformance to Standards.

The quality of design and conformance is also an important aspect that should be kept in mind while analyzing quality control in manufacturing industries. (Montgomery, 2009) [7].

According to the literature review and case studies conducted in [8], Economic, social, cultural, and technological factors) indicate that the company is subject to solid external influences; these external forces impact Toyota's organizational culture (Toyota Way), response to customers, responsibility, and the inability to accept that Toyota is no longer a Japanese company and is managed by the Toyota family. Executives, employee issues, and training, to name a few because they have had a significant impact on Toyota's failure or success. (Oyinkepreye Bebeteidoh, 2015) [8].

Statistical Process Control run charts are utilized to the utmost full potential and validity during a Quality Index project conducted by [9]. When utilized properly, a STATISTICAL PROCESS CONTROL chart serves as a statistical tool for analyzing a process's performance and a resource for alerting the team about which interventions are likely to be beneficial and worth executing on a bigger scale. (Patrick W. Brady, 2018) [9].

The article [10] they concluded that Most of the processes are within the control range, but the sample shows a rapid decline over time, and it is mentioned that the process parameter vibration alarm is not in a state of statistical control. (Patrik Flegner, 2020) [10].

The Weighted cumulative sum chart has potential uses for monitoring processes with different arithmetic means because any researcher/industry worker can dynamically adjust weights based on observations. Compared to other weighting functions, the wcom (et) - based weighting function can take full advantage of the sample size and observed number. (Qin Zhoua, 2014) [11].

The study's primary aim in [12] is to show how a systematic strategy to use Statistical Process Control across the many phases of production of a specific "X" item in a plant to improve its manufacturing processes may be applied. This strategy, which employs control charts, provides for detecting faults in the company's manufacturing process. Using the model made it possible to bring what had been taught and learned in theory into reality and develop knowledge and learning by overcoming the problems and challenges experienced at each phase of the Statistical Process Control application (Radu Godina, 2016).

In [13], the researchers hybridized the np chart and synthetic chart to produce a Syn-np chart. While the np control chart pointed out that np control charts are widely used to monitor non-compliance percentage processes in manufacturing to determine quality characteristics, they found that Syn np charts are more effective than standard NP control charts. (S. Haridy, 2011)

The paper presents the design algorithm optimizes the charting parameters of each np chart element cumulative sum chart element and the detection power allocation between the two elements, resulting in the best overall performance. As a result, it would be helpful to search for the optimal values of n and h of an np- cumulative sum scheme for such applications as a future project. (Salah Haridy, 2013) [14].

The study aimed to analyze the production process of buff sausages in the food manufacturing business using statistical control charts. All of the sample points are inside the two-sigma control boundaries, as shown in the range chart. (Shrestha, 2020) [15].
The primary purpose of [16] The authors evaluated and compared the performance of the Variable sampling Interval (VSI) np control chart with average time to signal (ATS) properties. After knowing the process parameters, they evaluated them, and the results showed that the performance of the VSI np table was slightly different when testing the number of samples in Phase 1, and it was minimal. (Shu Wu, 2019)
If a quality improvement program focuses on lowering variability in product performance characteristics over time, the quality control system should use the PSD improvement data to detect process mean shifts immediately. The - X -LC chart operates well even in cases with a slight process mean shift due to the employment of the time-varying, i.e., When the Shewhart-X-NLC chart is used without taking into account the learning curve, it has essentially little detection ability (Wang, n.d.) [17].
The findings in [18] this paper presents an optimal algorithm for designing the np control chart. The document states that the optimization algorithm can be applied to concentrated samples, which are more suitable for manufacturers with large production. They concluded by stating that the optimization design is much smaller than any traditional np chart generated. (Wu, 2000)
The locational np chart is proposed in [19], which uses a two-lever inspection system to determine process status. Position information for a non-compliant unit in a sample is examined during the second level inspection Position np chart overcomes both np table errors by default, which is close to the value established for a given sample size. It gives the QA engineer more control over the design of an NP diagram and allows them to get a near minimal sum of false positives and non-compliant costs (WU, 2004) [19].
In [20], the cumulative sum chart is used and put forward in four different scenarios to keep the quality under control and manage small process changes reasonably. It showed that the Shewhart control chart combined with the cumulative sum chart could prove to be highly useful, especially for the managers of the organization who aim to keep the production process and maintenance department under perfect control in terms of quality and standards. (Y. Li1, 2017)
The study in [21] examines the impact of changes in production and control chart factors on the optimal maintenance plan, which provides an integrated model of manufacturing policy, maintenance strategy, and quality control. They also provide a quality control guide that uses a statistical non-compliance checklist to select an appropriate maintenance plan. (Zied Hajej, 2021).

3. Methodology

Figure 3 below depicts the methodology implemented in this research using the data obtained from the authorities.

IOP Conf. Series: Materials Science and Engineering 1206 (2021) 012022 doi:10.1088/1757-899X/1206/1/012022

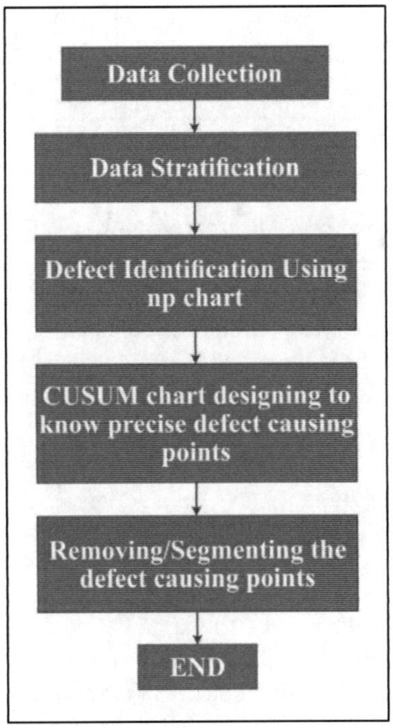

Figure 3. Methodology of the research.

3.1. Data collection

The data collection brings together data from all relevant sources to solve the research problem, test the hypothesis and evaluate the results. The data were collected to use various items from DAIMLER INDIA PVT LTD. Data were collected daily, with more than 200 samples per day, depending on production status, equipment status, and staff status. During the data collection, the data collected the day before was checked twice to rule out checking errors as far as possible. The data collected are shown in Table 1.

Quantitative data collection methods are used based on mathematical calculations and can be presented in different formats. Examples of quantitative data collection and analysis approaches include regression, mean, mode, and median methods. Open-ended systems, focus groups, observation, games or role play, case studies, and other qualitative data collection methods are used to ensure more profound understanding.

Table 1. Depicting the data of parts inspected and rejected on a particular day.

Part Name	Oil Filter Nipple		Circular Oblong		Chassis Brackets		Axe Bush		Centre Bolt Plate	
Time Slot ↓↓	Inspected	Accepted	Inspected	Accepted	Inspected	Accepted	Inspected	Accepted	Inspected	Accepted
8-9 AM	215	213	639	630	29	29	58	58	33	33
9-10 AM	202	199	620	610	21	20	63	62	35	39
10-11 AM	212	211	641	630	35	34	66	66	34	34
11-12 PM	206	214	618	615	22	22	51	51	30	30
12-01 PM	233	221	626	620	26	25	55	24	40	40
01-02 PM	211	205	636	630	25	24	67	67	35	35
02-03 PM	226	221	631	628	38	37	69	69	32	32
03-04 PM	234	201	629	620	31	30	57	57	31	31
04-05 PM	222	211	630	629	27	26	60	59	36	36
05-06 PM	231	210	604	599	23	22	54	54	37	37
06-07 PM	208	207	602	598	37	36	59	59	38	38
07-08 PM	204	197	603	595	33	32	64	64	35	35

08-09 PM	224	221	638	629	40	40	52	50	38	38
09-10 PM	213	207	634	629	32	31	70	70	38	38
10-11 PM	217	215	632	625	36	35	50	50	36	36
11-12 AM	229	220	616	600	39	38	56	56	37	37
12-01 AM	200	196	647	627	34	34	65	65	39	39

3.2. Data stratification

Classifying data, people, and objects into discrete groups or levels is known as stratification. It is a method for analyzing data that are utilized in conjunction with other technologies. The meaning of data can be challenging to see when data from several sources or classifications have been thrown together. One of the seven core quality tools, this data collecting, and analysis technique separates the data to see patterns. Typically, data is stratified in four ways. Who, what, where, and when are the four factors? Two elements are taken into account in this vehicle company's study to check their method.

3.3. Data identification using NP chart

Stratification is the process of categorizing data, people, and objects into discrete groups or tiers. It is a technique for analyzing data that works in tandem with other technologies. When data from multiple sources or categories are thrown together, the significance of the data can be challenging to discern. One of the seven core quality tools, this data collection, and analysis approach isolates the data to see patterns. Data is typically stratified in four ways. The four factors are who, what, where, and when. This automobile company's study considers two factors in order to validate its strategy.
Components that have been inspected vs. parts that have been accepted and time slots are the two variables. There were seventeen samples with three sample sizes recorded at different time windows in the data collection. MS-OFFICE 2017 edition was used to plot the charts and do the calculations in this research.
For np chart:

$$p = \text{Rejected parts / Inspected parts} \quad (1)$$

$$p^{-} = \sum[(\text{Rejected parts})/(\text{Inspected Parts})] \quad (2)$$

$$\sigma = \sqrt{p^{-} * (1 - p^{-})}/\sqrt{\text{Inspected parts}} \ ^{[5]} \quad (3)$$

$$UCL = p+3\sigma \quad (4)$$

$$LCL = p-3\sigma \quad (5)$$

3.4. CUSUM chart designing to know precise defect causing point

The primary purpose of the CUSUM control chart is to keep the process online. The "cumulative sum" for this type of graph is the sum of the deviations from the target of the results of individual samples or mean values of subgroups. The CUSUM control chart illustrates these cumulative deviations over time and shows when the process is "out of control" or, in this case, clearly deviating from the target.
For Cumulative Sum chart:

$$\text{Mean } (\mu) = \text{Accepted samples } (Xi) \text{ / Total number of Time slots.} \quad (6)$$

$$LCL = \mu - 3\,\sigma \quad (7)$$

$$UCL = \mu + 3\,\sigma \quad (8)$$

$$H = 5\sigma \quad (9)$$

$$\mu_1 = \mu + n\,\sigma \quad (10)$$

$$K = (\mu 1 - \mu)/2 \quad (11)$$

$$C_i^{+} = \max\,[0, Xi - (\mu + K) + C_{i-1}^{+}] \ ^{[6]} \quad (12)$$

$$C_i^{-} = \max\,[0, (\mu - K) - Xi + C_{i-1}^{-}] \ ^{[7]} \quad (13)$$

3.5. Removing & Segregating the defect causing point

Manufacturing flaws can occur themselves in quite a few methods in manufacturing, main to product recollects everywhere in the world. The automobile enterprise suffers from the identical defective problem. Human or operator adjustments in manufacturing settings, composition percent adjustments, fabric distribution, and environmental elements should all be reasons for those production failures.

Manufacturing flaws can occur themselves in quite a few methods in manufacturing, main to product recollects everywhere in the world. The automobile enterprise suffers from the identical defective problem. As a result, the manipulate chart has ended up our crucial tool for reinforcing the enterprise's pleasant methods. Table 1 suggests the results. Data stratification is the method of showing all the records acquired from the enterprise in a tabular format. There are quite a few elements produced there; however, it was checked out that the oil clear-out nipple for pleasant manipulates purposes during the research.

4. Results & Discussions

After taking the data of all the products manufactured on a particular day, only a singular product, i.e., oil filter nipple (Table 2), was considered for further calculation. The reason was simple to have a simplified answer within a given time frame, and the procedure to calculate the result for each procedure is the same only.

As stated in the literature review, to detect the defect in any production and remove it, control charts are one of the best tools, and for that exact purpose, np charts and cumulative sum charts were used here.

Table 2. Depicting the data of oil filter nipples that were inspected and rejected on a particular day.

Time Slot	No of Parts Inspected	No. of Parts Accepted
12-01 AM	200	196
9-10 AM	202	199
07-08 PM	204	197
11-12 PM	206	214
06-07 PM	208	207
01-02 PM	211	205
10-11 AM	212	211
09-10 PM	213	207
8-9 AM	215	213
10-11 PM	217	215
04-05 PM	222	211
08-09 PM	224	221
02-03 PM	226	221
11-12 AM	229	220
05-06 PM	231	210
12-01 PM	233	221
03-04 PM	234	201

After collecting the data, data stratification was done, designing a scatter diagram (Figure 4). This scatter diagram gave us an idea of how much variation was between the total number of parts inspected and the total number of parts inspected. The result, as shown in the figure, is quite clear that the variation was somewhat more significant as only 24% of the total parts produced were closer to the line proving that the precision of production versus acceptance is relatively low, and that is undoubtedly a factor that needs to be considered. Moreover, in the day, nearly 134 products were defective out of produced 3695 products.

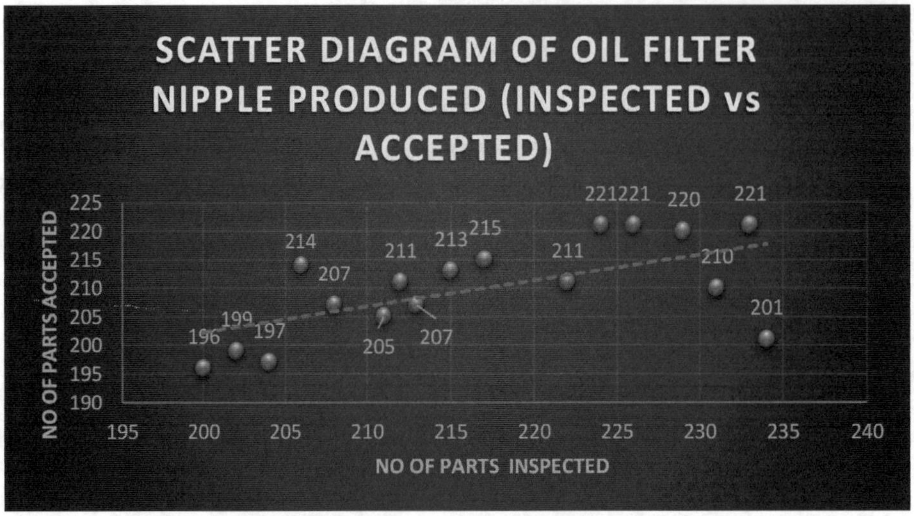

Figure 4. Figure depicting the scatter diagram of oil filter nipple

Table 3. Depicting the data of oil filter nipples inspected on a particular day with the lowest and highest slots.

Time Slot	No of Parts Inspected
8-9 AM	215
9-10 AM	202
10-11 AM	212
11-12 PM	206
12-01 PM	233
01-02 PM	211
02-03 PM	226
03-04 PM	234
04-05 PM	222
05-06 PM	231
06-07 PM	208
07-08 PM	204
08-09 PM	224
09-10 PM	213
10-11 PM	217
11-12 AM	229
12-01 AM	200

Then moving ahead to get an idea of at what time the production was at peak and when it was lowest frequency chart (Figure 5) was designed, another factor behind designing it was that most of the times it happens that when the production capacity is at peak chances are high that defects are produced at that particular time rather than focussing on the shift where fewer parts were produced. With this, the data stratification gets over. Then major work begins that is to find out defect causing shifts initially np chart is taken into consideration, the reason being np charts are the charts that show the points which are above the upper control limits and below lower control limit in a single go, and that gives us a rough idea of which process to keep and which not in order to improve the quality of production.

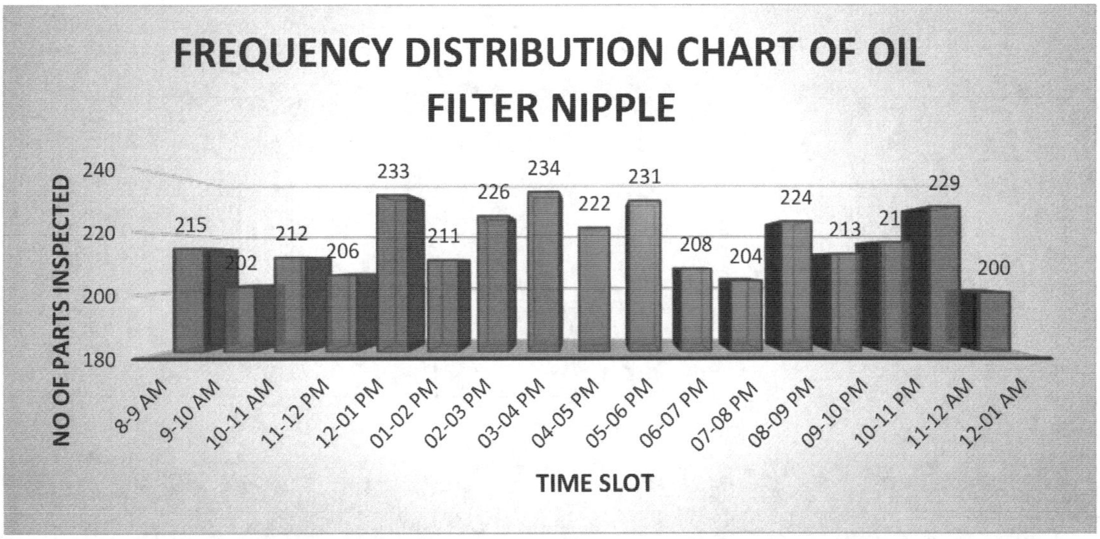

Figure 5. Depiction of the frequency distribution chart of the oil filter nipple.

Here for the np chart calculation, first, the ratio of total rejected part and inspected part was calculated *(refer to table 4),* and then its sum gave us p bar, which was further used for sigma calculation (standard deviation) the value of p- came to be 0.036265,
Then standard deviation was calculated for each shift individually with the formula:

$$\sigma = \sqrt{(\bar{p} * (1-\bar{p}))} / \sqrt{(\text{Inspected parts})}$$

Moreover, as one can see, the value came out to be either 0.012 or 0.013, the value of 3σ was calculated. This 3σ helps calculate the control limits of the particular process now for calculating the control limits' p,' i.e., the ratio of accepted parts and inspected parts need to be calculated individually and also this 'p' acts as the mean Control limit or only control limit in the process, moving ahead after calculating CL the upper control limit is calculated with the formula p+3σ for all process individually. The lower control limit is calculated by p-3σ now in np charts value should always be positive; therefore, the values of lower control limit were negative was converted by default to 0. From the calculated table, one can notice that the process that was highest among the LCL is highlighted in red, and the process with the highest UCL was highlighted in the yellow mark. After this, the np chart was designed using Microsoft Excel (Figure 6), and after creating it, the idea of which process is out of control was not visible as every process was confined accurately within the prescribed upper and lower control limits.

Table 4. Depicting the Np chart calculation along with the process that had negative **p-3σ** (marked in red) and process that has highest **p-3σ** (marked in yellow)

Time Slots	Inspected Parts	Accepted Parts	Rejected Parts	σ	3σ	p	p+3σ	p-3σ	p-3σ
8-9 AM	215	213	2	0.013	0.038	0.009	0.047	0.000	-0.029
9-10 AM	202	199	3	0.013	0.039	0.015	0.054	0.000	-0.024
10-11 AM	212	211	1	0.013	0.038	0.005	0.043	0.000	-0.034
11-12 PM	214	206	8	0.013	0.038	0.037	0.076	0.000	-0.001
12-1 PM	233	221	12	0.012	0.037	0.052	0.088	0.015	0.015
1-2 PM	211	205	6	0.013	0.038	0.028	0.067	0.000	-0.010
2-3 PM	226	221	5	0.012	0.037	0.022	0.059	0.000	-0.015

3-4 PM	234	201	33	0.012	0.036	0.141	0.178	0.105	0.105
4-5 PM	222	211	11	0.012	0.037	0.050	0.087	0.012	0.012
5-6 PM	231	210	21	0.012	0.037	0.091	0.128	0.054	0.054
6-7 PM	208	207	1	0.013	0.039	0.005	0.043	0.000	-0.034
7-8 PM	204	197	7	0.013	0.039	0.034	0.073	0.000	-0.005
8-9 PM	224	221	3	0.012	0.037	0.013	0.051	0.000	-0.024
9-10 PM	213	207	6	0.013	0.038	0.028	0.066	0.000	-0.010
10-11 PM	217	215	2	0.013	0.038	0.009	0.047	0.000	-0.029
11-12 AM	229	220	9	0.012	0.037	0.039	0.076	0.002	0.002
12-1 AM	200	196	4	0.013	0.039	0.020	0.059	0.000	-0.019
TOTAL	**3695**		**134**			**0.599**			

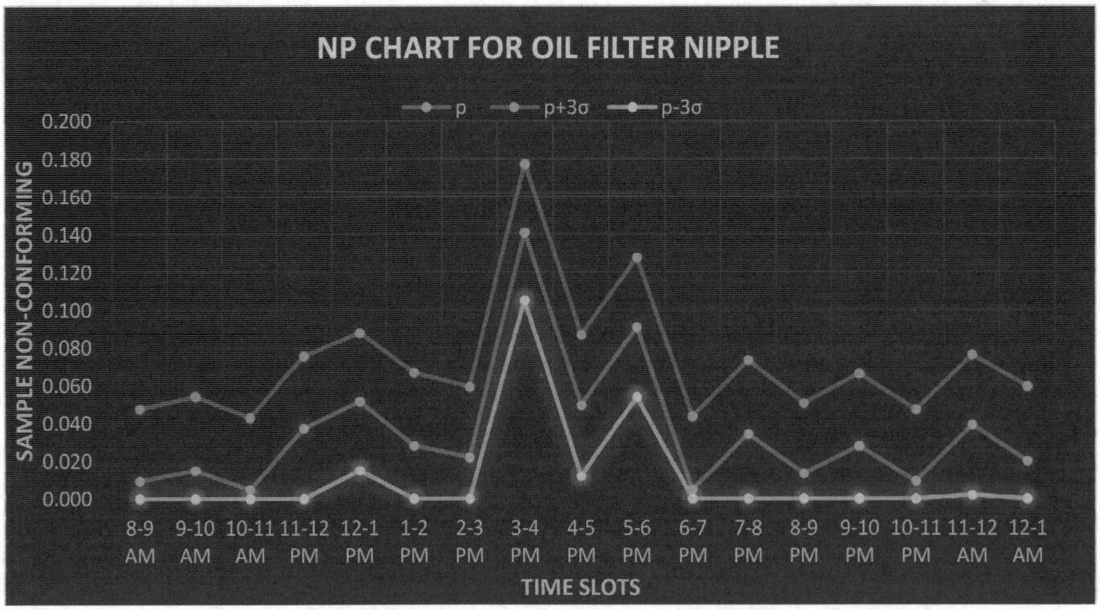

Figure 6. The NP chart of oil filter nipple

To get the exact picture, a cumulative sum chart or CUSUM chart was taken into consideration; CUSUM chart is generally used to detect the defects through small shifts, the primary mechanism behind it is that it utilizes the cumulative sum of deviations from a target then it plots the same for individual measurements so that the point or shift where the defect is there can be identified and can be separated to enhance the quality of produced goods. For that, initially, the mean of accepted samples was calculated that came around 209.5 (Table 5) that also acts as the control limit (CL), now the next thing is to subtract the calculated mean with the accepted sample data of each shift individually and then for Cumulative Index the first value is calculated by subtracting accepted samples with control limit after that the cumulative sum of the accepted sample minus mean is calculated. Based on the findings, the data is tabulated, and then the cumulative sum chart is constructed (Figure 7) with accepted sample and upper control limit, lower limit, and control limit.

Table 5. Depicting the calculated CUSUM chart calculations with processes that need to be eliminated

Time Slots	Accepted Samples (XI)	XI-209.5	CI	CL	UCL	LCL
8-9 AM	213	3.5	4.1	209.5	212.5	206.5
9-10 AM	199	-10.5	-7.0	209.5	212.5	206.5
10-11 AM	211	1.5	-5.5	209.5	212.5	206.5
11-12 PM	206	-3.5	-9.0	209.5	212.5	206.5
12-1 PM	221	11.5	2.5	209.5	212.5	206.5
1-2 PM	205	-4.5	-2.0	209.5	212.5	206.5
2-3 PM	221	11.5	9.5	209.5	212.5	206.5
3-4 PM	201	-8.5	1.0	209.5	212.5	206.5
4-5 PM	211	1.5	2.5	209.5	212.5	206.5
5-6 PM	210	0.5	3.0	209.5	212.5	206.5
6-7 PM	207	-2.5	0.5	209.5	212.5	206.5
7-8 PM	197	-12.5	-12.0	209.5	212.5	206.5
8-9 PM	221	11.5	-0.5	209.5	212.5	206.5
9-10 PM	207	-2.5	-3.0	209.5	212.5	206.5
10-11 PM	215	5.5	2.5	209.5	212.5	206.5
11-12 AM	220	10.5	13.0	209.5	212.5	206.5
12-1 AM	196	-13.5	-0.5	209.5	212.5	206.5
TOTAL	**3561**	**-0.5**				

PROCESS WHICH IS ABOVE THE UCL

PROCESS WHICH IS BELOW THE LCL

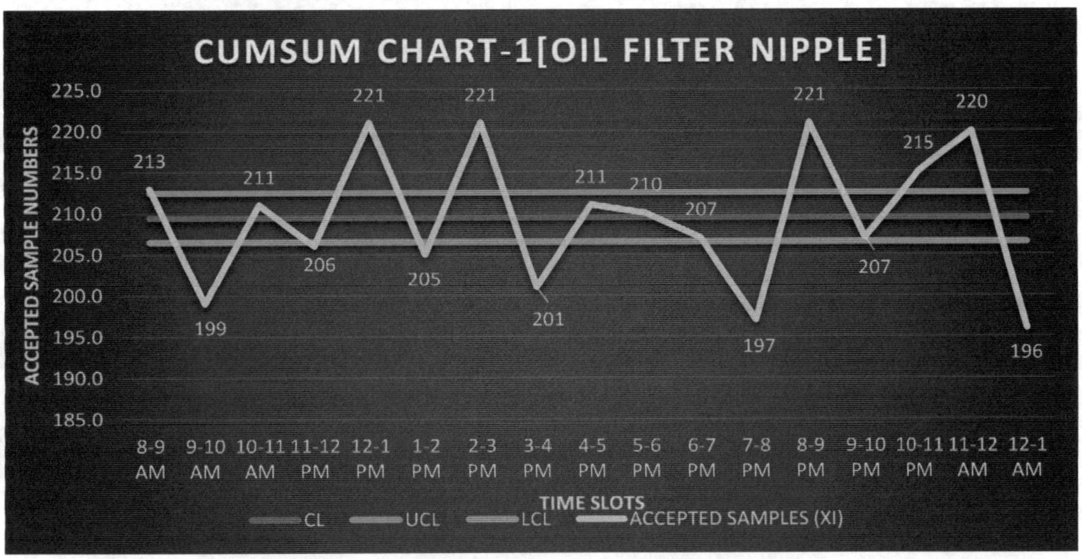

Mean (μ) = 209.5 = CL; σ = 1; UCL = $\mu + 3\sigma$ = 212.5; LCL = $\mu - 3\sigma$ = 206.5

Figure 7. The CUSUM chart of oil filter nipple

The values of shifts that lie outside the upper control limit are highlighted in yellow, and those below the lower control limit are highlighted in red. That process is removed, and then again, the cumulative sum chart is designed with modified data (Figure 8); the modified data is taken after removing the out-of-control limits (Table 6). Now the same procedure that includes calculating the mean that came around 208.7 and then subtracting the calculated mean with the accepted samples and then calculating the cumulative frequency and then calculating the upper control limits and lower control limits for individual shifts and then tabulating it and then finally designing a modified cumulative sum chart is done. It is applied once again to check whether the process is within the limit or not. The findings found that all the processes were within the control limits and that the process now was defect-free.

Table 6. Depicting the modified cumulative sum chart calculations

Time Slots	Accepted (XI)	XI-208.7	CI	CL	UCL	LCL
10-11 AM	211	2.3	3.0	208.7	211.7	205.7
11-12 PM	206	-2.7	0.3	208.7	211.7	205.7
4-5 PM	211	2.3	2.6	208.7	211.7	205.7
5-6 PM	210	1.3	3.9	208.7	211.7	205.7
6-7 PM	207	-1.7	2.2	208.7	211.7	205.7
9-10 PM	207	-1.7	0.5	208.7	211.7	205.7
TOTAL	**1252**	1043.3				
Mean(μ)=	208.7					
CL=	208.7					
UCL=μ+3σ	211.67			208.6667		
LCL=μ-3σ	205.7					

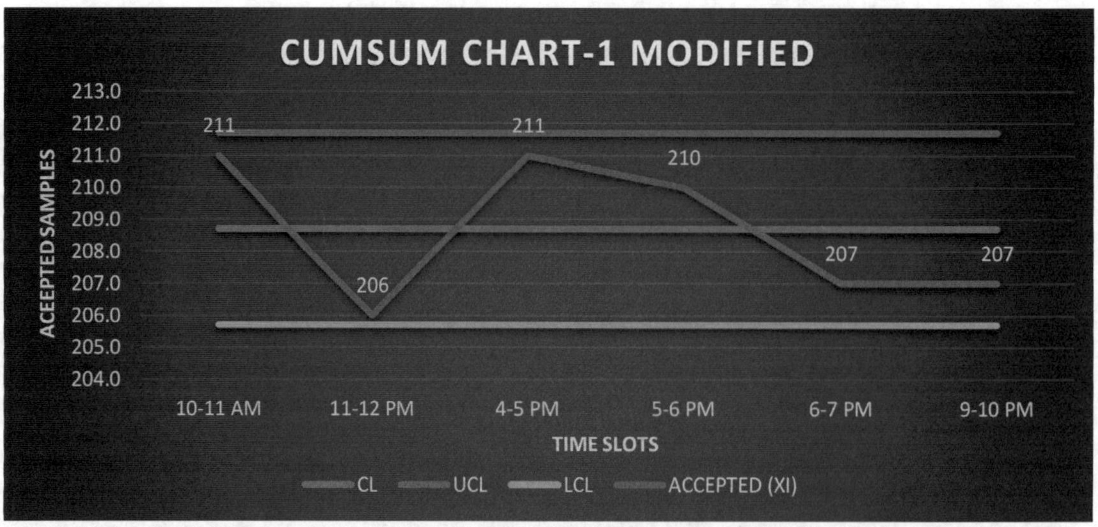

Figure 8. The modified CUSUM chart of oil filter nipple.

By chance, if the control limit had exceeded this process, this procedure would have continued until all the processes were within the control limits.

Now since all the out of control process shifts were separated, these given shifts are the process that would provide quality products with 100% defect-free and that if the out of control limits would have been forecasted in the np control chart itself, then the cumulative chart would not have been used extensively moreover in place of np chart, p chart could also have been used to detect non-conforming units, but then it would have just shown the proportion of non-conforming units rather than an actual process where non-conformities are happening.

5. Conclusion

The primary aim of this research is to remove the defect causing shifts after identifying it, and that was achieved completely using various statistical quality control tools like Np charts and cumulative sum charts. Moreover, this research analyzes only a single factor, i.e., Oil filter nipple, to get a rough idea of how effective the control charts are. The same can be applied to all the products manufactured in the industry for more precise and accurate results. Though one can argue that the np chart failed to confine the results due to which cumulative sum chart was considered, the cumulative sum chart required specific iterations and time to confine to an optimal result. That can be considered a disadvantage here, but that thing can be solved with more advancements and more accurate computerized methods in the future. Apart from that, one can doubt why the exponentially weighed Average chart (EWMA) was not used to detect the non-conformities in small shifts; EWMA cannot predict Cumulative Sum charts precisely. Also, the Shewhart chart has the same disadvantage: it cannot detect small shifts, and here the actual non-conformities were detected only through small shifts. Shewhart charts are way too high sensitive to significant process shifts and add to it the probability of precisely detecting small shifts non-conformities is relatively tiny, so cumulative sum chart was taken into consideration. Also, one must note that from a future perspective cumulative sum chart with computerized methods will be used more regularly than today as it provides instant and accurate results, and with the recent advancements in Cumulative sum charts like V- mask Cumulative sum chart, it obtains a more accurate, and new advancements like combining the Shewhart chart with cumulative sum chart to provide the advantages of both and providing better results will surely help to use cumulative sum charts more extensively indeed aims to increase the quality of production in the long run. [22-28]

Conflict of Interest

The authors declare no conflict of interest. Due permission was taken from the concerned authorities of Diamler India.

Contributions

TP, and SKr: conceptualization, methodology, software, and formal analysis. UC: validation, supervision, project administration. TP, and SKr: investigation. TP, SKr, and UC: resources and data curation. TP, SKr, and UC: writing—original draft preparation. TP, SKr, and UC: writing—review and editing and visualization. All authors have read and agreed to the published version of the manuscript.

References

[1] Aaradhya Sharma, Ananya Chouhan, Lakshmi Pavithran, et al., Implementation of LSS framework in automotive component manufacturing: A review, current scenario and future directions, Materials Today: Proceedings, 2021, ISSN 2214-7853, https://doi.org/10.1016/j.matpr.2021.02.374.

[2] Bo Yang, Y. H. (2020). Research on data analysis and Quality Control based on P control chart. 13th International Congress on Image and Signal Processing, BioMedical Engineering and Informatics.

[3] Dao PB. A CUSUM-Based Approach for Condition Monitoring and Fault Diagnosis of Wind Turbines. Energies. 2021; 14(11):3236. https://doi.org/10.3390/en14113236

[4] Huang, C. C. (2014). Max control chart with adaptive sample sizes for jointly monitoring process mean and standard deviation. Journal of the Operational Research Society, 1788-1799.

[5] M.A. Noman, A. A.-S.-A. (2020). A Model For Maintenance Planning and Process Quality Control Optimization Based on Ewma and Cusum Control Charts.
[6] Ming Ha Lee, M. B. (2020). Economic-statistical design of synthetic np chart with estimated process parameter. Plos One.
[7] Montgomery, D. (2009). Quality Improvement in the Modern Business Environment. In D. Montgomery, Introduction to Statistical Quality Control (pp. 6-9).
[8] Oyinkepreye Bebeteidoh, S. A. (2015). Manufacturing defects in the automobile industry, a case study of the remote causes and effects of Toyota's transmission malfunctions in cars. International Journal of Engineering and Applied Sciences (IJEAS) .
[9] Patrick W. Brady, M. J. (2018). Displaying and Analyzing Quality Improvement Data. Journal of the Pediatric Infectious Diseases Society.
[10] Patrik Flegner, J. K. (2020). Statistical Process Control Charts Applied to Rock Disintegration Quality Improvement. Applied Sciences, MDPI.
[11] Qin Zhoua, W. H. (2014). A comparison of weighted CUSUM procedures for monitoring process proportions with varying. International Journal of Production Research,.
[12] Radu Godina, J. C. (2016). Quality Improvement With Statistical Process Control in the Automotive Industry. International Journal of Industrial Engineering and Management (IJIEM), 1-8.
[13] S. Haridy, Z. W. (2011). Synthetic-np Chart for Attributes. IEEE.
[14] Salah Haridy, Z. W.-J. (2013). An optimisation design of the combined np-CUSUM scheme for attributes. European J. Industrial Engineering.
[15] Shrestha, N. (2020). Application of Statistical Process Control Chart in Food Manufacturing Industry. International journal of Engineering, Business and Management (IJEBM).
[16] Shu Wu, M. T. (2019). Optimal design of the variable sampling interval np control chart for monitoring service process . Science Direct.
[17] Wang, L. y.-R. (n.d.). Quality Control Scheme with Learning Effect Consideration. IEEE Explore.
[18] Wu, H. L. (2000). Optimal NP Control Charts for Concentrated. IEEE. ICMIT.
[19] WU, Z. (2004). A np Chart Using the Locational Information of Nonconforming Units. Electronics Packaging Technology Conference. IEEE.
[20] Y. Li, L. C. (2017). Jointly Optimal Design of Perfect Maintenance Policy and CUSUM Control Chart. IEEE, (pp. 1920-1924).
[21] Zied Hajej, A. C. (2021). An Integrated Model of Production, Maintenance, and Quality Control with Statistical Process Control Chart of a Supply Chain. Applied Sciences, MDPI.
[22] R. Sivasubramani, Aarya Verma, G. Rithvik, Utkarsh Chadha, Selvaraj Senthil Kumaran, Influence on nonhomogeneous microstructure formation and its role on tensile and fatigue performance of duplex stainless steel by a solid-state welding process, Materials Today: Proceedings, Volume 46, Part 17, 2021, Pages 7284-7296, ISSN 2214-7853, https://doi.org/10.1016/j.matpr.2020.12.983.
[23] Raj, A., Ram Kishore, S., Jose, L. et al. A survey of electromagnetic metal casting computation designs, present approaches, future possibilities, and practical issues. Eur. Phys. J. Plus 136, 704 (2021). https://doi.org/10.1140/epjp/s13360-021-01689-1
[24] Selvaraj S. K., Srinivasan K., Chadha U., et al. Contemporary Progresses in Ultrasonic Welding of Aluminum Metal Matrix Composites, Frontiers in Materials,(2021) 8(126), DOI: 10.3389/fmats.2021.647112
[25] Virmani, K., Deepak, C., Sharma, S. *et al.* Nanomaterials for automotive outer panel components: a review. *Eur. Phys. J. Plus* **136,** 921 (2021). https://doi.org/10.1140/epjp/s13360-021-01931-w
[26] Ghimire, T., Joshi, A., Sen, S. et al. Blockchain in Additive Manufacturing Processes: Recent Trends & its Future Possibilities. (2021), Mater. Today. Proc.

[27] Menon, A. P., Lahoti, V., Gunreddy, N. et al., Quality Control Tools and Digitalization of Real-Time Data in Sustainable Manufacturing. (2021), Mater. Today Proc.

[28] Gupta P., Krishna C., A. V., Patel S.S., Kapruan C., Chadha U., Selvaraj S.K. Industrial Internet of Things in Smart/Intelligent Manufacturing: A Review, Approaches, Opportunities, Open Challenges, and Future Directions (2021), Mater. Today Proc.

Economics-Based Payback and Life Cycle Cost Savings Assessment of Inverter Type Air Conditioners

Sunil Kumar Gupta*, B B Arora, Akhilesh Arora

Department of Mechanical Engineering, Delhi Technological University, Delhi-110042, India.
*Corresponding Author Email: sunilgupta_me@yahoo.co.in

Abstract. This paper investigates the options for a consumer to choose between different inverter and non-inverter type residential air conditioners (ACs) concerning their payback periods (PPs) and Life Cycle Cost (LCC) savings. The economics-based analysis is carried out to evaluate the PPs by which the costs associated with owning and operating 3-star, 4- star, and 5-star inverter AC models are recoverable compared to non-inverter type (baseline model). The product costs (C_P), the repair costs (C_R), the maintenance costs (C_M), and the energy costs (C_E)are taken as the decision parameters for evaluating the payback periods (PPs). It is shown that PP strongly depends on the energy cost. Estimates of PPs are calculated as the future value of present cost involved in buying, maintaining, and running the ACs. It is concluded that inverter technology can save electrical energy by 12-22.4% compared to a non-inverter air conditioner. PP of 3-star inverter AC is estimated 2.17 years while PPs of 4-star and 5-star inverter ACs are 2.42 years and 2.33 years for 10 hours operation in a day when a higher slab of unit energy cost is considered. Depending on PP and daily usage requirements, a consumer can choose either a less efficient AC priced at a lower initial cost but more running cost or a more efficient AC priced at a higher initial cost but lesser running cost.

Keywords. Inverter AC, life cycle cost, payback period

1. Introduction

New Delhi, India has a composite climate i.e., very hot-dry climate in summer (from April to June) and hot-humid climate in the rainy season (from July to September). Therefore, to counterbalance these extreme climatic situations, most people are switching over to own and use air conditioners (ACs). However, the high running cost of air conditioner is significant concern for consumers in general. To cut the electricity bills, they look for higher energy-efficient air conditioners.

The energy efficiency and performance of refrigeration cycle can be improved to a great extent by employing variable speed technology [1]. Ha & Jeong showed that a variable speed compressor contributed 21% reduction of the annual power consumption over a fixed speed compressor [2]. Huge electricity savings were obtained by replacing less energy efficient (e.g.,1- star ratings) air conditioners with more energy-efficient inverter ACs incorporating a variable frequency compressor [3].

Energy Efficiency Rating (EER) of the compressor was found to have much effect on system performance and economics [4]. In comparison to a base-case, a noticeable reduction in energy cost was obtained by investing more in equipment, primarily improving its electric motor, evaporator, and compressor [5]. Higher the cooling loads, the more was the product cost per unit of cooling capacity, while the energy cost per unit cooling load was decreased [6]. It was demonstrated that a higher value on more efficient components is justified when the electricity price is high [7]. Optimizing the energy savings require a variable-speed compressor's cost not to exceed 26% than that of a fixed speed compressor. When the aim is to optimize the cost savings, a variable-speed compressor could be justified for its price, not exceeding 20% than that of a fixed speed compressor [8].

The effects of energy reduction using the inverter air conditioner are mainly attributed to the daily or seasonal changes in the temperature and cooling load. Inverter technology saved energy by 18.3 to 51.7% in different climates. It was shown that the energy savings are enhanced during non-peak summer months and at part load operations.

It is possible to predict the payback period (PP) over the baseline model through economic modelling. Accordingly, a consumer can assess the cost-benefit through efficiency enhancement of AC [9].

From the open literature, it is understood that very little information is available on the payback periods (PPs) and life cycle cost savings of inverter ACs over non-inverter ones. The author aims to determine the PPs and life cycle cost savings through economics analysis of inverter ACs with different star ratings against non-inverter AC and investigate the effects of different variables on the PPs.

2. Materials and Methods

Residential air conditioners meet the cooling requirements by employing a vapor compression refrigeration system due to its suitability to almost all types of applications with a wide range of cooling capacities. The input to the vapor compression refrigeration system is in the form of mechanical energy required to run the compressor obtained by an electric motor.

An inverter is a device that can be used to control electric voltage, current, and frequency. The inverter air-conditioner's function is to vary the compressor's power supply frequency to adjust its cooling/heating capacity. The inverter continuously adjusts the compressor speed by changing the frequency to maintain a specific desired room temperature by controlling the refrigerant flow rate. After attaining the preset temperature, the unit annihilates any temperature variations by adjusting its capacity. If the load increases, air-conditioner works at a higher capacity, and if there is a reduction in load, the unit switches to lower capacity.

On the other hand, non-inverter air conditioners operate at a fixed cooling capacity only. The compressor of the non-inverter air-conditioner repeatedly stops and starts to control the indoor temperature.

Energy Star rating is a ready reference to compare and choose from among the best options available. In India, the Bureau of Energy Efficiency develop these ratings. A higher star rating corresponds to the higher energy efficiency of AC. As the cost increase is associated with the technological development of components (mainly compressor) to improve efficiency, the product's cost increases with an increase in its energy efficiency or star rating.

2.1 Mathematical Modelling

Various costs involved in purchasing and operating an air conditioner are product cost (C_P), installation cost (C_{INS}), repair cost (C_R), maintenance/service cost (C_M), and energy cost (C_E).

Thus, total cost comprises the sum of all the above costs as:

$$C_{Total} = Initial\ Cost\ (IC) + Operating\ Cost\ (OC)$$

$$\mathrm{C_{Total}} = \text{Initial Cost (IC)} + \text{Operating Cost (OC)}$$

where,

$$IC = Product\ Cost\ (C_P + \ Installation\ Cost\ (\ C_{INS})$$

and

$$OC = Energy\ Cost\ (C_E) + \ Repair\ Cost\ (C_R) + \ Maintenance\ Cost\ (C_M)$$

$$C_{Total} = C_P + C_{INS} + C_E + C_R + C_M \quad (1)$$

2.1.1 Product cost (C_P). The product cost varies according to the inclusion of some unique features (such as smart eye) or more energy-efficient or both. More efficient products employ components with improved technology, which results in an exponential hike in the product cost. Therefore, a higher star (e.g.5-star) rated product of the same capacity is costlier than a low star (e.g.3-star) rated product.
Installation cost (C_{INS}). Generally, the installation charge is not included in the product cost. For a standard installation, the charge is usually at par irrespective of the air conditioner's model or brand. Also, as a marketing strategy to attract consumers, manufacturers provide free installation of air conditioners or include this cost in the product price. Therefore, for all selected models of ACs this component (C_{INS}) has been excluded in the calculations.

2.1.2 Energy cost (C_E). Energy cost is directly proportional to the energy usage (EU) of AC. Unit capacity, energy efficiency ratio (EER) and operating time are the three factors which quantify the energy usage [10] where

$$EU = Unit\ capacity * operating\ time/EER \tag{2}$$

The following expression can find the energy cost,

$$C_E = EU * C_{UE} \tag{3}$$

Where C_{UE} is the cost per unit of energy consumed.
In India, to encourage consumers for efficient use of electrical energy, many state governments have prescribed different slabs of the tariff. Electricity tariff in New Delhi (India) for domestic consumers varies from ₹ 3 up to 200 units (kWh) to ₹ 8 above 1200 units. The use of AC generally falls in a higher consumption slab [9] .

2.1.3 Repair cost (C_R). Usually, the manufacturers provide warranty for carrying out any repair free of cost in the first year of purchase. It is unpredictable to state when the air conditioner will undergo some breakdown and what price will be applicable to carry out the appliance repair. However, a simple correlation between repair cost (C_R), product cost (C_P) and unit's lifetime has been proposed [10]as follows,

$$C_R = 0.5 * \frac{C_P}{Unit's\ lifetime} \tag{4}$$

The air conditioner's useful lifetime is assumed ten years (assumption based on compressor warranty by most manufacturers).

2.1.4 Maintenance cost (C_M). The air conditioner's working is affected by prolonged use and surrounding conditions that differ from place to place. Maintenance cost includes the general checks and works carried out for maintaining the AC in good working condition.

After the first year of free maintenance/service, the annual maintenance/service for subsequent years is generally provided by the manufacturer on a yearly payment basis. The maintenance/ service cost (C_M) has been estimated to 2.5% of the product cost (C_P) [10], i.e.

$$C_M = 0.025 * C_P \tag{5}$$

2.2 Payback Period (*PP*) *and Life Cycle Cost* (*LCC*)

Payback period (PP) is the shortest time in which higher initial investment in buying a more efficient AC is received back in terms of its lower running cost i.e., energy cost (C_E) [11]. The payback period can be used as a tool by a consumer to select among costly but more energy efficient and economical but less efficient models.

2.2.1 Simple Cost. PP with the simple cost is calculated using the following equation [11]:

$$PP = \frac{\Delta Initial\ Cost(IC)}{\Delta Operating\ cost\ (OC)} \tag{6}$$

For the first year, annual maintenance cost is zero since the product is under warranty by the manufacturer, and the repair cost has been assumed to incur at the end of the fifth year. Therefore, the first year's operating cost covers only the energy cost, which is assumed to incur for six months (operating period of AC) in a year.

2.2.2 Compounding Cost. Estimating the future value of money associated with owning and operating the AC unit provides the consumer's criteria to justify its selection. Since the consumer always spends the money either for purchase or operating the AC therefore, the future value (FV) will reflect the compounding effect of money spent during the lifetime of AC. With compounding interest rate, FV of non-recurring present value (PVNR) has been calculated as

$$FV_{NR} = PV_{NR} * (1+i)^n \quad (7)$$

and FV for the recurring expenditure amounts (A_R),

$$FV_R = \frac{A_R(1+i)^n - 1}{i} \quad (8)$$

where '*i*' is the rate of interest per annum compounded monthly, and '*n*' is the number of months for which future value was calculated.

It may be noted here that costs (C_P *and* C_R) are assumed to incur only once in the unit's lifetime. The energy cost is paid at the end of every month. The maintenance costs are born at the beginning of every year, excluding the year of purchase.

Average operating time per day has been considered as 8 hours, 10 hours & 12 hours. Annual operation of air conditioner has been assumed for 180 days. Interest rate (i) is taken as 3% & 6% per annum and unit's lifetime (n) = 120 months (10 years).

Salvage value at the end of the product's life was not considered as it was assumed to be same irrespective of models.

2.3 Selection of Air Conditioners

Three different inverter air conditioners, i.e., 3-star, 4-star, and 5-star (Daikin make), have been selected with a non-inverter AC as a baseline model for this study. The chosen models have similar capacities, with comparable differences within 5%.

In this study, we assess the compounding cost-based payback periods (PPs) for the higher efficiency models of inverter air conditioners by evaluating additional prices paid for their efficiency enhancement (including costs for their non-efficient features) against the price of baseline efficiency product. Selected models of the split air conditioner and product costs with some critical specifications have been shown in Table 1. In the present analysis, 1.5 ton (5.2 kW) non-inverter AC is chosen as a baseline model with its product cost as ₹ 31,990.

Table 1. Product cost and specifications of selected models of Daikin Air Conditioners (as on 1st August,2021)

Parameter	3-Star Non Inverter AC (Baseline)	3-Star Inverter AC	4-Star Inverter AC	5-Star Inverter AC
Retail Price (₹)	31990	37100	39990	42990
Cooling capacity Rated Full/Half (min.~max.)(kW)	5.2	5.0/2.5 (2.5~5.0)	5.0/2.5 (2.5~5.0)	5.2(2.5-5.7)
Power consumption Rated Full/Half (min.~max.) (W)	1425	1752/615 (615-1752)	1680/510 (510-1680)	1415 / 475 (375~1650)
ISEER(Wh/Wh)	3.65	3.99	4.17	4.7
Operating current (A)	6.5	7.71	7.4	6.37

Figure 1 display cash flows for 5-star inverter AC model. The initial cost is paid at the beginning of the investment or purchase of AC. The energy cost (C_E) is paid at the end of months 1 to 6. The pattern repeats for the month 13 to 18 and so on. The final payment of energy cost is made at the end of month 114. The annual maintenance cost (C_M) is paid either at the beginning of the 2nd year or at the end of the first year and after that, C_M is paid at the annual intervals. The repair cost (C_R) has been taken as one time cost incurred in the middle of the product's life, i.e., at the end of 5 years. The compound costs are calculated using equations 7 and 8 to estimate the payback periods (PPs) of all three selected models.

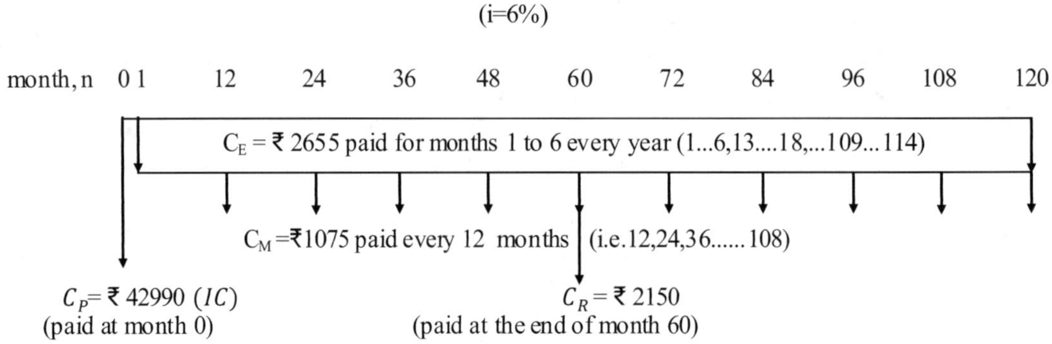

Figure 1. Cash flow for 5-star inverter AC

Changing input variables that affect the compounding cost in arriving at PP are; energy usage, operating hours, and interest rate. To estimate the energy cost, we first need to determine energy usage or electricity consumption. In this paper, we compare the per day energy usage based on 8 hours, 10 hours, and 12 hours of operation time.

3. Results and Discussion

It is expected that the higher efficient model ACs shall payback in terms of energy savings accumulated every month to compensate for the initial higher cost. The results in this section have been discussed with unit energy cost ₹ 8.00 and interest rate 6%.

3.1 Effect of unit cost of energy

A 5-star inverter AC has lower PP than 3 or 4-star models with lower usage and more economical energy tariff (Table 2).

Table 2. PPs (in Years) at different tariff and interest rates (t=10 hour)

	Simple cost			*i*=3 %			*i*=6%	
C_E (₹)	4.5	6.5	8	4.5	6.5	8	4.5	6.5	8
3-Star Inverter AC	5.33	3.59	2.08	6.42	4.08	2.17	7.33	4.25	2.17
4-Star Inverter AC	5.58	3.85	2.46	7.25	4.08	2.33	8.17	4.42	2.42
5-Star Inverter AC	4.25	2.95	2.25	4.5	3.17	2.33	5.33	3.25	2.33

It is evident from figure 2 that for lower electricity tariff, the payback periods are more and PPs decrease at a higher unit price of energy.

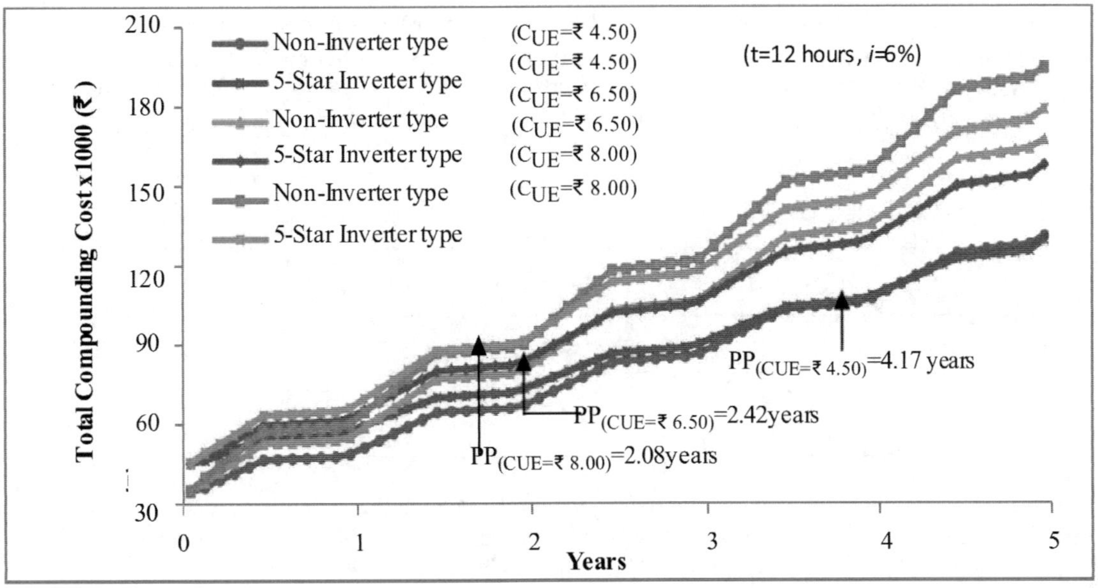

Figure 2. Payback periods of inverter type (5-Star) AC with different unit energy costs

3.2 Effect of operating time

As shown in figure 3, payback periods for 5-star inverter AC are 3.25 years, 2.33 years, and 2.17 years. Inverter type 4-star and 5-star ACs show higher *PPs* than a 3-star inverter AC corresponding to their operating hours. The steeper curves show the AC's energy consumption phases, and flatter portions are the non-operational slots. The payback periods (*PPs*) for all the models decrease with higher usage of the air conditioner, i.e., increase in running time. Thus, more running hours help to recover initial costs to consumers quickly.

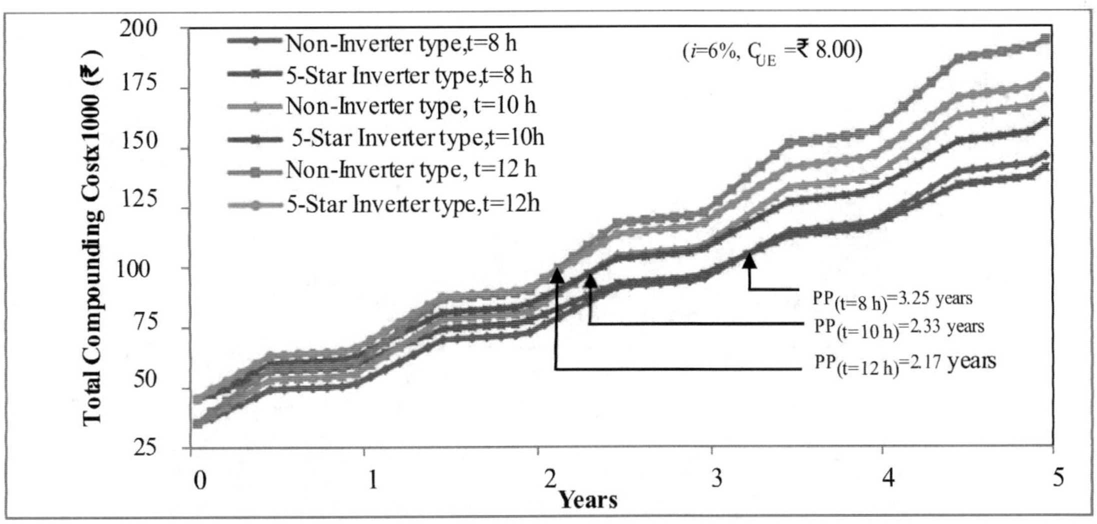

Figure 3. Effect of operating hours on payback periods for 5-star AC

3.3 Effect of interest rate

According to Table 3, the higher interest rate has a negative influence on the PPs of the inverter type ACs. Simple costing presents the lowest PP. For a 3-star inverter AC with 8 hours average daily operating time, PP increases from 2.33 years with simple cost to 2.42 years (3.9% increase) at the interest rate of 3%. PP moves to 2.5 years (3.3% increase) if the interest rate is changed from 3% to 6%.

Table 3. Effect of interest rate on payback periods of inverter ACs (C_{UE} =₹8.00)

Usage time	simple cost	*i*=3%	*i*=6%	simple cost	*i*=3%	*i*=6%	simple cost	*i*=3%	*i*=6%
8 hr	2.33	2.42	2.5	3.08	3.25	3.33	3.08	3.25	3.25
10 hr	2.08	2.17	2.17	2.46	2.33	2.42	2.25	2.33	2.33
12 hr	1.42	1.42	1.42	2.05	2.17	2.17	2.08	2.08	2.17

The adverse influence of interest rate owes to a relatively higher C_P difference compared to the monthly variation in C_E between non- inverter and inverter types. It is also noted that the effect of interest rate on PP vanishes at longer operating time (12 hours daily).

Figure 4 shows PPs at different interest rates with ₹ 4.50 unit energy cost. The results establish that simple cost fetches the PP far earlier than compounding interest rates.

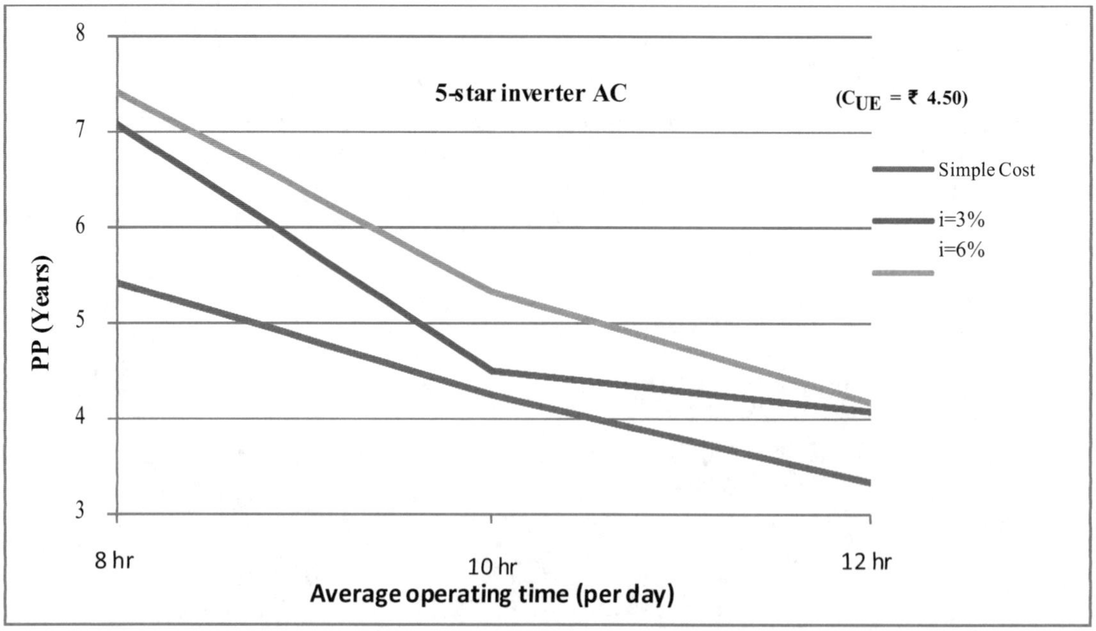

Figure 4. PPs at different interest rates and C_{UE} =₹4.50 for 5-star inverter AC

3.4 Cost-benefits of efficiency improvement

The difference in the initial cost of a more efficient and a baseline product is supposed to transfer cost-benefits of efficiency improvement to the consumer in energy cost savings. The cost to the energy-saving ratio for 5-star AC is higher than that of 3-star AC (Table 4). These models offer consumers ease of operation (such as mobile app controller) and aesthetic appearance etc. but do not contribute to energy efficiency improvement. Due to higher cost margins for these additional features, the initial cost is significantly higher than that anticipated for just improving efficiency.

Table 4. Energy Savings by more efficient ACs with simple costing and 6% interest rate

	Operation Time	3-Star Inverter AC		4-Star Inverter AC		5-Star Inverter AC	
		C_{UE}=8.00	C_{UE}=4.50	C_{UE}=8.00	C_{UE}=4.50	C_{UE}=8.00	C_{UE}=4.50
simple cost	8 h	19740	7860	25980	11520	36660	20640
	10 h	24660	9780	32460	14340	45840	25740
	12 h	29640	11820	39000	17280	55020	30960
i=6%	8 h	27361	10895	36011	15968	50814	28609
	10 h	34181	13556	44993	19877	63538	35679
	12 h	41084	16383	54058	23951	76263	42913

3.5 *Life-cycle cost (LCC) savings*

Comparisons of simple cost and compounding cost on LCC savings for 3-star and 5-star inverter ACs with energy tariff ₹ 8.00 are shown in figure 5.

The LCC savings for a 5-star, 4-star and 3-star inverter ACs running for 12 hours daily (C_E=₹ 8.00 and i= 6%) are ₹ 52,229, ₹ 36,578 and ₹ 29,916 respectively. Corresponding LCC savings on a simple cost basis are worth ₹ 43,195, ₹ 30,400, and ₹ 24,147.

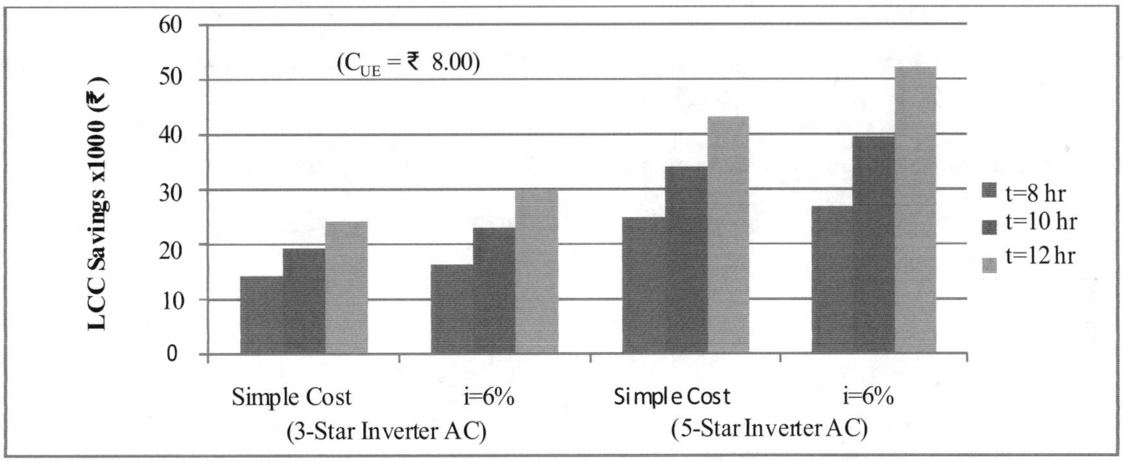

Figure 5. Life Cycle Cost Savings for 3-Star and 5-Star Inverter ACs (C_{UE}=₹8.00)

4. Conclusions

The following conclusions are derived from the present study:

1. The energy savings have a remarkable effect on the *PPs*. Energy cost reductions of about 12%, 15.8%, and 22.4% are achieved with 3-star, 4-star, and 5-star inverter ACs respectively, over the baseline non-inverter AC.
2. The 3-star inverter AC provides the lowest payback period. With unit cost in the highest slab of ₹ 8.00, *PPs* of 3-star inverter AC are 2.5 years, 2.17 years, and 1.42 years for a daily average operating time of 8 hours, 10 hours, and 12 hours respectively. Corresponding *PPs* for 5-star inverter AC are 3.25 years, 2.33 years, and 2.08 years.
3. The LCC savings are highest for a 5-star inverter AC. Running for 12 hours daily (C_E=₹ 8.00), the 5-star inverter AC saves ₹ 52,229 at 6% compounding interest and ₹ 43,195 with simple costing.
4. Increasing trends are shown for *PPs* with an increase in interest rate.
5. Higher usage time does not increase the percentage energy savings but considerably lowers *PP* up

to about 40%.

6. The effect of C_R on PP is not noticeable in the higher tariff slabs. C_R reduces the PP generally at lower unit energy cost.

Finally, it is concluded that selecting the best model based on compounding cost analysis depends mainly on more significant energy savings and higher operating hours. The results show that substantial efficiency improvement can be achieved at a modest increase in the retail price of 3-star inverter AC which will be paid back relatively quickly through energy bill savings. Higher expenditure on more efficient AC models are justified only when electricity prices are high, and AC operates for a longer duration.

5. References

[1] A. O. Elsayed and T. S. Kayed 2020 Dynamic performance analysis of inverter-driven split air conditioner Int. J. Refrig. vol. 118 pp. 443–452

[2] D. Ha and J. H. Jeong 2015 Performance characteristics of a combined air conditioner and refrigerator system interconnected via an intercooler Int. J. Refrig. vol. 49 pp. 57–68

[3] R. Opoku, E. A. Adjei, D. K. Ahadzie, and K. A. Agyarko 2020 Energy efficiency , solar energy and cost saving opportunities in public tertiary institutions in developing countries : The case of KNUST , Ghana Alexandria Eng. J. vol. 59 no. 1 pp. 417–428

[4] C. E. Vincent and M. K. Heun 2006 Thermoeconomic Analysis & Design of Domestic Refrigeration Systems

[5] M. D. D'Accadia and F. De Rossi 1998 Thermoeconomic optimization of a refrigeration plant Int. J. Refrig. vol. 21 no. 1 pp. 42–54

[6] S. Sanaye and H. R. Malekmohammadi 2004 Thermal and economical optimization of air conditioning units with vapor compression refrigeration system Appl. Therm. Eng. vol. 24 no. 13 pp. 1807–1825

[7] Y. Sahebi and A. Motallebi 2011 Economical and Exergical Analysis of Refrigeration Systems Middle-East J. Sci. Res. vol. 9 no. 3 pp. 285–290

[8] R. S. Mitishita, E. M. Barreira, C. O. R. Negrão, and C. J. L. Hermes 2013 Thermoeconomic design and optimization of frost-free refrigerators Appl. Therm. Eng. vol. 50 no. 1 pp. 1376–1385

[9] N. Shah et al. 2016 Cost-Benefit of Improving the Efficiency of Room Air Conditioners (Inverter and Fixed Speed) in India no. June https://cloudfront.escholarship.org/ dist/prd/content/qt2v6646cf/qt2v6646cf.pdf

[10] K. Almutairi, G. Thoma, J. Burek, S. Algarni, and D. Nutter 2015 Life cycle assessment and economic analysis of residential air conditioning in Saudi Arabia Energy Build. vol. 102 pp. 370–379

[11] G. Rosenquist, K. Coughlin, L. Dale, J. McMahon, and S. Meyes 2004 Life-cycle Cost and Payback Period Analysis for Commercial Unitary Air Conditioners no. March p. 60 [Online]. Available: http://docplayer.net/1085499-Life-cycle-cost-and-payback-period-analysis-for-commercial-unitary-air-conditioners.html%0Ahttp://files/116/1085499-Life-cycle-cost-and-payback-period-analysis-for-commercial-unitary-air-conditioners.html.

Stress-strain distribution in intact L4-L5 vertebrae under the influence of physiological movements: A finite element (FE) investigation

Devismita Sanjay[1], Neeraj Kumar[1] and Souptick Chanda[1]

[1] Department of Biosciences and Bioengineering, Indian Institute of Technology Guwahati, North Guwahati, 781 039, Assam, India

E-mail: devismitasanjay@iitg.ac.in

Abstract: This study is aimed at finding the stress and strain distribution in functional spinal unit of L4-L5 occurring due to physiological body movements under five loading conditions, namely compression, flexion, extension, lateral bending and torsion. To this purpose, 3D finite element (FE) model has been generated using 4-noded unstructured tetrahedral elements considered both for bones and intervertebral disc, and 1D tension-only spring elements for ligaments. The analyses were performed for a compression load of 500 N and for other load cases, a moment of 10 N-m along with a preload of 500 N was applied. The model was validated against *in-vitro* experimental data obtained from literature and FE analysis data for a range of motion (RoM) corresponding to various loading conditions. The highest stress was predicted in the case of torsion though the angular deformation was highest in case of flexion.

Keywords. Lumbar spine, finite element analysis, ligaments, stress development, strain, range of motion, functional spinal unit, intervertebral disc.

1. Introduction

The lumbar spine plays a vital role in transferring the weight to the pelvis. It also permits various types of physical movements during daily activities. The physiological movements may include the loading conditions of compression, flexion, extension, lateral bending and torsion. A major portion of the population, especially the elderly people, suffer from lower back pain and may need surgical help depending on severity of the ailment. It is, therefore, essential to gain insights into the biomechanics of lower vertebrae. In the past, many *in-vitro* studies were performed to describe the mechanical behaviour of lumber spine [1, 2, 3, 4]. Moreover, finite element (FE) based *in-silico* analyses were also performed for different physiological movements for functional spinal unit (FSU) of L4-L5 [5, 6, 7]. During these experimental and FE analyses, the measurements or predictions, respectively, were carried out with regard to range of motions, displacements, intradiscal pressure and disc bulges. In the FE study by Tsouknidas et al [8], stress development in lumbar spine was investigated for two mobility scenarios of walking and running. In another FE study by Talukdar et al [9], the stress-strain distribution was investigated for the physiological movement load influences on L3-L4 FSU. In another FE studies, for physiological movements of compression and torsion, the stress-strain distribution was investigated on pediatric femur bone [10, 11]. There is, however, hardly any study on the stress and strain distribution on the bones and intradiscal segments of L4-L5 FSU, occurring due to physiological loading conditions of body movement like compression, flexion, extension, lateral bending and torsion. In the present study, FE investigation was carried out in order to understand the load influence of daily movements for five loading conditions mentioned above, on L4-L5 vertebrae.

2. Materials and method:

The effects of physiological loading conditions were analysed using FE models of the functional spinal unit of intact L4-L5 vertebrae. The L4-L5 vertebrae was assessed for range of motions (RoM) and stress-strain distribution, and the validation of the *in-silico* scheme was conducted based on available literatures. The materials and method starting from preparation of virtual osteotomy till the biomechanical analyses have been described stepwise in the following subsections.

2.1 Development of model

Anatomically representative CAD models of L4 (model #3902) and L5 (model #3901) vertebra were procured from the manufacturer (Sawbones, Europe AB, Malmo, Sweden) [12] and the FSU was anatomically placed in NURBS modelling platform of Rhinoceros v5.0 (Rhinoceros, Robert McNeel & Associates Seattle, USA). The cancellous and cortical bone were virtually segregated by performing Boolean operations in Rhinoceros. The average thickness of cortical bone was kept as 1 mm. The intervertebral disc (IVD) consisting of annulus fibrosus and nucleus pulposus was prepared thereafter. The nucleus pulposus occupies almost 43% of the entire disc volume [9]. A thickness of 0.5 mm was assigned to both bony endplates and cartilaginous endplates. Facet cartilage was developed with an initial gap of 0.1 mm.

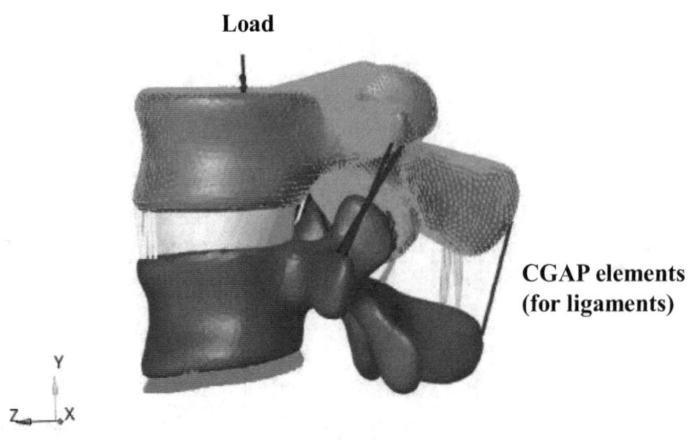

Figure 1: FE model of intact L4-L5 with loading and boundary condition

The 3D FE model for the FSU of L4-L5 was generated using Hypermesh v2019.1 (Altair Engineering Inc., Troy, Michigan, USA). The volumetric mesh for the model was generated using 4-noded unstructured tetrahedral elements (Figure 1). The average edge length of the elements was kept as 1 mm. Six types of ligaments were modelled using 1D tension-only spring elements, called CGAP elements. These elements were ideally suited to mimic soft tissue properties since they imparted stiffness only in case of tension whereas in compression the stiffness was zero.

2.2 Material properties

The material properties of the FE models were considered to be linear, elastic and isotropic and are presented in table 1. Low elastic modulus (E = 0.1 MPa) was considered for nucleus pulposus in order to simulate the incompressible fluid like behaviour [13].

Table 1. Material properties

Components	**Young's modulus (MPa)**	**Poisson's ratio**	**References**
Cortical bone	12000	0.3	Li et al. [13], Goel et al. [14]
Cancellous bone	100	0.3	Li et al. [13], Goel et al. [14]
Bony endplate	1200	0.29	Li et al. [13]
Cartilaginous endplate	23.8	0.28	Wang et al. [15]
Nucleus pulposus	0.1	0.49	Li et al. [13]
Annulus pulposus	9	0.4	Talukdar et al. [9]

The lengths of the ligaments were curated based on data from available literature [16, 17]. Stiffness values of the ligaments were calculated from the length, anatomical cross-section area and Young's moduli of ligaments taken from the literature [6, 7, 13]. For calculating the stiffness, axial stiffness formula, $K=AE/L$, was employed, where K is the stiffness, A is the cross-sectional area, E is the Young's modulus and L is the length of the ligament. The calculated stiffness values were found to corroborate those from existing literature [16, 17]. The stiffness values of the ligaments are presented in table 2.

Table 2. Ligament stiffness

Ligaments	**Stiffness (N/mm)**
Anterior longitudinal	45.2
Posterior longitudinal	26.49
Flavum	43.71
Intertransverse	2.769
Interspinous	35.5
Supraspinous	12.72

2.3 Loading and Boundary Condition

The analysis was performed for five loading conditions namely compression, flexion, extension, lateral bending and torsion. A rigid body element (RBE3) was created, where a 'dependent' node was selected roughly at the centre of top surface of L4 cortical bone, and all surface nodes of L4 cortical bone were selected as 'independent' nodes. The loads were applied at central node of RBE3 which is at the top surface of L4 and calculated based on the weighted average of motions for all the surface nodes of L4 cortical (Figure 1). Thus, the loads were assumed to be acting through the center of gravity of L4. For the compression loading condition, a load of 500N was applied. For flexion, extension, lateral bending and torsion, 10 N-m moment was applied along with a preload of 500N [5]. All the bottom surface nodes of L5 cortical of FSU L4-L5 was constrained for all six degrees of freedom (Figure 1).

2.4 Contact Analysis

Surface-to-surface contact elements were employed (coefficient of friction, $\mu = 0.2$) [18], on both the left and right side at the facet cartilage interface of L4-L5. The contact problem was solved using Augmented Lagrangian method, where L4 cartilage was considered as slave and L5 cartilage as master body.

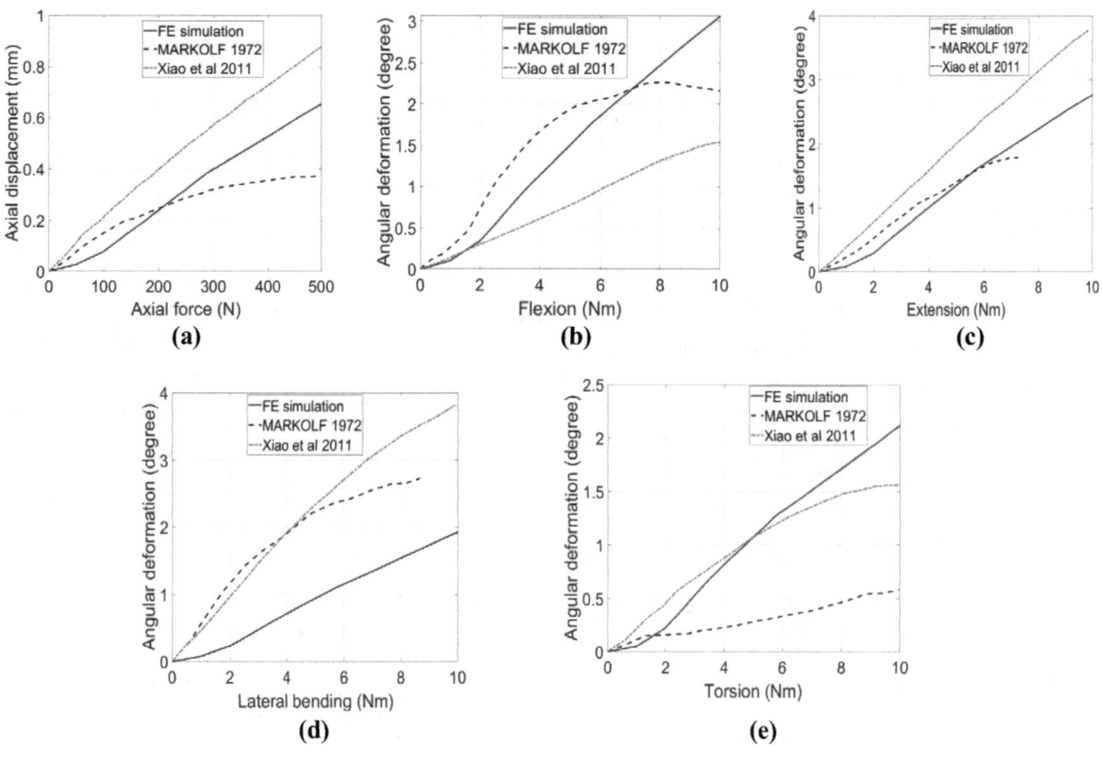

Figure 2: Comparison of RoM in present study with existing literture [1, 5]: (a) axial compression, (b) flexion, (c) extension, (d) lateral bending and (e) torsion.

2.5 Validation and analysis

The FE analysis was performed using 'Optistruct' solver of Hypermesh 2019.1. For the analysis, geometric nonlinearity was employed, wherein the stiffness is a function of displacement contrary to linear analysis where stiffness remains constant. The nonlinear analysis was implemented using piecewise linearity such that load was split into small increments and stiffness matrix was updated after each incremental load application. The iterations were performed to get equilibrium condition after every increment and Newton's Raphson method was used to solve the equations [19]. The validation of the model was done considering the RoM under various loading conditions.

3. Result

In order to have an insight into the influence of different physiological movements on lower vertebra, stress-strain characteristics of the intact L4-L5 model were investigated in the present study. The results for range of motion (RoM) versus load were validated against the experimental data [1] and FE analysis data [5], as shown in figure 2. The contour plots of stress-strain distribution for different loading conditions were obtained and reported.

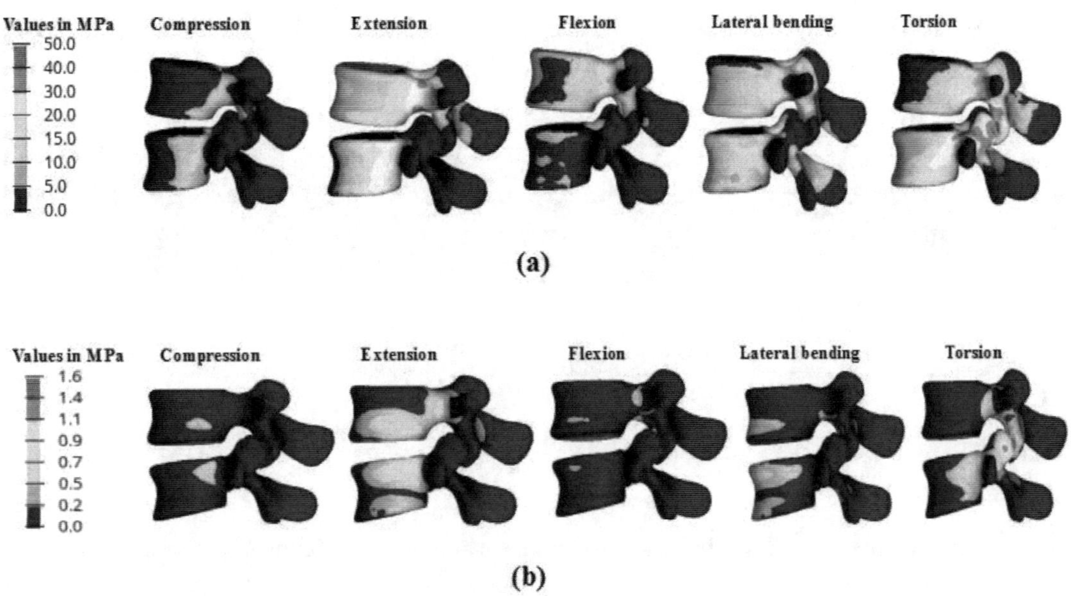

Figure 3: Von Mises stress contours: (a) cortical bone and (b) cancellous bone

3.1 Stress analysis

Von Mises stress contour plots for five loading condition are presented in figure 3. Both the cortical bone (Figure 3a) and cancellous bone (Figure 3b) were found to be following identical trend in stress distribution. The peak stress values in case of cortical and cancellous bone (Figure 3) were predicted to be ~50 MPa and ~1.6 MPa, respectively. The torsional movement predicted maximum stress followed by the extension. On the other hand, minimum stress was obtained in case of simple compression. In case of compression, the stress value for the cortical bone was found to be under 15 MPa (Figure 3a) while the value was less than 0.7 MPa for cancellous bone (Figure 3b). Under all loading condition, except lateral bending, the maximum stress region was found towards the posterior side. The lateral bending condition was found to induce the maximum stress towards the anterior side of the FSU.

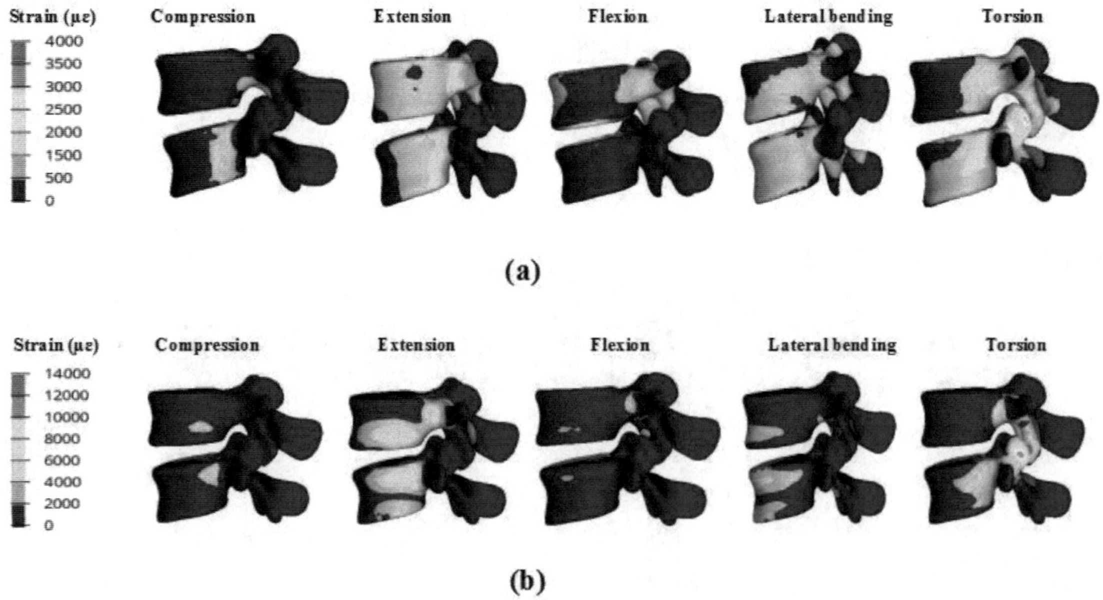

Figure 4: Von Mises strain contours: (a) cortical bone and (b) cancellous bone

3.2 Strain analysis

Von Mises strain contour plot for the five loading conditions are shown in figure 4. The strain distribution was found to be following similar trend as stress distribution and also across the two types of bone – cortical and cancellous (Figure 4). The torsion was found to register peak strain for both cortical (~0.004) and the cancellous bone (~0.014) (Figure 4). The value of strain in case of simple compression for the cortical bone (Figure 4a) was found below 0.0015 and for the cancellous bone the predicted value was below 0.004.

4. Discussion

An intact L4-L5 model was assessed under different physiological movements to gain insights into biomechanical behaviour of lumber vertebra. This study is specifically helpful to know how the load transfer occurs during daily physiological activities. Previous studies suggested that proper loading and boundary condition plays a significant role for stress-strain analysis in lumbar FSU [9]. The validation of the model was done considering the RoM under various loading conditions. The loading conditions were obtained from the available literatures [1, 5]. Axial compression loading conditions considered as vertical force of 500N applied axially at the top surface of cortical of L4. This type of loading may arise while standing. In all the other four loading conditions, the axial force of 500N was considered along with the moment of 10 Nm. The magnitude of moment was kept same in all four cases of physiological movements though it differed in direction depending on the load case.

The FE results of RoM were found to be closely associated with the experimental results [1], as shown in figure 2. A maximum relative displacement of 0.38 mm was reported in the experimental study [1] whereas Xiao et al [5] predicted the value to be 0.88 mm under axial compression. In the present study, the predicted displacement was ~0.65 mm which was an average of the previous two findings. The maximum difference between predicted and measured angular deformation, however, was observed under torsion though the value was close to that reported by Xiao et al. [5]. In all other cases, a reasonable agreement with values reported in literature was found.

Earlier studies with implants (pedicle screw) [20-22] reported that the stress was found to be higher in neck region of pedicle for loading conditions such as flexion, extension, lateral bending and torsion. The stress distribution pattern in case of intact model of L3-L4 FSU by Talukdar et al. 2021 [9] and our FE result for L4-L5 FSU depicted a similar trend in the physiological movements. However, depending on the type of load, the peak stress was found to be in different locations, i.e. for extension it was highest towards posterior side whilst for lateral bending, it was in the lateral part of the vertebra. The von Misses strain distribution pattern for intact L4-L5 model was found to be similar in case of compression load of 500 N for both numerical value and the reported pattern [9]. Furthermore, the peak strain was found to follow a similar trend as that in the stress distribution.

5. Conclusion

An improved FE model of lumber vertebra was developed to understand the load transfer during physiological movements. From the study, it was evident that loading and boundary condition have significant influence on stress-strain distribution in L4-L5 vertebra. This study gave an insight into the stress-strain distribution corresponding to different loading conditions. The maximum stress obtained in case of torsion through our FE analysis was found near to the value reported by Biswas et al FE investigation performed on L3-L5. In their study, the maximum values of stress were reported but it did not display the contour plot of stress distribution. It is evident from previous study that the stress distribution pattern can help in better design of implants. From the present study it was found that, the highest stress was obtained in the case of torsion but the angular deformation was highest in case of flexion. Therefore, it can be concluded that only range of motions (RoM) values may not give a wholesome picture with regards to estimation of the effect of loading. Rather, a stress and strain distribution pattern can help understand the effect of loading and failure criteria in a better way, and serve as a reference for improved design of implant.

Acknowledgements

The authors would like to acknowledge the computational facilities available at the Biomechanics laboratory of the Department of Biosciences and Bioengineering, Indian Institute of Technology Guwahati, which has helped to carry out this research study. The study has been partially supported by SERB, India (Grant no. SRG/2019/000235).

Reference:

[1] Markolf KL. Deformation of the thoracolumbar intervertebral joints in response to external loads: a biomechanical study using autopsy material. *J Bone Joint Surg Am.* 1972 Apr; **54**(3):511-33. PMID: 5055150.

[2] Markolf KL, Morris JM. The structural components of the intervertebral disc. A study of their contributions to the ability of the disc to withstand compressive forces. *J Bone Joint Surg Am.* 1974 Jun; **56**(4): 675-87. PMID: 4835815.

[3] Virgin WJ. Experimental investigations into the physical properties of the intervertebral disc. *J Bone Joint Surg Br.* 1951 Nov;**33**-B (4):607-11. doi: 10.1302/0301-620X.33B4.607. PMID: 14880588.

[4] Brown T, Hansen RJ, Yorra AJ. Some mechanical tests on the lumbosacral spine with particular reference to the intervertebral discs; a preliminary report. *J Bone Joint Surg Am.* 1957 Oct; **39**-A (5):1135-64. PMID: 13475413.

[5] Xiao Z, Wang L, Gong H, Zhu D, Zhang X. A non-linear finite element model of human L4-L5 lumbar spinal segment with three-dimensional solid element ligaments. *Theoretical and Applied Mechanics Letters* 2011; Volume **1**, Issue 6, 064001.

[6] Vena P, Franzoso G, Gastaldi D, Contro R, Dallolio V. A finite element model of the L4-L5 spinal motion segment: biomechanical compatability of an interspinous device, *Computer Methods in Biomechanics and Biomedical Engineering.* 2005; **8**:1,7-16, DOI: 10.1080/10255840500062914.

[7] Boccaccio A, Vena P, Gastaldi D, Franzoso G, Pietrabissa R, Pappalettere C. Finite element analysis of cancellous bone failure in the vertebral body of healthy and osteoporotic subjects. *Proc Inst Mech Eng H.* 2008 Oct; **222**(7):1023-36. doi: 10.1243/09544119JEIM296. PMID: 19024151.

[8] Tsouknidas, A.; Savvakis, S.; Tsirelis, N.; Lontos, A. and Michailidis, N. (2013). A Non-linear Finite Element Model for Assessment of Lumbar Spinal Injury Due to Dynamic Loading. In *Proceedings of the*

International Conference on Bioinformatics Models, Methods and Algorithms - BIOINFORMATICS, (BIOSTEC 2013) ISBN **978**-989-8565-35-8; ISSN 2184-4305, pages 292-295. DOI: 10.5220/0004236902920295

[9] Talukdar RG, Mukhopadhyay KK, Dhara S, Gupta S. Numerical analysis of the mechanical behaviour of intact and implanted lumbar functional spinal units: Effects of loading and boundary conditions. *Proc Inst Mech Eng H.* 2021 Jul; **235**(7):792-804. doi: 10.1177/09544119211008343. Epub 2021 Apr 9. PMID: 33832355.

[10] Kanu NJ, Patwardhan D, Gupta E, Vates UK, Singh GK. Finite element analysis of mechanical response of fracture fixation functionally graded bone plate at paediatric femur bone fracture site under compressive and torsional loadings. *Materials today: Proceedings* 2021, Volume **38,** Part 5, https://doi.org/10.1016/j.matpr.2020.08.740

[11] Kanu NJ, Patwardhan D, Gupta E, Vates UK, Singh GK. Numerical investigations of stress-deformation responses in fractured paediatric bones with prosthetic bone plates. 2020 *IOP Conf. Ser.: Mater. Sci. Eng.* **814** 012038

[12] *www.sawbones.com*

[13] Li J, Shang J, Zhou Y, Li C, Liu H. Finite Element Analysis of a New Pedicle Screw-Plate System for Minimally Invasive Transforaminal Lumbar Interbody Fusion. *PLoS One.* 2015 Dec 9; **10**(12): e0144637. doi: 10.1371/journal.pone.0144637. PMID: 26649749; PMCID: PMC4674154.

[14] Goel VK, Kiapour A, Faizan A, Krishna M, Friesem T. Finite element study of matched paired posterior disc implant and dynamic stabilizer (360° motion preservation system). *SAS J.* 2007 Feb 1; **1**(1):55-61. doi: 10.1016/SASJ-2006-0008-RR. PMID: 25802579; PMCID: PMC4365571.

[15] Wang W, Zhang H, Sadeghipour K, Baran G. Effect of posterolateral disc replacement on kinematics and stress distribution in the lumbar spine: a finite element study. *Med Eng Phys* 2013; **35**:357–64. https://doi.org/10.1016/j.medengphy.2012.05.013.

[16] Yoganandan, N., Kumaresan, S., and Pintar, F. A. "Geometric and Mechanical Properties of Human Cervical Spine Ligaments." ASME. *J Biomech Eng.* 2000 December; **122**(6): 623–629. https://doi.org/10.1115/1.1322034.

[17] *www.wheelessonline.com/issls/lumbar-vertebrae-ligaments/*

[18] Zhou Q, Zeng F, Tu J, Dong Z, Ding ZH. Influence of cement augmented pedicle screw instrumentation in an osteoporotic lumbosacral spine over the adjacent segments: a 3d finite element study. *J Orthop Surg Res* 2020; **15**(1):132.

[19] Gokhale NS, Deshpande SS, Bedekar SV, Thite AN, Book- *practical finite element analysis.*,(2008)445.

[20] Kim K, Park WM, Kim YH, Lee S. Stress analysis in a pedicle screw fixation system with flexible rods in the lumbar spine. *Proc Inst Mech Eng H.* 2010; **224**(3):477-85. DOI: 10.1243/09544119JEIM611. PMID: 20408492.

[21] Biswas JK, Rana M, Roy S, Majumder S, Karmakar SK, Roychowdhury A. Effect of range of motion (ROM) for pedicle-screw fixation on lumbar spine with rigid and semi-rigid rod materials: A finite element study *2018 IOP Conf. Ser.: Mater. Sci. Eng.* **402** 012146

[22] Kumbhalkar MA, Rambhad KS, Kanu NJ. An insight into biomechanical study for replacement of knee joint. *Materials today: Proceedings,* Volume 47, Part 11, 2021, Pages 2957-2965, https://doi.org/10.1016/j.matpr.2021.05.202

Electromagnetic shielding effectiveness for Al6061 metal matrix composite based mesh wire reinforced with Flyash for oblique incidence of EM wave

Srinu Budumuru[1], **M. Satya Anuradha**[2]
[1]Assistant Professor, Department of Electrical Electronics and Communication Engineering, GITAM Institute of Technology, GITAM Deemed to be University, Visakhapatnam, Andhra Pradesh, India,530045
[2]Professor Department of Electrical Electronics and Communication Engineering, Andhra university, Visakhapatnam, Andhra Pradesh, India,530045
E-mail - srinubudumuru@gmail.com[1] , radhamsa@gmail.com[2]

Abstract. Nowadays, flywire is used exclusively in aeronautical applications. A plane's complete control is dependent on electronic technology, yet it suffers from high-intensity radiated fields. An electromagnetic shield may be necessary to protect this equipment from external electromagnetic pollution. The current project attempts to create a protective barrier around the operating equipment to enhance its efficiency. AL6061 composite material was used to create a metal matrix mesh shield. It is reinforced with fly ash in various volume fractions, and the electrical characteristics and Shielding Effectiveness are determined (SE). The maximum SE is 45.36dB obtained, which can be effectively used as a shield for aerospace and other applications.

Keywords. Shielding effectiveness, Reflection loss, reinforcement, Flyash, oblique incidence.

1. Introduction

With the promotion of electronic communication equipment in contemporary society, the subsequent electromagnetic (EM) radiation has become genuine contamination that can't be overlooked. Thusly, it is of extraordinary importance to foster electromagnetic interference (EMI) shielding materials. Lightweight and productive electromagnetic interference (EMI) shielding composites are of incredible significance for improving cutting-edge correspondence innovation, wearable gear, and high-power electronic hardware. The impact of high-intensity radiated fields on aerospace applications and wind power generating units is growing. Every commercial flight experiences a lightning strike effect once every 1000 hours [1]. Lightning contact generates electromagnetic interference, which disrupts the normal operation of electrical devices. This also harms the airplane's physical structure [2][3]. Massive currents created by a lightning strike on the plane's surface induce delamination, embrittlement, and structural collapse [4]. Adding lightning strike protection (LSP) is a critical job for protecting aircraft from electrical and physical harm.

When lightning strikes the plane's surface, the resulting currents attempt to travel along the least resistive regions of the surface [5]. To save weight, metal was often transformed into fiber-reinforced materials (FRP) to create the aircraft surface. In most cases, metal was converted into fiber-reinforced materials (FRP) to construct the aeroplane surface in order to reduce weight. Unfortunately, FRP lacks the conductivity required to efficiently flow the enormous currents produced by a lightning strike, resulting in electromagnetic interference and physical damage [6]. To address the poor conductivity issue in FRP composite materials, a new material was created. Electrical and mechanical characteristics of composite materials are adequate. The composites have high conductivity, which may be enhanced by strengthening them with various materials [7]. The Al6061 metal matrix material was used as a composite material in this study, and the shielding characteristics can be enhanced by strengthening the flyash [8][9]. The electromagnetic wave's incidence is assumed oblique rather than normal, with obliques incidence approximation extremely close to the real scenario [10]. The addition of the mesh structure reduces the total weight, which reduces the operation's fuel usage [11]. The Al6061 metal matrix is employed as a composite material in this study, and it is reinforced with fly ash to increase both shielding efficacy and density.

The total work is organized as section 1 selection of material, section 2 preparation of the material, section 3 shielding mechanism, section 4 results analysis, section 5 conclusion.

2. Selection of material:

Because of their improved strength, hardness, and weight ratio, as well as their superior wear resistance, Al6061-based hybrid metal matrix composites outperform traditional aluminium alloys [12][13][14]. Because the Fe_2O_3 in the fly ash is linked to improving electromagnetic shielding effectiveness, it is included as a mandatory reinforcing material in the preparation of all specimens of the proposed material. The fly ash (Siliceous fly ash) used in this study was obtained from a thermal power plant in Visakhapatnam, and its density and specific surface area were 2.1 g/cm3 and 3449 cm2/g, respectively [15][16].

3. Preparation of the material:

The stir casting procedure was used to create the three separate sets of Composites with variable reinforcing materials since it is a cost-effective way for creating particle reinforced composites. A stir casting system comprises of a furnace, a feeder for reinforcing ingredients, and a stirrer. A stirring rod and blades are included with the stirrer. One kilogramme of Al6061 alloy is placed in a ceramic crucible and heated to 720°C in a resistance-heated muffle furnace. In the furnace, the Al 6061 alloy is heated to a molten condition before being mixed with warmed fly ash particles [17][18]. The experiment conducted for stir casting of flyash with Al6061and shown in figure 1.

Figure 1: Stir casting equipment for reinforcement of flyash with Al6061.

After the composite material has been prepared, it may be examined for electrical properties, and the values of Permeability, Permittivity, and Conductivity (μ, ϵ, σ) can be acquired using a vector network analyzer. Shielding may be quantified using these electrical characteristics.

4. Shielding mechanism:

The major EMI shielding techniques are reflection and absorption. Shielding material should have a reasonably high electrical conductivity in order to achieve considerable electromagnetic radiation reflection. Metals are the most often used materials for EMI shielding, and their primary function is reflection owing to free electrons in them. Absorption is the secondary process of EMI shielding [19]. When electromagnetic waves strike the surface of a protective

electromagnetic shield, the shield's capacity to reflect the waves without crossing the sheet offers superior EM shielding for external electromagnetic fields.
The mesh wire's shielding efficiency is determined by reflection, absorption loss, and internal multiple reflections. However, if it is low, numerous reflections cannot be considered. The loss due to reflection and absorption is determined by the material characteristics [20][21].

$$SE_{dB} = (Absorption\ loss)A_{dB} + (Refelction\ loss)R_{dB} + (Multiple\ refelction\ loss)M_{dB} \quad (1)$$

We are calculating the SE for the wire mesh in the oblique case. Let us consider the rectangular mesh with size a_s. For our convenience, we are estimating the wires are skinny. The radius of the mesh is taken as r_w. The mesh structure is shown in figure 3.
The sheet inductances of the mesh are represented as L_s [22][23].

$$L_s = \frac{\mu_0 . a_s}{2\pi} \ln(1 - e^{-\frac{2\pi r_s}{a_s}})^{-1} \quad (2)$$

The internal impedance of the wire is represented as z'_w

$$z'_w = r'_w \frac{(\sqrt{jw\tau_w}).I_0.(\sqrt{jw\tau_w})}{2.I_1.(\sqrt{jw\tau_w})} \quad (3)$$

From equation 3 we are considering the resistance per unit length is r'_w *and* τ_w is considered as the time constant. $I_n(.)$ Represent the first kind of the Bessel function.

$$r'_w = (\pi r_w^2 \sigma_w)^{-1} \quad (4)$$

$$\tau_w = \mu_w \sigma_w r_w^2 \quad (5)$$

The sheet impedances are represented as z_{s1} and z_{s2}

$$z_{s1} = z'_w a_s + j\omega L_s \quad (6)$$

$$z_{s2} = z_s - \frac{j\omega L_s}{2} sin\theta^2 \quad (7)$$

The reflection(R) and transmission (T) are given as per the angle of incidences (***θ***) as given below. The sheet impedance of perpendicularly polarized plane waves is z_s= z_{s1} and similarly the sheet impedance of parallel-polarized plane waves is z_{s2}

$$R_1 = \frac{-1}{1+2\left(\frac{z_{s1}}{z_0}\right)cos\theta} \quad (8)$$

$$T_1 = \frac{2\left(\frac{z_{s1}}{z_0}\right)cos\theta}{1+2\left(\frac{z_{s1}}{z_0}\right)cos\,\theta} \quad (9)$$

$$R_2 = \frac{cos\theta}{1+2\left(\frac{z_{s2}}{z_0}\right)cos\theta} \quad (10)$$

$$T_2 = \frac{2\left(\frac{z_{s2}}{z_0}\right)cos\theta}{1+2\left(\frac{z_{s2}}{z_0}\right)cos\theta} \quad (11)$$

Equation 9 represents the TE polarization, and equation 11 represents the TM polarization. From these two equations, the shielding effectiveness is determined as given below

$$SE = -10log_{10}\left(\left(\frac{1}{2}.T_1^2\right) + \left(\frac{1}{2}.T_2^2\right)\right) \quad (12)$$

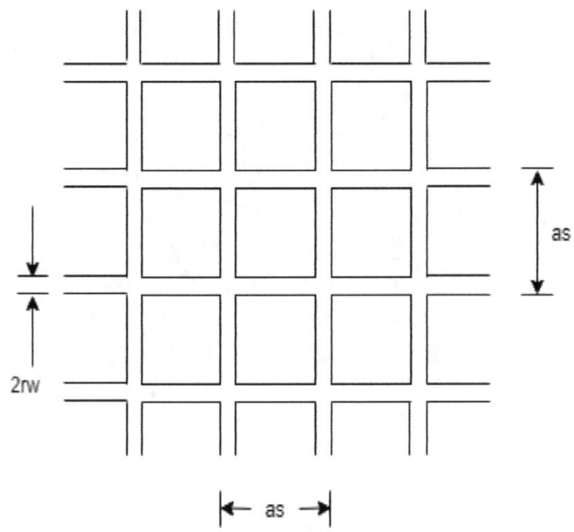

Figure 2: Composite mesh representation with the rectangular mesh with size a_s and the radius of the mesh is taken as r_w.

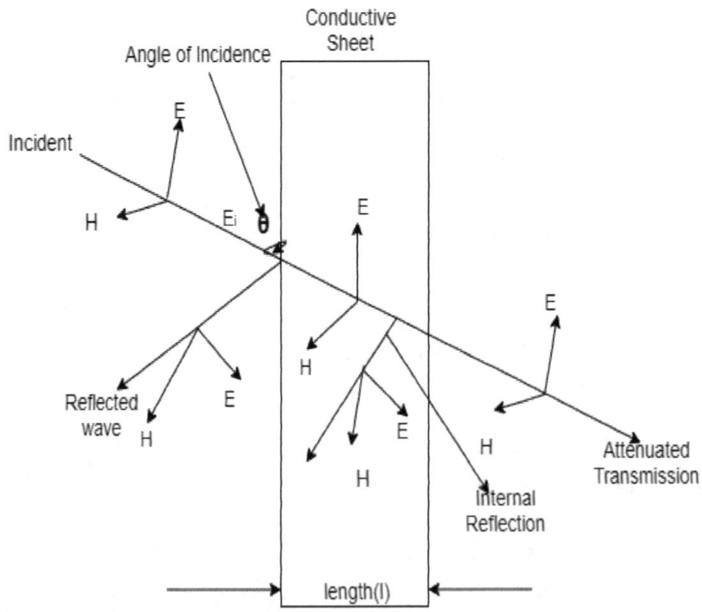

Figure 3: Oblique incidence of EM wave on Mesh sheet.

5. Results and Analysis:

Shielding effectiveness of Al6061 reinforced with flyash at different percentages executed and corresponding results shown from figure 4 to 7. Comparison has made between all the combinations of the composite materials and shown in figure 8.

Figure 4: SE of Pure Al6061 w.r.t angle of incidence.

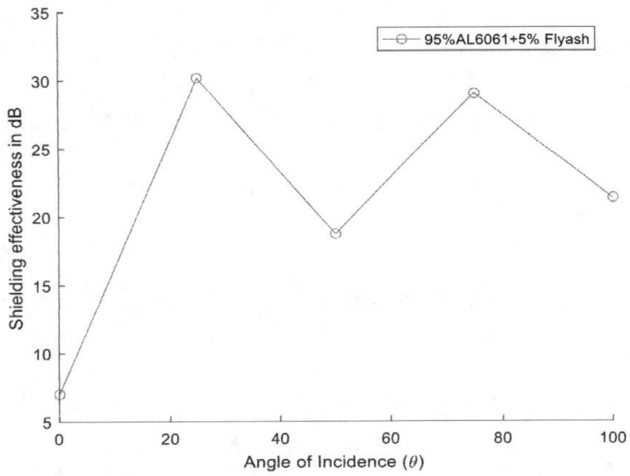

Figure 5: SE of Al6061 reinforced with 5% of fly w.r.t angle of incidence.

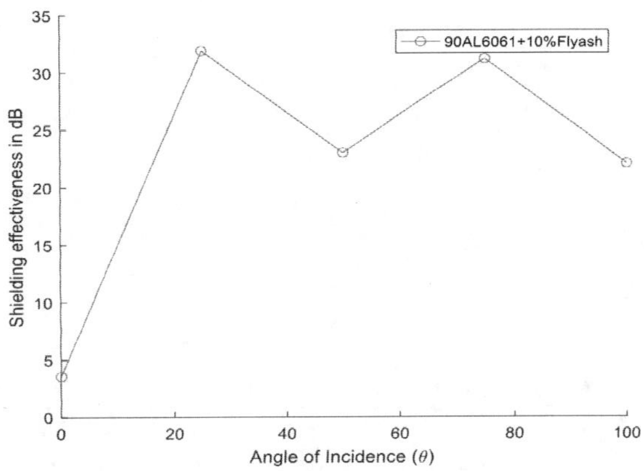

Figure 6: SE of Al6061 reinforced with 10% of fly w.r.t angle of incidence.

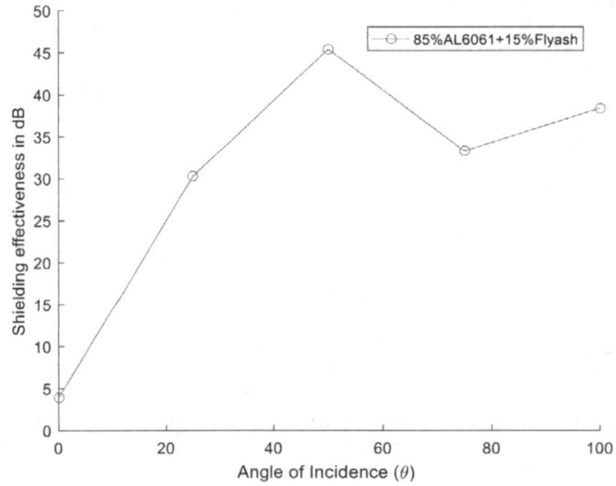

Figure 7: SE of Al6061 reinforced with 15% of fly w.r.t angle of incidence.

Table 1: Maximum SE of Al6061 with different volume fractions of flyash.

S.no	Material	Maximum Shielding Effectiveness (dB)
1	Pure AL6061	21.35
2	(95%) AL6061+ (5%) Fly ash	30.17
3	(90%) AL6061+ (10%) Fly ash	31.89
4	(85%) AL6061+ (15%) Fly ash	45.36

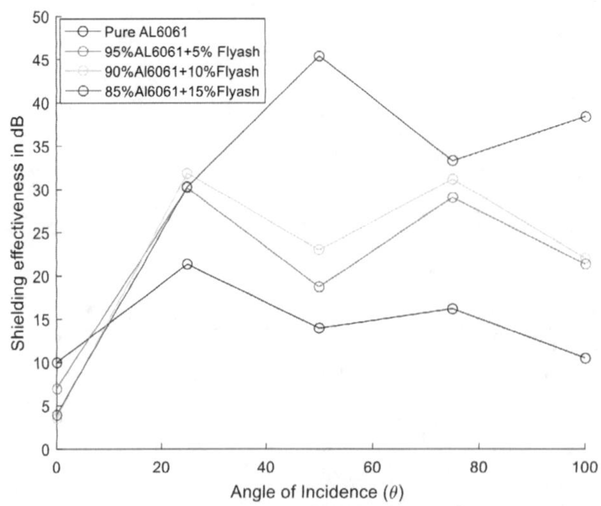

Figure 8: Shielding effectiveness of Al6061 reinforced with different combinations of flyash variation w.r.t to angle of incidence.

Table 1 represents the total SE of all composites reinfored with fly ash at different volumes fractions. The shielding effectiveness of the composite mesh calculated at different volume fraction of flyash. Analysis was done for oblique incidence of the electromagnetic wave to approximate practical case. For pure Al6061 the shielding effectiveness of 21.35dB was obtained. For Al6061 with a volume fraction of 95% and 5% of fly ash gives SE of 30.17dB obtained. Whereas for Al6061 with a volume fraction of

90% and 10% of fly ash gives SE of 31.89dB obtained. Finally, Al6061 with a volume fraction of 85% and 15% fly ash gives SE of 45.36dB.

6. **Conclusion:** The primary goal of switching from metallic sheet to mesh is to minimize weight. The current study created the electromagnetic protective layer using a mesh structure, and a good shield was achieved. The composite material Al6061 was utilised as the base material and fly ash was supplemented with to enhance the SE. The highest shielding effectiveness of 45.36dB is recognized and can be employed in lightning protection systems.

7. **References:**

1. Larsson A (2002) The interaction between a lightning flash and an aircraft in flight. C R Phys 3:1423–1444.
2. Guo Y, Xu Y, Wang Q, Dong Q, Yi X, Jia Y (2019) Eliminating lightning strike damage to carbon fiber composite structures in Zone 2 of aircraft by Ni-coated carbon fiber nonwoven veils. Compos Sci Technol 169:95–102.
3. Kumar, V., Sharma, S., Pathak, A., Singh, B.P., Dhakate, S.R., Yokozeki, T., Okada, T., Ogasawara, T., Interleaved MWCNT buckypaper between CFRP laminates to improve through-thickness electrical conductivity and reducing lightning strike damage, Composite Structures (2018), doi: https://doi.org/10.1016/ j.compstruct.2018.11.088.
4. Kawakami, H. and Feraboli, P., 2011. Lightning strike damage resistance and tolerance of scarf-repaired mesh-protected carbon fiber composites. *Composites Part A: Applied Science and Manufacturing*, *42*(9), pp.1247-1262.
5. Wu, N., Du, W., Hu, Q. and Jiang, S.V.Q (2020) Recent development in fabrication of Co nanostructures and their carbon nanocomposites for electromagnetic wave absorption. Engineered Science 13(13):11-23.
6. Gagné, M. and Therriault, D (2014) Lightning strike protection of composites. Progress in Aerospace Sciences 64:1-16
7. Zhao, B., Deng, J., Zhang, R., Liang, L., Fan, B., Bai, Z., Shao, G. and Park, C.B., 2018. Recent advances on the electromagnetic wave absorption properties of Ni based materials. *Engineered Science*, *3*(28), pp.5-40.
8. Yinsuo Dai, Jianhua Wu, DerongWang, Effect of Mineralogical Phase and Chemical Composition of Fly Ashon Electromagnetic Wave-Absorbing Properties, Materials Transactions, Vol. 59, No. 6 (2018) pp. 876 to 882.
9. Reddy, T.P., Kishore, S.J., Theja, P.C., Rao, P.P. (2020) Development and wear behavior investigation on aluminum-7075/B 4 C/fly ash metal matrix composites. Advanced Composites and Hybrid Materials. p.1- 11.
10. Han, F. and Zhang (1996) August. Degeneration of shielding effectiveness of planar shields due to oblique incident plane waves. Proceedings of Symposium on Electromagnetic Compatibility : 82-86.
11. Lan Cheng, Tonghua Zhang, Min Guo, Jiali Li, Shu Wang & Hua Tang (2015) Electromagnetic shielding effectiveness and mathematical model of stainless steel composite fabric, The Journal of The Textile Institute, 106:6, 577-586, DOI: 10.1080/00405000.2014.929275.
12. Reddy, T.P., Kishore, S.J., Theja, P.C., Rao, P.P. (2020) Development and wear behavior investigation on aluminum-7075/B 4 C/fly ash metal matrix composites. Advanced Composites and Hybrid Materials. p.1- 11.
13. Kanu, N.J., Mangalam, A., Gupta, E., Vates, U.K., Singh, G.K. and Sinha, D.K., 2021. Investigation on secondary deformation of ultrafine SiC particles reinforced LM25 metal matrix composites. *Materials Today: Proceedings*.

14. Kanu, N.J., Vates, U.K., Singh, G.K. and Chavan, S., 2019. Fracture problems, vibration, buckling, and bending analyses of functionally graded materials: A state-of-the-art review including smart FGMS. *Particulate Science and Technology*, *37*(5), pp.583-608.
15. Lingvay, D., Velciu, G., Bors¸, A.-M., Moant¸a, A., Caramitu, A.-R. and Radermacher, L. (2018), “Fly ash for increased electromagnetic protection of constructions”, Electrotehnica, Electronica, Automatica: EEA, Vol. 66 No. 2, pp. 76-80.
16. Budumuru, S. and Anuradha, M.S., 2021. Analysis of Electromagnetic Shielding Effectiveness Properties of Al6061 Metal Matrix Composites at X-Band for Aerospace Applications. In *Microelectronics, Electromagnetics and Telecommunications* (pp. 467-476). Springer, Singapore.
17. Manimaran R, Jayakumar I (2018) Mechanical properties of fly ash composites—A review. Energy Sour Part A: Recovery Util Environ Effects.
18. Chauhan, A., Vates, U.K., Kanu, N.J., Gupta, E., Singh, G.K., Sharma, B.P. and Gorrepati, S.R., 2021. Fabrication and characterization of novel nitinol particulate reinforced aluminium alloy metal matrix composites (NiTip/AA6061 MMCs). *Materials Today: Proceedings*, *38*, pp.3027-3034
19. Dharma Raj C, Sasibhushana Rao G, Jayasree PVY, Srinu, Lakshman P (2010) Analysis of reflectivity and shielding effectiveness of absorbing material–conductor laminate for electromagnetic compatibility. J Electromag Anal App.
20. Raj, C.D., Rao, G.S., Jayasree, P.V., Srinu, B. and Lakshman, P., 2010. Analysis of reflectivity and shielding effectiveness of absorbing material–conductor laminate for electromagnetic compatibility. *Journal of Electromagnetic Analysis and Applications*, *2010*.
21. Budumuru, S. and Mosa, S.A., 2021. Analysis of shielding effectiveness and mechanical properties of metal matrix composite AL6061 reinforced with Al2O3 and fly ash for oblique incidence of EM wave. *International Journal of Intelligent Computing and Cybernetics*.
22. Casey, Kendall F. "Electromagnetic shielding behavior of wire-mesh screens." *IEEE transactions on Electromagnetic Compatibility* 30.3 (1988): 298-306.
23. Lovat, Giampiero, Paolo Burghignoli, and Salvatore Celozzi. "Shielding properties of a wire-medium screen." *IEEE transactions on electromagnetic compatibility* 50.1 (2008).

IOP Conf. Series: Materials Science and Engineering 1206 (2021) 012026 doi:10.1088/1757-899X/1206/1/012026

Reliability Failure in Microelectronic Interconnects by Electric Current Induced Chemical Reaction

S Kumar[1,*], P Kumar[2], R Pratap[3]

[1,3]Center for Nano-Science and Engineering, Indian Institute of Science, Bangalore
[2]Materials Engineering, Indian Institute of Science, Bangalore
[*]Corresponding Author Email: sumitk@iisc.ac.in

Abstract. The electric field-induced chemical reaction in Cr thin film by a micro/ nano-probe has been recently reported with detailed characterization. Although the phenomenon is employed for micro-nano fabrication, this can act as a reliability failure, where Cr is used as an adhesion layer or main interconnects in microelectronic circuits. Here, we present an investigation on the role of electric current density for such failure using a specifically designed sample. A 100 μm width and 100 nm thin Cr film is deposited perpendicular to the Pt film of similar dimensions. The anode probe (20 μm diameter) is positioned onto the Pt film, whereas the cathode probe onto the Cr film. It is observed that the chemical reaction, for an applied voltage, initiates at the edge of the Pt film and not at the cathode probe. The localized chemical reaction causes to damage the interconnection line. The analysis based on the COMSOL multiphysics simulation illustrates that the chemical reaction evolves at the high current density locations. The study also builds a fundamental understanding of the mechanism of evolution of patterning by electric field-induced chemical reactions.

Keywords. Reliability failure, Electric field-induced chemical reaction, Electromigration, Micro-nano fabrication

1. Introduction

Reliability failure due to high current density in metallic interconnections has acted as a villain in microelectronics [1]. Miniaturization of electronic devices necessitates the utilization of thin metallic films (nanometre thickness) as interconnects. Therefore, even a tiny current flowing through the circuit results in a high current density (~10^4 A/cm^2) [1]. Such a high current density leads to electromigration, where mass migrates from cathode to anode or vice-versa and the interconnection is permanently damaged [2]. The study of solid and liquid electromigration has been conducted by various researchers. Huntington presented the theoretical model of electromigration responsible for mass flow under an applied electric field [2]. K.N Tu has reported the measurement of electromigration force in Au, Ag, Cu, Al, and Pb [1]. Yang et al. have demonstrated the interconnection failure due to electromigration in Pd, Sc and Y strips [3]. They reported the current density of the order of 1 MA/cm^2 responsible for electromigration. Kumar et al. reported electromigration in liquid metal, wherein Cu lines are coated with a long-range continuous flow of Bi liquid [4]. We introduced a theory of liquid electromigration that explains the dichotomy (flow direction dependent on metal parameters) in the flow behaviour of liquid metals [5]. Although electromigration is a damaging phenomenon, it can be utilized in positive aspects, for example, coating, nano-patterning, and electrical nano-contacts formation [6]. More detail

on electromigration in various aspects can be found in the literatures [7–9]. Here, we report a different mechanism (other than electromigration) of reliability failure.
Before going to the actual scenario, we would like to revisit the phenomenon of patterning by scanning probe tip based on electric field-induced chemical reaction [10]. Herein, a scanning probe microscope, e.g., atomic force microscope (AFM), is utilized for direct writing in metallic/ non-metallic films [11]. Figure 1 shows a ring pattern formation in a 100 nm thin Cr film by tungsten electrode probes [12]. In order to pattern the structure, first, the anode is positioned on Cr film. A potential difference of 10-20 V is applied across the probes. Now, the cathode probe is slowly touched to the Cr film in order to create the pattern. The moment, cathode tip touches the film, the high electric field at the tip-apex causes to initiate a chemical reaction at the cathode location, and chromium trioxide (CrO_3) is formed as a chemical product [10]. If the cathode probe is scanned on a given path (for example, straight lines), line or curved patterns of various dimensions can be generated [13].

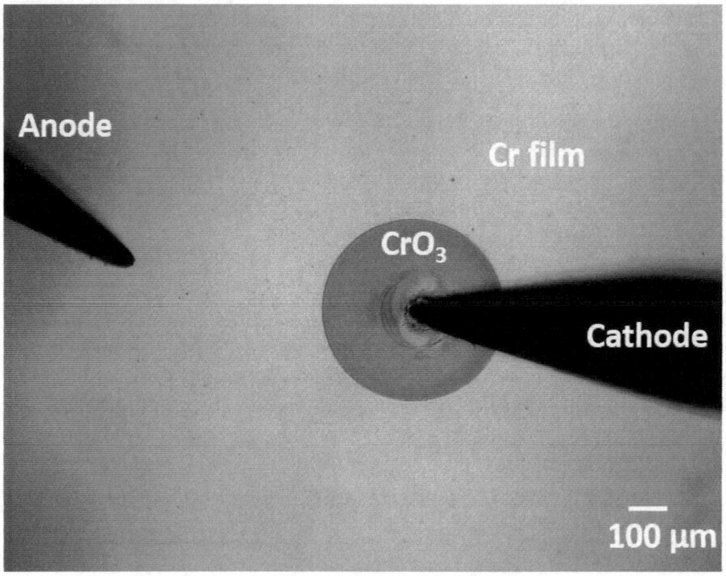

Figure 1. A ring pattern formation in Cr film by electric field-induced chemical reaction with a scanning probe tip. The chemical reaction is initiated at the cathode tip.

The final developed pattern of dimensions up to 40 nm has been demonstrated by this technique [11]. Nano beads in the range of 10-40 nm have been fabricated by atomic force microscope (AFM) [14,15]. In this paper, we report an investigation to understand the role of electric current density on chemical reaction at cathode-probe in a Cr film. A simple question arises that why reaction starts at the cathode only and not elsewhere in the film. The answer lies in the current density. The configuration shown in Figure 1 doesn't explain clearly the effect of current density. Therefore, an experiment is conducted on a specifically designed sample to understand the contribution of current density. A COMSOL simulation is also performed to verify the experimental observation.

2. Experiment

A typical sample for patterning is prepared by depositing Cr film (10-100 nm) on an insulator substrate (usually Si/ SiO_2). Here, to observe the current density contribution, we fabricated a sample by cross-depositing Cr film on a Pt line. First, a Pt line of 100 µm width and 100 nm thickness is fabricated by standard lift-off process on a Si/SiO_2 substrate. Second, the Cr film of 200 µm width and 100 nm thickness is sputter deposited on and perpendicular to the Pt line (Figure 2).

For the experiment, the anode probe is placed on the Pt line, whereas the cathode is touched on the Cr line. A constant voltage of 10 V is applied across the probes. Surprisingly, the reaction starts at the edge of Pt-film and not at the cathode point. Figure 2 shows the time-lapse images of the chemical reaction. The video of the phenomenon is recorded by a digital microscope mounted above the sample (see supplemental material).

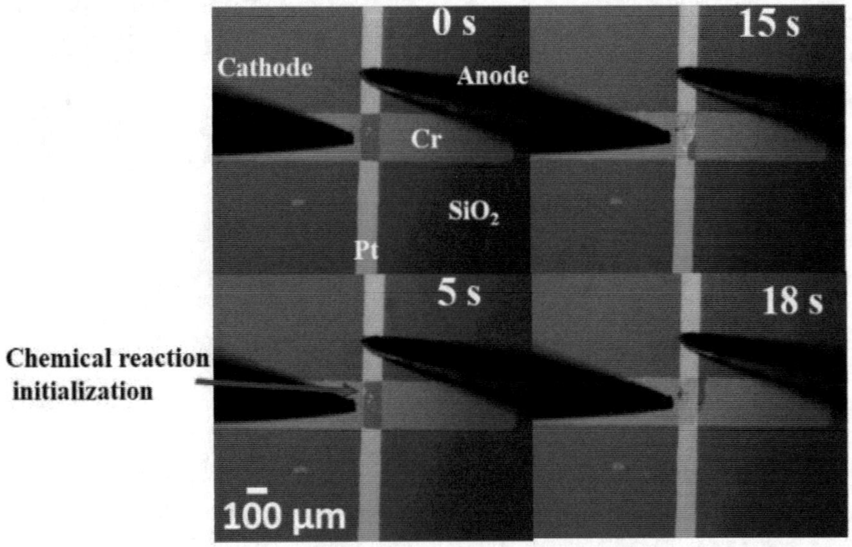

Figure 2. Reaction initialization check with the cathode probe on a Cr film deposited on and perpendicular to the Pt line. The reaction starts (see at 5 s) at the edge of the Pt line. The Cr over the Pt-region is completely consumed and converted to CrO_3 at 18 s.

3. Results and discussion

COMSOL multiphysics is used to perform FEA simulations to get the distribution of the current density. A 3D FEA model having a similar geometry as the sample is constructed. The thickness of the Pt and Cr films are taken to be 1 µm for each for easier meshing. Small circular discs were chosen as the anode and the cathode points on the Pt and Cr film, respectively; the diameters of these discs are 10 µm. Free tetrahedral mesh of sizes in the range of 100-500 nm is used for the meshing. The electrical conductivity of Pt and Cr were taken to be 8.9×10^6 S/m and 7.9×10^6 S/m, respectively. Electric current module is used for the simulation.

Figure 3 shows the FEA simulation results. The electric current density is the highest at the edge of the Pt film. This is also the point where the chemical reaction is initiated. In other words, the chemical reaction is most probable at the high current density locations. The high current density location serves as a source of the electric field induced chemical reaction in Cr film.

Microelectronic utilizes the Au, Cu, and other metallic films as interconnects where Cr is also used as an adhesion layer. The reliability failure can be caused due to the chemical reaction at a high current density location. Therefore, it is suggested to avoid the use of Cr film for interconnections as it may cause chemical reaction-based reliability failure. Titanium metal could be a better choice instead for adhesion purposes.

(a) (b)

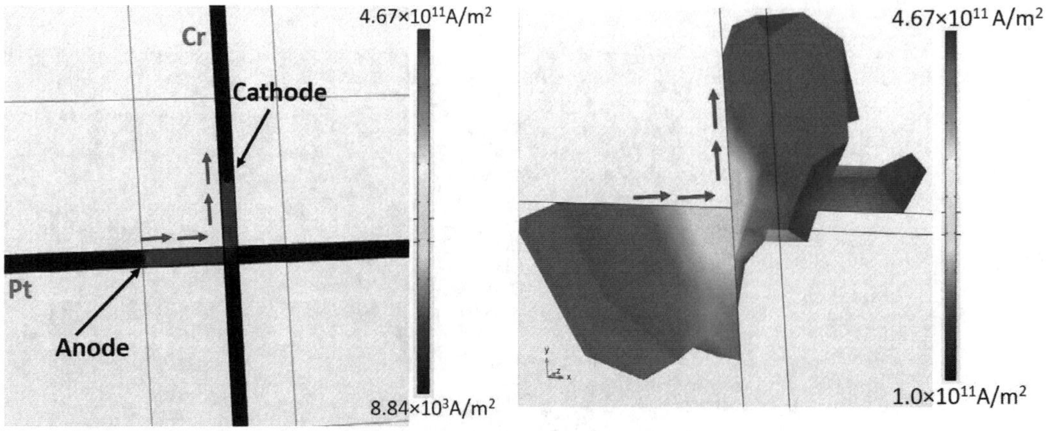

Figure 3. FEA simulation showing the (a) distribution of current density in the sample shown in Figure 2. The red arrows indicate the current direction. The current density is high in the region between the anode and cathode. (b) Isometric views of the model, especially focusing on the edge of Pt and Cr interaction region shown in Figure 3(a). The maximum current density is observed at the turning edge of the Cr-Pt configuration. The red arrows show the current flow direction.

4. Summary

The reliability failure in microelectronic circuits caused by an electric field-induced chemical reaction in Cr film (other than electromigration) is introduced in this work. The investigation is done on a Cr film, as an interconnection, deposited perpendicular to the Pt film. An electric current is made to pass from Pt to Cr by applying a potential difference through metallic probes. The chemical reaction in Cr film is observed to be initiated at the high current density location (at the edge Pt film), and it damages the Cr film. The experimental observation is verified by COMSOL simulation, and it is identified that the high current density points act as the source of the chemical reaction. Based on the study presented in this paper, it is recommended to avoid the use of Cr film for adhesion or as a direct interconnection in micro/ nano-electronic circuits.

Acknowledgements

We would like to acknowledge the financial support by the funding agency, DSTO 1526 (Technology System Development Program of Department of Science and Technology, India) and DSTO 1759 (Indo-Russia Joint Research Project of Program of Department of Science and Technology, India).

References

[1] Tu, N. Electromigration in stressed thin films.(1992). *Physical Review B* **45**, 1409–1413.

[2] Huntington, H. B. Current-induced marker motion in gold wires(1961). *J.Phys.Chem.Solids* **20**, 76–87.

[3] Yang, Y., Xu, S., Xie, S. & Peng, L. M. Current sustainability and electromigration of Pd, Sc and Y thin-films as potential interconnects(2010). *Nano-Micro Letters* **2**, 184–189.

[4] Dutta, I. & Kumar, P. Electric current induced liquid metal flow: Application to coating of micropatterned structures(2009). *Applied Physics Letters* **94**, 14–17.

[5] Kumar, S., Kumar, P. & Pratap, R. A model for electromigration induced flow in liquid metals(2017). *Journal of Physics D: Applied Physics* **50**, 1–9.

[6] Kumar, P. *et al.* Electrical Jointing at Micro- and Nanoscale by Electromigration and Mechanical Nano-Manipulation for Bottom-Up Nano-Assembling(2018). in *2018 IEEE International Conference on Manipulation, Manufacturing and Measurement on the Nanoscale, 3M-NANO 2018 - Proceedings* 167–170 (IEEE, 2018). doi:10.1109/3M-NANO.2018.8552248.

[7] Pierce, D. G. & Brusius, P. G. Electromigration: A review(1997). *Microelectronics Reliability* **37**, 1053–1072.

[8] Gupta, E., Kumar Sinha, S., Kumar Vates, U. & Chavan, S. S. A high performance solid state transformer for an efficient electric grid application: Design and analysis for advanced nanocrystalline core material(2021). *Materials Today: Proceedings* **46**, 2146–2149.

[9] Ho, P. S. & Kwok, T. Electromigration in metals(1989). *Rep. Prog. Phys.* **52**, 301.

[10] Kumar, S., Suresh, H., Sethuraman, V. A., Kumar, P. & Pratap, R. Electric field induced patterning in Cr film under ambient conditions : A chemical reaction based perspective(2020). *SN Applied Sciences* **2**, 1–9.

[11] Talukder, S., Kumar, P. & Pratap, R. Electrolithography- A New and Versatile Process for Nano Patterning.(2015). *Scientific reports* **5**, 17753.

[12] Kumar, S., Abraham, E., Talukder, S., Kumar, P. & Pratap, R. Micro Electrolithography System Development(2018). *2018 4th IEEE International Conference on Emerging Electronics, ICEE 2018* 1–4 doi:10.1109/ICEE44586.2018.8937922.

[13] Kumar, P. *et al.* Patterning Cr film by passing electric current through a traversing pointy stylus: Introduction to electrolithography and its prospects(2018). in *2018 IEEE International Conference on Manipulation, Manufacturing and Measurement on the Nanoscale, 3M-NANO 2018 - Proceedings* 181–185 (IEEE, 2018). doi:10.1109/3M-NANO.2018.8552166.

[14] Sharma, M. *et al.* Melting and electromigration in thin chromium films(2020). *Physics of the Solid State* **62**, 988–992.

[15] Sharma, M. *et al.* Electric current driven formation of micro-and nano-sized beads in thin Cr films(2019). in *2019 IEEE International Conference on Manipulation, Manufacturing and Measurement on the Nanoscale, 3M-NANO 2019 - Proceedings* 86–89 (IEEE, 2019). doi:10.1109/3M-NANO46308.2019.8947369.

Numerical Simulation and Parametric Analysis of Fatigue Crack in UIC60 Rail Thermite Welded Joint

Prakash Kumar Sen[1], Mahesh Bhiwapurkar[2] and S P Harsha[3]

[1] Mechanical Engineering Department, O.P. Jindal University Raigarh, India
[2] Mechanical Engineering Department, O.P. Jindal University Raigarh, India
[3] Department of Mechanical & Industrial Engineering, IIT Roorkee, India

prakashkumarsen@gmail.com

Abstract. The mechanism of rail-wheel contact is the most essential field of study in railway engineering since it requires extensive application expertise, diagnostic skills, and a trustworthy analysis technique. In this research the fatigue life of a UIC60 rail AT weld under vertical load and its parametric effect has been studied, and for that a three-dimensional elastic-plastic finite element model is created using ANSYS space-claim software, and then finite element method is employed to analyse the full-scale model of wheel-track and weld system with realistic three-dimensional solution. Model assembly components include axle, wheel, and thermite-welded rail. Simulation of contact between wheel and UIC60 rail weld with crack on weld at angles of 30 and 60 degrees with different coefficients of friction between the weld wheel contact and between crack surfaces was carried out under vertical loadings. In general, the Hertz contact theory assumptions are taken into consideration throughout the analysis, and the impacts on fatigue life are given by using damage mechanics method. The results of the wheel/weld fatigue crack analysis have been displayed to demonstrate the influence of different parameters on the fatigue life of cracks. The purpose of this study is to identify and safeguard the rail against failure, as well as to ensure the safety of passengers and to reduce the cost of maintaining the rail system.

Keywords. Railway, fatigue, crack, FEM, ANSYS.

1. Introduction

Rails are now connected together by welding whenever it is possible. Welding is, an average, a less expensive process that generates a higher-quality connection than other methods. Every year, a considerable quantity of AT weldments is installed in rail pathways, applying customary processes and ingredients which are established primarily via field practises over the years, and a very high proportion of them found reliable.

Lengths of standard rails are welded together using flash welding methods to form long sections, which are normally six times in length of a standard rail. These sections are then brought to the construction site and joined together to form a continuous track. Alumino-thermic welding is perhaps the most often used technology for joining rails in the work field, for multiple of motives, like easy to align the tracks and other field parameters, as well as cost and time savings. Aluminium-thermite weld on rails are exactly positioned having joint-gap of approximately 35 mm between the rail ends as a regular practice [1]. For joining through gape, an exclusive ceramic mould is placed around it, which is then packed with cementitious sand, and a combine mix powder having iron oxide and its alloys with aluminium is poured through hopper in it. As the combination of this mixture is burnt, molten metal rushes downward in the cavity, forming the joint over there. The method is also said as casting joint. After the weld-joint cooled, the mould frame is removed, and any excess metal on joint area is chipped out and by grinding operation the joint has provided surface finish so it can match the shape and profile of the parent rail.

The rail-wheel assembly is considered as very complex geometry who rely on friction to convert energy. As a consequence of this rolling contact friction and frictional slip or sticky kind of action a damage mechanism comes in to picture which manifest the structure to initiate wear and growth of crack in it. This phenomenon now becomes a challenge for railway engineers to identify and solve such problems as early as possible as it arises. That's why it's essential to understand the fundamentals of contact theory and generation of contact stress, and its distribution to predict the fracture behaviour and propagation. In addition, for above mentioned terms the 3-D stresses parameters and locations are estimated correctly in [2, 3], so the wheel-rail wear/damage can be accurately predicted. In the literature [4-6], for some circumstances, it was difficult to find the direct solutions for stresses on the contact point hense new methods were developed for those cases and presented. These solutions were based on specific elastoplastic material properties where their qualities were improved in physically and do not consider the friction coefficient normally for wheel rail contact. It's also found that these closed forms solutions are with very basic geometries and full of errors thus rail-wheel contact becomes quite sophisticated.

(a) (b)

Figure 1. (a) Rail weld with wheel contact (b) AT weldment with crack [18]

Many scholars have observed the contact between a rail and a wheel in various contexts and with various rail and wheel geometries in order to find the contact and its behaviour, the generated stress and its distribution. Result shown by the clarifies about surface fatigue fracture and damage mechanism and plastic deformation on structure and other vibrational parameters. The proper study and simulation of these systems allowed us to understand thorough comprehension of the specific knowledge about real time interaction of wheel and the rail. These study have been investigated in the literature [7-11] through the use of some laboratory data, and finite element simulations with analytical assumptions under variety of scenarios [12-15]. This research will not provide any new insight into the literature review or the fundamentals of the area of interest because the objective of this work is to merely illustrate a well-known finite element analysis approach on an assembly of 3-D rail–weld. Following are the steps involved in the 3-D contact study of rail weld–wheel contact that were taken into account in this work, and they are depicted in the Figure 2 below in the following order:

- Rail-weld assembly layout and design with Elastic-plastic material model.
- Fixing the boundary and loading conditions.
- Parameters in which the varied coefficients of friction on contact are taken into consideration.
- Solve the problem and post-process the results

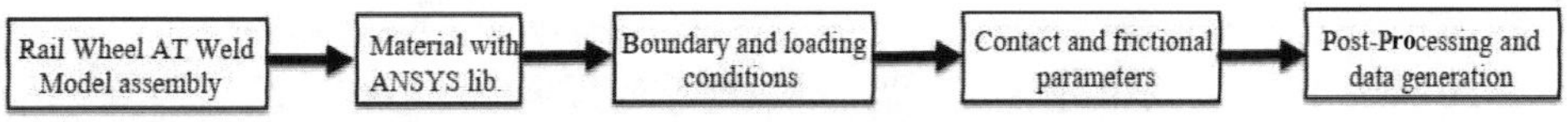

Figure 2. Flow diagram for simulation process

2. FEA model, material and methodology: Pre-Processing

Specifically, a mono-block wheel with a diameter of 1150 mm and the matching rail UIC 60 [16] and with its axle and a connection of realistic AT weld with side embossed and grinded surface have been modelled in accordance with the Indian Railways criteria indicated in Figure 3(a) for this study. This assembly model is shown in Figure 3(b) after the geometries have been produced in ANSYS space-claim 2020 student edition [17].

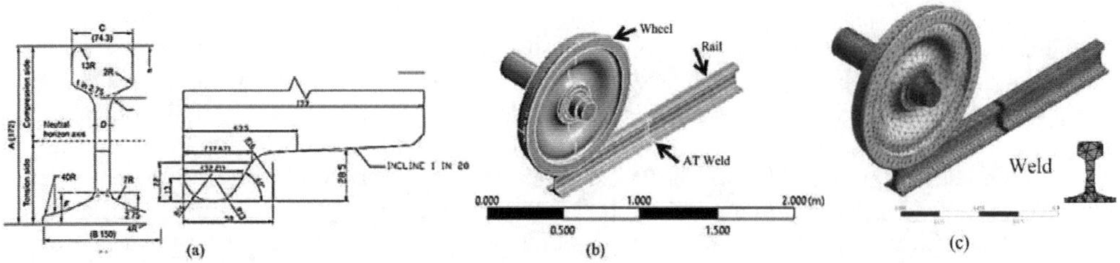

Figure 3. (a) Rail and wheel geometry (b) Rail weld with wheel assembly model and (c) Mesh

The 3-D model assembly was imported into the ANSYS mechanical workbench and then mesh was provided with 8 node hexagonal elements type of solid185, as in Figure 3(c), along with weld zone meshing as shown in Figure 3(d). The frictional surface contact is defined using conta174 type elements, which are on trade surface of wheel, and other surface as the targe170 elements, which are placed on the rail and its weld. There were a total of 190572 nodes and 204846 elements in the FEA model mesh for all geometry. The loading conditions are depicted in Figure 4, where a vertical force of 55 kN is applied to the centre of the wheel at its centre of rotation. The rail and weld bottom are anchored at the transverse site in all directions to prevent the body from moving. The effects of the rotating wheel and lateral force were not taken into consideration. The FE model is anticipated to reflect the weld in a more realistic manner by including a 1mm-wide side extruded portion of the weld and the assumption that the head of the weld has been grounded.

Table 1. Parameters of elastic plastic weld model

Mechanical property	Value
Poisons Ratio	0.3
Young's Modulus (GPa)	207 GPa
Ultimate tensile strength (MPa)	996.7 MPa
Yield strength (MPa)	675.7 MPa
Percentage reduction of area	4.22
Percentage Elongation	3.09

Table 2. CAD parts material properties

Part name	Modulus of elasticity(GPa)	Modulus of plasticity (GPa)	Yield stress (MPa)	Poisson's ratio
Rail	206.9	22.7	483	0.295
Wheel	205	22.7	640	0.3
Axle	205	-	-	0.3

Both Tables 1 and 2 represent the relevant material attributes that were used in the ANSYS material library's, material database and then assigned to the CAD assembly model in workbench.
Figure 4 depicts the assembly with a view of the AT weld on rail as a joint, where two fractures are postulated for the purpose of analysis under vertical loading, as shown in the assembly. The position of the fracture is thought to be the region where the greatest amount of stress distribution may generate. For the purpose of individual analysis, the orientation of cracks has been considered to be at an angle of 30° and 60°. The coefficients of friction used in this calculation of stress between contacts are 0.25, and the coefficients of friction used in the parametric analysis of crack are 0.1 to 0.4, respectively, with vertical loads of 92, 120, and 140 kN in the case of the contact.

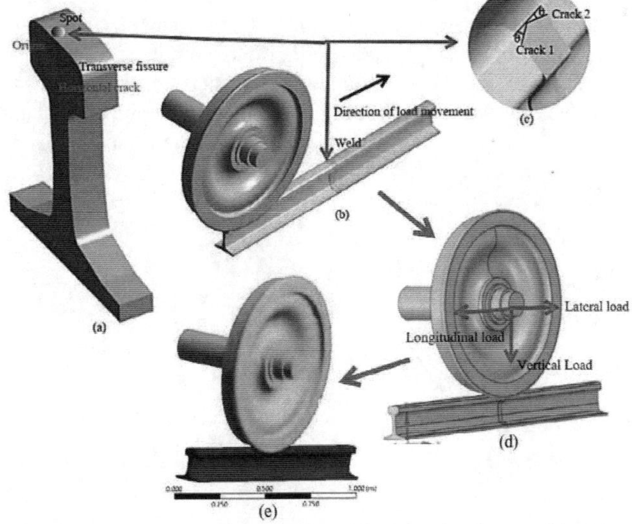

Figure 4. (a) AT weldment crack position (b) Rail weld with wheel assembly (c) weld part having crack positions (d) Loading conditions (f) simulation view of wheel contact on weldment

3. Simulation results: Post-Processing

The assembly is applied with boundary conditions with loading and simulated as post-processing, allowing for the visualization of stress distributions. Figure 5 depicts the maximum stress distribution at various points in contact. Figure 5(a) shows the pictorial view of simulation having a maximum Von-Mises stress value as 597 MPa under the frictional coefficient of 0.25 just beneath the head surface about 2.89 mm as expected to be occurred. It may be observed that the induced stress is above the value of yield strength of parent rail material and quite below the yield strength of the rail weld material. For both weld part and contact region stress contour may be seen in Figure 5(b).

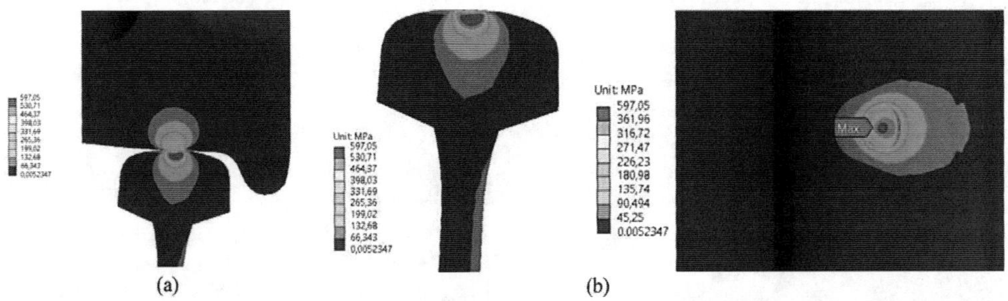

Figure 5. (a)Stress AT weldment contact (b) Stress showing at weld and at contact point

It has been observed in many studies that the majority of the distortion due to plastic deformation happens on the wheel body, and then rail and its weldment experience essentially no specific plastic deformation. According to Figure 5, stress concentration where the greatest equivalent deformable stress value occurs just having same node, is approx. 462 MPa. An examination of the Von-Mises stress contours of the wheel and rail-weld reveals the presence of a fracture that develops over the running cycle exactly in the area depicted in Figure 4(a). The result has been plotted between frictional coefficient and equivalent stress is in Figure 6(f).

For the study of effect of different parameters on weld life due to fatigue is explored using the proposed damage mechanics approach and carried out by numerical simulation, and their plots are drawn. The effective parameters considered here are loads which are vertically applied and coefficients of friction variations having two condition first in between the contact surfaces of wheel and weld and second in between the crack surfaces. Here two cracks are named as crack 1 and crack 2 as discussed previously. Now for parametric analysis with varying loading conditions i.e. 92, 120, 140 kN through wheel has applied on weld and observations have been made on weld crack 1 with a constant opening angle 60° and coefficient of friction 0.3. The welds cracks fatigue life for crack 1 and crack 2 under various vertical loads is shown in Figure 6 (a-b) respectively for different crack lengths and coefficients of friction.

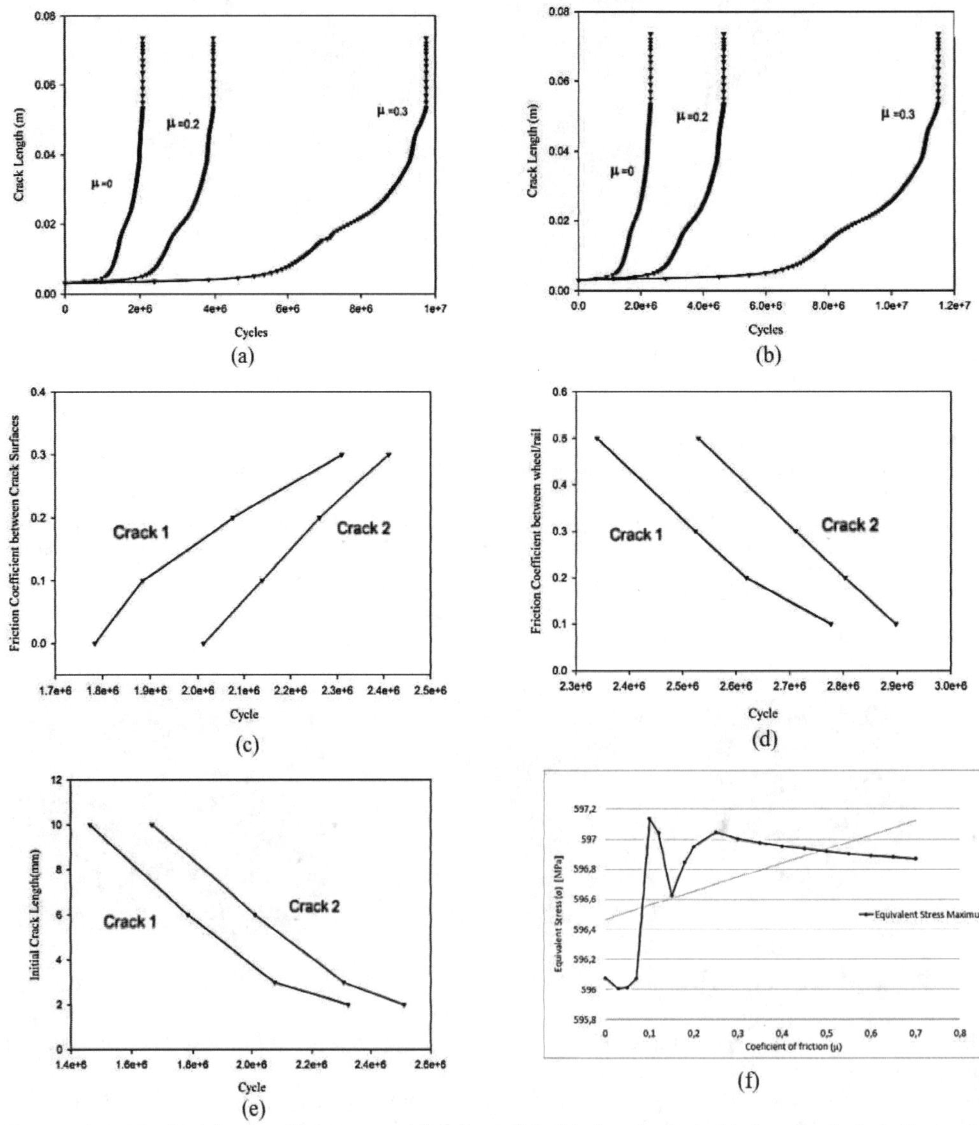

Figure 6. Vertical loads effects on weld fatigue life (a) for crack 1 (b) for crack 2, Influence of Coefficient of friction on fatigue life (c) between wheel and weld contact (d) between crack surfaces, Influence of initial crack length on fatigue life (e) for both crack 1 and 2 (f) Von-Mises Stress on contact

When the weld is subjected to a vertical load on a crack having fixed initial length of 1 mm and at an angle 30° for both crack 1 and crack 2 under coefficients of friction 0.1, 0.2, 0.3, and 0.5, the weld crack fatigue life is estimated for both crack 1 and crack 2. Graph in Figure 6 (c) represent the weld life for

different coefficient of friction. Figure 6 (d) represents the estimation of welds crack life under fatigue for coefficients of friction between the surfaces of crack taking in to account as 0, 0.1, 0.2 and 0.3 respectively, considering a fixed vertical load of 92 kN and an opening angle of 60° and fixed initial crack length of 6 mm with wheel-weld surface friction as 0.3, and plotted for both crack 1 and crack 2. Figure 6 (e) illustrate the life of fatigue for weld with initial crack lengths of 1, 3, 6 and 10 mm respectively.
A good agreement between results of experimental fatigue test and FEM based solution on two different surface crack growth, performed by Yang Liu et al. [19] can be co-related with this wok for satisfactory and concluding remarks.

4. Conclusions
At wheel weld contact the Von- Mises equivalent stress condition has been achieved in this work, using the elastic-plastic modelling and interaction has been developed between wheel and UIC60 weldment were with different coefficients of friction. The Von Mises equivalent stress state at the wheel-rail-weld contact has been achieved using the elastic-plastic model in this work. The numerical method and finite element software ANSYS were used to obtain the results, and then analysed. The presence of frictional forces on the contact zone results as following:

- Von Mises stress is distributed in an asymmetric manner.
- Increment in Von Mises stress value at narrow depths on weldment occurs.

It has been shown that increasing value of Von Mises equivalent stress close to the contact surface have a negative impact on wheel and weld fatigue endurance when operating under the rolling contact phenomenon.
The linear elastic fracture mechanics used to perform the parametric analysis of the cracks, which represents operational factors such as vertical load, coefficient of friction between the wheel and weld contact, and surface friction for two types of surface cracks. On observational basis it is possible to draw the following conclusions:

- The greater the coefficients of friction at the contract surface between weld and wheel, results in higher growth of crack 1 and, as a result, the reduction in the growth of crack 2.
- The friction between the fracture surface and the friction between the weld and wheel have dramatically opposed effects on life of fatigue. The fatigue life of a fracture grows as the friction between the crack surface and the surrounding material increases. When the frictional constant between wheel and the weld increases, the fracture has a tendency to grow in a progressive manner. As a result, the duration of wear can be minimised.
- Crack 1 expands toward the rail depth as friction increases, and eventually transforms into transverse cracks as the rail depth decreases.

Results for novel parametric variables like stress due to manufacturing process, bending loads, and rail/weld hardening, may be obtained through future research for crack formation.

Acknowledgment
Authors appreciate the support from ANSYS for using the student edition for the simulation work and also express our gratitude for RDSO, Lucknow from the Indian railways to make use of the available resources.

References

[1] Indian Railways year-book 2020 , www.indianrailways.org.
[2] Cui X L, Chen G X, Yang H G, Zhang Q, Ouyang H and Zhu M H 2015 Effect of the wheel/rail contact angle and the direction of the saturated creep force on rail corrugation (Elsevier B. V.)
[3] Xin Z, Xiaogang Z, Chao L, Zefeng W and Xuesong Jin 2016 A study on dynamic stress intensity factors of rail cracks at high speeds by a 3D explicit finite element model of rolling contact (Elsevier B. V.)

[4] Crețu S 2009 Contactul concentrat elastic-plastic" (Editura „Politehnium", Iași) [2] Vo K D, Tieu A K, Zhu H T and Kosasih P B 2014 A tool to estimate the wheel/rail contact and temperature rising under dry, wet and oily conditions COMPRAIL 14th Conference on Railway Engineering Design and Optimization United Kingdom: WIT Press pp. 191-201
[5] Zhang J, Shouguang S and Xuesong J 2009 Numerical simulation of two-point contact between wheel and rail (AMSS Press, Wuhan, China).
[6] Wang W J, Lewis R, Yang B, Guo L C, Liu Q Y, and Zhu M H 2016 Wear and damage transitions of wheel and rail materials under various contact conditions (Elsevier B. V.)
[7] Peixoto D F C et al. 2007 Non-linear analysis of the wheel/rail contact (Departamento de Engenharia Mecânica, Faculdade de Engenharia da Universidade do Porto Portugal)
[8] Busquet M et al. 2003 3D Finite element investigation on the plastic flows of railheads. Correlation with micro structural observations. In: 6th International conference on contact mechanics and wear of rail/wheel systems CM2003 in Gothenburg, Sweden.
[9] Zili L, Xin Z, Coenraad E, Rolf D and Molodova M 2008 An investigation into the causes of squats—Correlation analysis and numerical modelling (Elsevier B.V.)
[10] Wen-jian W, Wen-juan J, Heng-yu W, Qi-yue L, Min-hao Z and Xue-song J 2016. Experimental study on the wear and damage behavior of different wheel/rail materials (Journal of rail and rapid transit)
[11] Masoudi Nejad, R., Shariati, M., and Farhangdoost, Kh. (2016), "Effect of Wear on Rolling Contact Fatigue Crack Growth in Rails,"Tribology International, 94, pp 118–125.
[12] Liu, Y., Liu, L., and Mahadevan, S. (2007), "Analysis of Subsurface Crack Propagation under Rolling Contact Loading in Railroad Wheels Using FEM," Engineering Fracture Mechanics, 74, pp 2659–2674.
[13] Liu, Y., Liu, L., Stratman, B., and Mahadevan, S. (2008), "Multiaxial Fatigue Reliability Analysis of Railroad Wheels," Reliability Engineering and System Safety, 93, pp 456–467.
[14] Liu, Y., Stratman, B., and Mahadevan, S. (2006), "Fatigue Crack Initiation Life Prediction of Railroad Wheels," International Journal of Fatigue, 28, pp 747–756.
[15] Cornell Fracture Group, Cornell University. Available at: http://www. cfg.cornell.edu (accessed February 16, 2014).
[16] Anderson, T. L. (1994), Fracture Mechanics, Fundamentals and Applications, 2nd ed. CRC Press: New York, NY.
[17] ANSYS Workbench help.
[18] http://www.railroadpart.com/news/fishplate-joint-and-rail-joint-welding-in-indian-railway.html
[19] Yang Liu, Kin Shun Tsang, Hsin Jen Hoh, Xiaopeng Shi, John Hock Lye Pang, Structural fatigue investigation of transverse surface crack growth in rail steels and thermite welds subjected to in-plane and out-of-plane loading, Engineering Structures, Volume 204,2020,110076, ISSN 0141-0296, https://doi.org/10.1016/j.engstruct.2019.110076.

Enhancing electrical properties of carbon nanotubes thin films by silicon incorporation

Sk Faruque Ahmed[1], Mohibul Khan[1] and Nillohit Mukherjee[2]

[1] Nanoscience Laboratory, Department of Physics, Aliah University, IIA/27, New Town, Kolkata - 700160, India.
[2] Center of Excellence for Green Energy and Sensor Systems, Indian Institute of Engineering Science and Technology, Howrah 711103, India.

E-mail: fahmed.phys@aliah.ac.in

Abstract. Silicon incorporated carbon nanotube (Si-CNTs) thin films was prepared by radio frequency plasma enhanced chemical vapor deposition technique. Tetraethyl orthosilicate solution was used for incorporation of silicon in CNTs thin films. Energy dispersive X-ray analysis shows that the silicon atomic percentage was varied from 0 % to 6.1 %. The chemical binding energies of carbon and silicon were analyzed from X-ray photoelectron spectroscopy data. The various peaks at ~531 eV, ~ 285 eV, ~155 eV and ~104 eV was observed in the XPS spectra due to the oxygen, carbon and silicon respectively. Surface morphologies of Si-CNTs thin films have been analyzed by field emission scanning electron microscopy, which revels that the length of the silicon incorporated carbon nanotubes ~500 nm and corresponding diameter ~80 nm. The room temperature electrical conductivity was increased whereas the activation energy was decreased with the increase of atomic percentage of silicon in Si-CNTs thin films. The room temperature electrical conductivity was increased from 4.3 x 10^3 to 7.1 x 10^4 S cm^{-1} as the silicon atomic percentage in Si-CNTs thin films increases from 0 to 6.1 % respectively.

Keywords. Carbon nanotubes, RF-PECVD Technique, Silicon incorporation, XPS, Electrical conductivity.

1. Introduction

Nanostructures graphene planes in rolled form are the building block of carbon nanotubes (CNTs). CNTs are composed of one or more closed hollow graphite cylinders, with few hundred nm in diameter and a few microns in length. Generally CNTs are two types; single- walled when the number of graphene sheet is one and multiwalled, if more than one graphene sheet formed tube [1,2]. CNTs have been widely used for many applications, such as device making for optical communication, probe tips for electron microscope, different electronic circuits, water purification, water disinfection, hydrogen storage, field emission electron guns, nanowires, gas sensors, biomedical energy and air purification [3-11]. Depending on the synthesis technique, single walled carbon nanotube and multiwalled carbon nanotubes with different surface structure can be prepared. The various preparation methods and characterizations of CNTs have been reported by various researchers because of their incredible properties and different prime applications in the nanotechnology. The basic procedure for preparation of CNTs are RF magnetron sputtering, solvothermal technique, AC-arc discharge, laser evaporation, Electrolysis, hot filament-assisted sputtering, catalytic chemical vapour deposition (CCVD), electrospinning and plasma enhanced chemical vapor deposition (PECVD) [11-21]. PECVD technique is one of the best root for the preparation of nanostructured carbon because of high yield and controllable technique for the production of CNTs at lower substrate temperature.

Doping of CNTs is a promising approach to control the electronic structure and is of high technological importance for the practical device manufacturing. Study of doped CNTs useful in understanding the perturbed physical properties caused by the dopants in one-dimensional materials and gives an opportunity to attainment their remarkable properties in innovative and relatively fast growing technology. The opportunity of alternate element such as iron, gold, boron, phosphorus, sulfur and nitrogen doping in CNTs has generated intense interest due to the feasibility of tailoring structural and electronic properties of CNTs by different doping elements [22-29].

In this paper, we have reported the synthesis of Si-CNTs thin films via radio frequency plasma enhanced chemical vapor deposition (RF-PECVD) technique in thin film form. Nanostructure surface morphology characterization was done by field emission scanning electron microscope (FESEM). The details chemical bonding information was analysed by X-ray photoelectron spectroscopy (XPS) data. Temperature dependence electrical conductivities of Si-CNTs thin films were studied for various atomic percentage of silicon in the CNTs thin films. The activation energy was calculated and has explained.

2. Experimental Procedure

2.1. Preparation of silicon incorporated CNTs thin films

Different transition metal such as nickel, iron, cobalt has been used as catalyst in the growth of carbon nanotubes by chemical vapour deposition technique. Nickel (Ni) thin films were deposited on alumina substrates was used as a catalyst for Si-CNTs preparation. The deposition of Ni catalyst thin film on the substrates was done by direct current (DC) sputtering technique. Before starting the deposition of Ni thin films, all alumina substrates were ultrasonically cleaned in methanol for 10 minutes at 30 ^{o}C temperature and rinsed thoroughly with double distilled water 2-3 times continuously then soaked with hot air gun. A Ni target of 99.99 % pure was used for DC sputtering technique. The DC sputtering performed at 0.15 mbar chamber pressure using Ar as a sputter gas. During DC sputtering the voltage and current density were 2.5 KV and 20 mA cm^{-2} respectively. A ~20 nm thick (measured by Inficon make digital thickness monitor Model: SQM-160) Ni catalyst thin film deposited on alumina substrates. After preparation of Ni catalyst thin film, the alumina substrates were kept within the RF-PECVD unit. The RF-PECVD chamber was evacuated with a base pressure of 10^{-6} mbar. A transformer was used for heating the substrate during Si-CNTs thin films preparation and the temperature of the substrate was maintained at 450 ^{o}C. The mass flow controller (MFC) key was used for entering the acetylene (C_2H_2) gas into the evacuated chamber with 12 sccm and was maintained the RF-PECVD chamber pressure 1.5 mbar during CNTs thin films growth. The preparation of Si-CNTs thin films has been done at 170 watt RF (13.56 MHz) power for 20 min. The precursor solution for silicon incorporation in CNT thin films made by tetraethyl orthosilicate (TEOS) mixed with methanol with different concentration. The solution was kept in a bubbler and Ar gas (10 sccm) was send to the RF-PECVD unit through the bubbler, which controlled by MFC key. The different concentration of TEOS solution were used for the variation of silicon atomic percentage in Si-CNTs thin films. The complete synthesis parameters for preparation of Si-CNTs thin films are described in table 1.

Table 1. Synthesis parameters for preparation of Si-CNTs thin films.

Synthesis parameters	Corresponding values
Acetylene gas	12 sccm
Ar gas	10 sccm
Base pressure	10^{-6} mbar
Working pressure	1.5 mbar
Substrates used	Alumina coated with nickel thin film
Deposition temperature	450 ^{o}C
Deposition time	20 min.
RF power	170 watt

2.2. Characterization of Si-CNTs thin films

The nanostructure surface morphology of Si-CNTs thin films were investigated by using a scanning electron microscope (FESEM, JEOL-JSM-6360) and the silicon, carbon composition were measured by energy dispersive X-ray analysis (EDX, Oxford, model-7582). The binding energy of different electronic states of carbon and silicon of the Si-CNTs thin films were analyzed by X-ray photoelectron spectroscopy (XPS) (Perkin-Elmer-1257). The hemispherical energy analyzer was used for the XPS spectra analysis. The background correction and analysis of the XPS data were done by Shirley technique. The Mg Kα non-monochromatic X-ray was used as the excitation source with energy 1253.6 eV. The XPS system pressure was ~10^{-9} mbar during measurement. A standard four-probe instrument used to study the electrical conductivity measurement of the Si-CNTs thin films with different temperature. To make the ohmic contact for electrical conductivity measurement, gold electrodes were deposited on Si-CNTs thin films by thermal evaporation technique.

3. Results and discussion

3.1. Analysis of composition for Si-CNTs thin films

The atomic percentage of silicon and carbon in the Si-CNTs thin films was analyzed from the EDX spectra. The EDX spectra of the Si-CNTs thin films synthesized on alumina substrate is shown in figure 1. It was observed that the atomic percentage of silicon in the CNTs thin films within range of the concentration of TEOS solution. The atomic percentage of silicon in the Si-CNTs thin films was varied by using different concentration of TEOS solution which shown in In table 2. In EDX spectra, the peak due to carbon, aluminium and oxygen was observed in both pure CNTs and silicon incorporated CNTs thin films, whereas silicon peak was detected for doped CNTs thin films. In the EDX spectra, the peak of the aluminium and oxygen comes from alumina substrate.

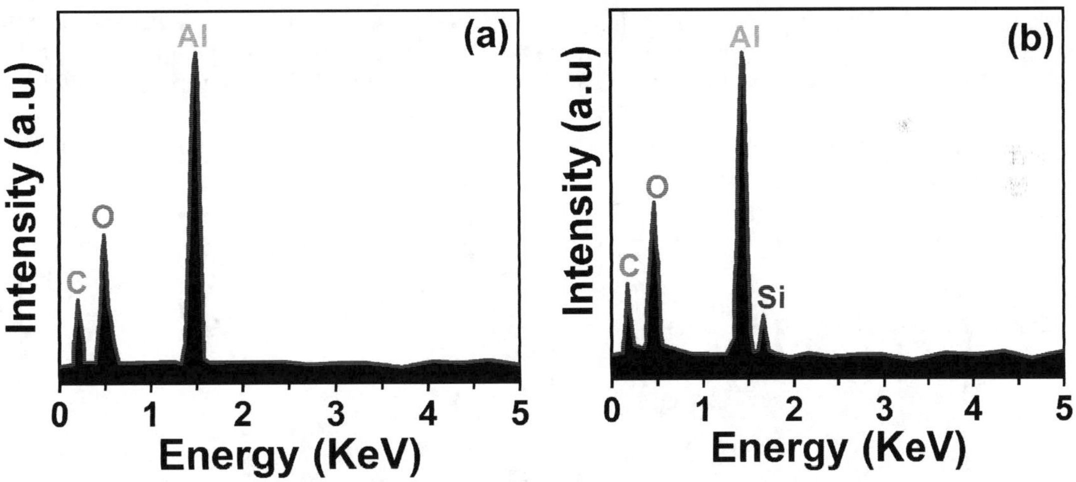

Figure 1. EDX spectra for (a) pure CNTs thin films and (b) silicon incorporated CNTs thin films was prepared using 8 % TEOS solution.

3.2. Surface morphology studies

The topographical image of Si-CNTs thin films was taken using a FESEM as shown in figure 2. The FESEM image showed the CNTs are randomly oriented on substrate. The length and diameter of the CNTs and Si-CNTs ~ 500 nm and ~ 80 nm respectively.

IOP Conf. Series: Materials Science and Engineering 1206 (2021) 012028 doi:10.1088/1757-899X/1206/1/012028

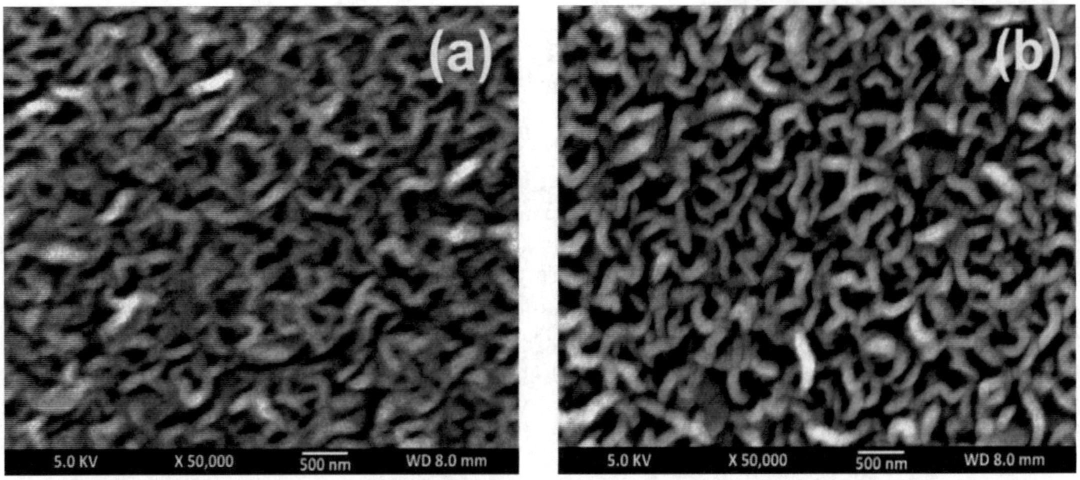

Figure 2. FESEM micrograph of (a) pure CNTs thin films and (b) 6.1 atomic percent silicon incorporated CNTs thin films.

3.3. XPS analysis

Figure 3(a) and (b) shows the XPS spectra of pure CNTs thin films and 6.1 atomic percent Si content CNTs thin films respectively. The XPS spectra of the Si-CNTs thin films shows the different peaks. In the XPS spectra, the peaks at ~531 eV, ~ 285 eV, ~155 eV and ~104 eV were observed due to O 1s, C 1s, Si 2s and Si 2p respectively [30,31]. The Si core level XPS spectrum was shown in the inset of figure 3(b). Two binding energy peaks corresponding to Si 2s at 104 eV and Si 2p at 155 eV respectively was found in core level spectra, which justify that silicon has been incorporated into the CNTs. XPS is a quite surface sensitive analysis technique. In XPS spectra, the peak at ~531 eV was appeared due to oxygen. In figure 3(a), there is a peak corresponds to oxygen for the pure CNTs thin films synthesized without using of TEOS solution. The oxygen peak in the both XPS spectra confirms that Si-CNTs surface contaminated by oxygen from different sources [32-34].

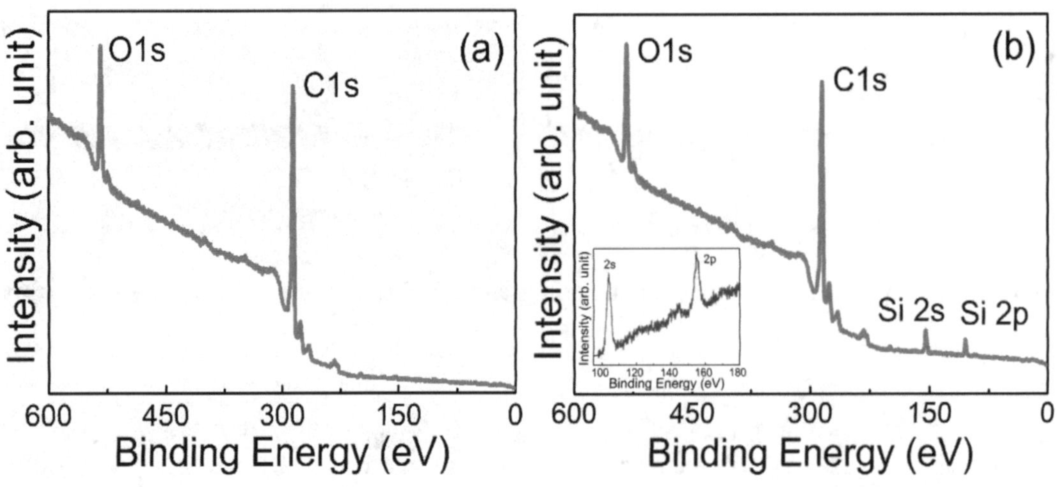

Figure 3. (a) Full XPS graph of the pure CNT thin films and (b) Si-CNTs thin films with 6.1 atomic percentage silicon content and the Si core level spectra of 6.1 at. % Si content CNF thin films in inset.

3.4. Electrical Conductivity analysis

The well-known equation for the study of electrical conductivity (σ) of semiconductor thin films is expressed by equation 1 as follows [35]:

$$\sigma = \sigma_o \exp[-\frac{E_a}{kT}] \quad (1)$$

Where k is Boltzmann constant and its value is 1.38×10^{-23} joule/ kelvin, σ_0 is a temperature independent factor and E_a is the activation energy of the material respectively. The activation energy known as the energy difference between top of the valence band and the acceptor level in case of p-type semiconductor and similarly the energy difference between bottom of the conduction band and the donner level in case of n-type semiconductor respectively. The equation (1) is a straight line equation and the plot between lnσ vs. 1/T should be a straight line. The electrical conductivity measurement with the variation of temperature for the Si-CNTs thin films was carried out. Several set of measurements was performed for different silicon content Si-CNTs thin films. The change in the electrical conductivities of Si-CNTs thin films ware observed from the conductivity measurement analysis. The lnσ vs. 1000/T plot for Si-CNTs thin films with different atomic percentage of silicon is shown in figure 4(a). The room temperature electrical conductivity of the Si-CNT thin films was increased from 4.3 x10^3 to 7.1x10^4 as the silicon atomic percentage in the CNTs thin films increase from 0 to 6.1 respectively as shown in figure 4(b).

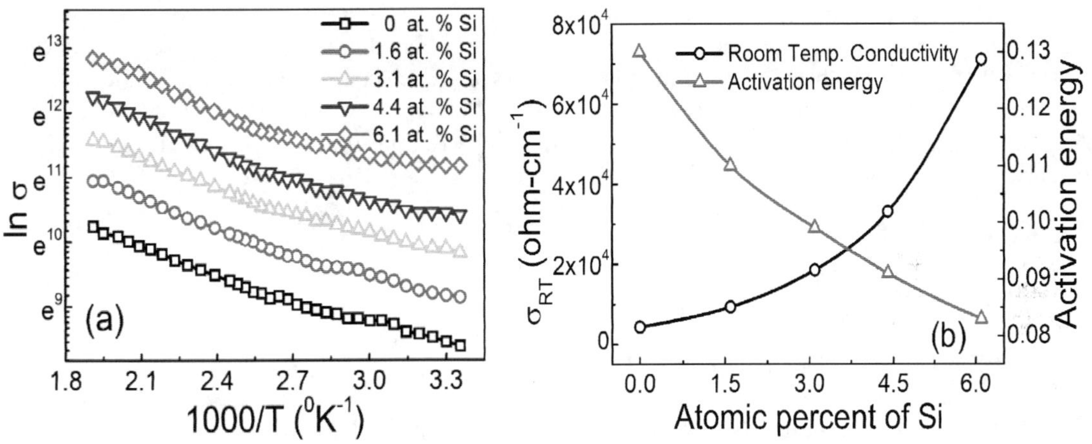

Figure 4. (a) Temperature dependent electrical conductivity of Si-CNTs thin films for different atomic percent of silicon and (b) variation of room temperature electrical conductivity and activation energy with various atomic percent of silicon in Si-CNTs thin films.

Table 2. The room temperature electrical conductivity for Si-CNTs thin films with various atomic percentage silicon as measured by EDX.

Nominal Si in TEOS solution (at. %)	Si from EDX (at. %)	(σ_{RT}) S cm^{-1}
0	0	4.3x10^3
2.5	1.6	9.3x10^3
4.0	3.1	1.8x10^4
6.0	4.4	3.3x10^4
7.5	6.1	7.1 x10^4

The increased of temperature, electrical conductivity (σ_{RT}) of the Si-CNTs thin films was increased, which similar as for semiconductor. The activation energy(E_a) for Si-CNTs thin films was calculated from the slope of the graph of lnσ vs. 1000/T, which decrease from 0.13 to 0.08 eV as the silicon content in the CNTs thin films increase from 0 to 6.1 atomic percent respectively. The electron transport phenomenon can be explain with the diffusion process [36,37]. There is a barrier potential exits in the doped CNTs, which accountable for the electron transport considering diffusive transport phenomenon. The height of the potential energy barrier is given by equation 2

$$E_{barrier} = E_F - E_V \quad (2)$$

where E_F is the Fermi energy of CNTs and E_V is the top energy of valence band. In electron transport, the outer wall of CNTs impact is maximum. As a result of doping in the outer wall of the CNTs, the Fermi level perturbed and reduce the energy gap difference between valence and conduction band [37]. As the barrier height decreases, the large number of electrons smoothly tunnel the potential and helps in conductance compare to the pure CNTs. The diffusive transport mechanism highly influence in the electrical conductivity of Si-CNTs thin films, which was prepared by RF-PECVD technique.

4. Conclusions

Si-CNTs thin films was prepared on alumina substrate by using RF-PECVD technique. The atomic percentage of silicon in the CNTs thin films has been varied from 0 % to 6.1 %, which confirm from EDX analysis. FESEM image showed that the CNTs are randomly oriented and the average length of the Si-CNTs ~500 nm and corresponding diameter ~80 nm. Different peaks were observed in XPS spectra due to carbon, silicon and oxygen respectively. The analysis of silicon core level spectra confirms that silicon has been incorporated into the CNTs. The electrical conductivity of the Si-CNTs thin films was increased with increase of the temperature whereas the activation energy was decreased. The room temperature electrical conductivity (σ_{RT}) of Si-CNTs thin films was increased from 4.3 x 10^3 to 7.1 x 10^4 S cm^{-1} with increase of silicon atomic percentage 0 to 6.1 respectively.

5. References

[1] Ijima S and Ichiashi T 1993 *Nature* **363** 603
[2] Ijima S 1991 *Nature* **354** 56
[3] Rinzler A G, Hafner J H, Nikolaev P, Lou L, Kim S G, Tomanek D, Nordlander P, Colbert D T and Smalley R E 1995 *Science* **269** 1550
[4] Tatsuura S, Furuki M, Sato Y, Iwasa I, Tian M and Mitsu H 2003 *Adv. Mater* **15** 534
[5] Liu K, Avouris P, Martel R and Hsu W K 2001 *Phys. Rev. B* **63** 161404
[6] Simonyan V V and Johnson J K 2002 *J. Alloys Compd* **330-332** 659
[7] Wei Y Y and Eres G 2000 *Appl. Phys. Lett* **76** 3759
[8] Jain N and Kanu N J 2021 *Mater. Today: Proceed* **43** 2998
[9] Lal A and Kanu N J 2020 *IOP Conf. Ser.: Mater. Sci. Eng* **814** 012033
[10] Jain N, Gupta E and Kanu N J 2021 *Smart Science DOI: 10.1080/23080477.2021.1940752*
[11] Robertson J, Milne W I, Teo K B K and Chhowalla M 2002 *AIP Conf. Proc* **633** 537
[12] Cornejo J CR, Sebastian D and Lazaro M J 2020 *Rev. Chem. Eng* **36** 493
[13] Honda S, Lee K Y, Fujimoto K, Tsuji K, Ohkura S, Katayama M, Hirao T and Oura K 2002 *Physica B* **323** 347
[14] Mishra S, Nguyen H, Adusei P K, Hsieh Y Y and Shanov V 2019 *Surf. Interface Anal* **51** 290
[15] Ahmed S F, Mitra M K and Chattopadhyay K K 2008 *Ind. J. Phys* **82** 223
[16] Zhang X and Shneider M N 2019 *J. Appl. Phys* **125** 203304
[17] Zhang L, Xia G L, Guo Z P, Li X G, Sun D L and Yu X B 2016 *Int. J. Hydrogen Energ* **41** 14252
[18] Ahmed S F, Das S, Mitra M K and Chattopadhyay K K 2006 *Ind. J. Pure & Appl. Phys* **44** 700.
[19] Ahmed S F, Mitra M K and Chattopadhyay K K 2008 *Appl. Phys. A* **91** 429
[20] Chena C M, Dai Y M, Huan J G and Jehng J M 2006 *Carbon* **44** 1808
[21] Ahmed S F, Mitra M K and Chattopadhyay K K 2007 *Appl. Surf. Sci* **254** 610
[22] Gupta N, Gupta S M and Sharma S K 2019 *Carbon Lett* **29** 419

[23] Fujisawa K, Hayashi T, Endo M, Terrones M, Kim J H and Kim Y A 2018 *Nanoscale* **10** 12723
[24] Wang L, Yang C L, Dou S, Wang S Y, Zhang J T, Gao X, Ma J M and Yu Y 2016 *Electrochim. Acta* **219** 592
[25] Bermudez N B, Avendano M G, Tourino I L, Naranjo A C and Cervantes M L R 2020 *Nanomaterials* **10** 749
[26] Iizumi Y, Yudasaka M, Kim J, Sakakita H, Takeuchi T and Okazaki T 2018 *Sci. Reports* **8** 6272
[27] Liu S, Li G, Gao Y, Xiao Z, Zhang J, Wang Q, Zhang X and Wang L 2017 *Catal. Sci. Technol* **7** 4007
[28] Muhulet A, Miculescu F, Voicu S I, Schutt F, Thakur V K and Mishra Y K 2018 *Mater. Today Energy* **9** 154
[29] Silva W, Ghica M E and Brett C M A 2018 *Anal. Methods* **10** 5634
[30] Ahmed M H, Byrne J A and Ahmed W 2015 *Diam. Relat. Mater* **55** 108
[31] Aghaei R and Eshaghi A 2017 *J. Alloys Compd* **699** 112
[32] Ahmed S F, Mitra M K and Chattopadhyay K K 2007 *J. Phys. Condens. Matt* **19** 346233
[33] Ahmed S F, Moon M W and Lee K R 2009 *Thin Solid Films* **502** 4035
[34] Ahmed S F, Alam M S and Mukherjee N 2018 *Physica E: Low-dimensional Sys. Nanostructures* **97** 120
[35] Chopra K L 1969 *Thin Film Phenomena* (New York: McGraw-Hill)
[36] Liu K, Avouris P, Martel R, Hsu W K 2001 *Phys. Rev. B* **63** 161404
[37] Likodimos V, Glenis S, Guskos N and Lin C L 2003 *Phys. Rev. B* **68** 45417

Acknowledgments
The authors wish to acknowledge the financial support from Aliah University during the execution of the work and one of the authors (M. Khan) wish to thank the University Grand Commission, New Delhi, India for providing the fellowship under Maulana Azad National Fellowship (MANF) scheme.

AUTHOR INDEX